AF780675

Optical and Quantum Electronics: Physics and Materials

Optical and Quantum Electronics: Physics and Materials

Guest Editor

Sergio J. Jiménez Sandoval

MDPI

Basel • Beijing • Wuhan • Barcelona • Belgrade • Novi Sad • Cluj • Manchester

Guest Editor
Sergio J. Jiménez Sandoval
Materials Science
Centro de Investigación y de
Estudios Avanzados del IPN
Querétaro
Mexico

Editorial Office
MDPI AG
Grosspeteranlage 5
4052 Basel, Switzerland

This is a reprint of the Special Issue, published open access by the journal *Inorganics* (ISSN 2304-6740), freely accessible at: https://www.mdpi.com/journal/inorganics/special_issues/UP4GOP5K5A.

For citation purposes, cite each article independently as indicated on the article page online and as indicated below:

Lastname, A.A.; Lastname, B.B. Article Title. *Journal Name* **Year**, *Volume Number*, Page Range.

ISBN 978-3-7258-6232-0 (Hbk)
ISBN 978-3-7258-6233-7 (PDF)
https://doi.org/10.3390/books978-3-7258-6233-7

© 2026 by the authors. Articles in this book are Open Access and distributed under the Creative Commons Attribution (CC BY) license. The book as a whole is distributed by MDPI under the terms and conditions of the Creative Commons Attribution-NonCommercial-NoDerivs (CC BY-NC-ND) license (https://creativecommons.org/licenses/by-nc-nd/4.0/).

Contents

About the Editor

Sergio J. Jiménez Sandoval

Sergio J. Jiménez Sandoval, Centro de Investigación y de Estudios Avanzados del IPN Unidad Querétaro, Mexico. His current interests include the doping and codoping of semiconductor films, novel materials and architectures for inorganic solar cells, energy generation through quantum mechanics principles, and the application of Raman and photoluminescence micro-spectroscopies for optical and photoelectronic properties. A novel method to improve and control the electrical and photoelectronic properties of semiconductors, through so-called Adjacent Compensated Codoping, was recently developed by his group.

Preface

The field of optical and quantum electronics (OQE) is at the forefront of current scientific and technological developments. The scope of OQE includes quantum mechanical modeling not only of the photon-matter interactions, but also of the electronic performance in devices. Photonic energy may be described, in general, by submillimeter wavelengths (i.e., terahertz, infrared, visible, and ultraviolet). Quantum electronic devices apply various kinds of materials such as semiconductors, metals, insulators, and semimetals at scale lengths, where quantum (or semiclassical) descriptions play a vital role. The aim of this Reprint is to provide interested readers with the latest results presented by various research groups working on OQE, who present their groundbreaking theoretical and experimental developments. It is indisputable that technology will continue to evolve as the progress in this field thrives, and this Reprint represents a contribution to the state of the art of the physics and materials related to Optical and Quantum Electronics.

Sergio J. Jiménez Sandoval
Guest Editor

inorganics MDPI

Editorial

Optical and Quantum Electronics: Physics and Materials

Sergio Jiménez-Sandoval

Unidad Querétaro, Centro de Investigación y de Estudios Avanzados del IPN, Querétaro 76230, Mexico; sergio.jimenez@cinvestav.mx

The field of optical and quantum electronics (OQE) is a pillar of current technology and scientific development. The generation, control, and detection of electromagnetic radiation in the submillimeter regime (terahertz, infrared, visible, and ultraviolet) have become ubiquitous in devices used every day and in research laboratories. The interaction of electromagnetic radiation with matter at semiclassical and quantum level is a founding block on which our current understanding and development of OQE rely on [1–3]. To a large extent, the technological evolution shall evolve depending on the progress on this field, which comprises an ample portfolio not only about the physical phenomena of 3D and lower dimensionality materials [4–6] but also about the applications involving the generation and detection of electromagnetic radiation, in addition to the study of light-probed physical properties, where quantum phenomena play a central role.

This Special Issue of *Inorganics* collects research articles and reviews on optical and quantum electronics comprising advances in the properties and performance of a wide range of materials. These include theoretical calculations of electronic and phonon properties, experimental results on low-dimension materials such as nanowires, atomic clusters, Langmuir–Blodgett films, heat and charge transport including hot carriers, applications of photonic upconversion, and results on devices such as transparent RRAMs and perovskite solar cells.

Currently, perovskites are a class of materials highly relevant for a good number of applications. Elucidating their properties has become a major task for device design. In particular, surface and interface phenomena can determine their performance. In ABO_3 perovskites, there are serious issues related to surfaces. R.I. Eglitis et al. [7] present a comparison of density functional theory calculations for B3PW and B3LYP exchange-correlation functionals for the neutral (001) and polar (111) surfaces of ABO_3 surfaces (A = Ba, Sr, Pb, Ca; B = Sn, Ti, Zr). They found different types of relaxations for the near-surface atomic layers and specific terminations. It was found that the calculated band gaps for the bulk were generally reduced near the neutral (001) and polar (111) surfaces.

Zincblende group-IV binary XC and ternary $X_xY_{1-x}C$ alloys (X, Y $\equiv$ Si, Ge, and Sn) are of scientific and technological interest as promising alternatives to silicon for high-temperature and high-power applications. Efforts have been made to calculate the structural, electronic and vibrational properties of binary alloys, although no vibrational and/or thermodynamic studies have been conducted on ternary alloys. D.N. Talwar [8] employed a rigid-ion model to account for the lattice dynamics and thermodynamic properties of both binary and ternary alloys. Positive values of the acoustic modes in the entire Brillouin zone implied structural stability of the XC binaries. The case of ternary alloys was addressed through Green's function theory in the virtual crystal approximation to calculate composition-dependent phonon frequencies ω_j ($\mathbf{q}$), including the one-phonon density of states. It was concluded that devices based on GeC, SnC, and/or $Ge_{1-x}Sn_xC$ may not be appropriate for radiation detection in nuclear reactors or high-temperature, high-power settings. However, ultrathin XC binary and $X_{1-x}Y_xC$ ternary alloys can be suitable for heterostructures in MQW and SL-based micro/nanodevices for different strategic and civil applications.

The search for efficient transparent conducting oxides is one of the main research areas in materials science. The applications of these materials depend on the high demand for present and future applications. These include high-resolution screens for portable computers, flat-screen high-definition televisions, low-emissivity and electrochromic windows, the manufacture of thin-film photovoltaics, and an increasing demand of smart displays for hand-held devices [9–14]. Silver nanowires (AgNWs) are considered a potential alternative to conventional transparent conductive materials for a wide range of applications. H. Ha et al. [15] review the incorporation of polymeric materials to AgNW electrodes with an emphasis on their protective performance as well as on their applications. This review includes an evaluation of relevant factors that may affect compatibility with AgNWs and a perspective of challenges and opportunities to overcome current drawbacks with emerging technology.

Finally, all contributing authors are kindly acknowledged, as well as the *Inorganics* editorial team for their indispensable, diligent, and efficient support.

Funding: This research received no external funding.

Conflicts of Interest: The author declares no conflict of interest.

References

1. Zhai, Y.; Gong, P.; Hasaien, J.; Zhou, F.; Zhao, J. Coherent phonons in correlated quantum materials. *Prog. Surf. Sci.* **2024**, *99*, 100761. [CrossRef]
2. Xu, Y.; Vergniory, M.G.; Ma, D.-S.; Mañes, J.L.; Song, Z.-D.; Bernevig, B.A.; Regnault, N.; Elcoro, L. Catalog of topological phonon materials. *Science* **2024**, *384*, seadf8458. [CrossRef] [PubMed]
3. Charlier, J.; Eklund, P.C.; Zhu, J.; Ferrari, A.C. Electron and Phonon Properties of Graphene: Their Relationship with Carbon Nanotubes. In *Carbon Nanotubes*; Topics in Applied Physics; Jorio, A., Dresselhaus, G., Dresselhaus, M.S., Eds.; Springer: Berlin/Heidelberg, Germany, 2007; p. 111. [CrossRef]
4. Razeghi, M. Low Dimensional Quantum Structures. In *Fundamentals of Solid State Engineering*; Springer: Boston, MA, USA, 2002. [CrossRef]
5. Ganguly, A.; Goswami, R. Low dimensional materials in nanoelectronics. In *Nanoelectronics: Physics, Materials and Devices, Micro and Nano Technologies*; Elsevier: Amsterdam, The Netherlands, 2023; pp. 173–192.
6. Kim, J.M.; Haque, M.F.; Ezekiel, H.Y.; Nahid, S.M.; Zarin, I.; Jeong, K.-Y.; So, J.-P.; Park, H.-G.; Nam, S. Strain Engineering of Low-Dimensional Materials for Emerging Quantum Phenomena and Functionalities. *Adv. Mater.* **2023**, *35*, 2107362. [CrossRef] [PubMed]
7. Eglitis, R.I.; Purans, J.; Jia, R.; Kruchinin, S.P.; Wirth, S. Comparative B3PW and B3LYP calculations of ABO_3 (A = Ba, Sr, Pb, Ca; B = Sn, Ti, Zr) neutral (100) and polar (111) surfaces. *Inorganics* **2025**, *13*, 100. [CrossRef]
8. Talwar, D.N. Composition-dependent phonon and thermodynamic characteristics of C-based $X_xY_{1-x}C$ (X, Y = Si, Ge, Sn) alloys. *Inorganics* **2024**, *12*, 100. [CrossRef]
9. Chopra, K.L.; Major, S.; Pandya, D.K. Transparent conductors—A status review. *Thin Solid Films* **1983**, *102*, 1–46. [CrossRef]
10. Ginley, D.S.; Bright, C. Transparent Conducting Oxides. *MRS Bull.* **2000**, *25*, 15–18. [CrossRef]
11. Granqvist, G.; Azens, C.G.; Hjelm, A.; Kullman, A.; Niklasson, L.; Ronnow, G.A.; Mattsson, D.; Stromme, M.; Veszelei, M.; Vaivars, G. Recent advances in electrochromics for smart windows applications. *Sol. Energy* **1998**, *63*, 199. [CrossRef]
12. Kammler, D.R.; Edwards, D.D.; Ingram, B.J.; Mason, T.O.; Palmer, G.B.; Ambrosini, A.; Poeppelmeier, K.R. Photovoltaics for the 21st Century. In Proceedings of the Electrochemical Society, Pennington, NJ, USA, 17–22 October 1999; Kapur, V.K., McConnel, R.D., Carlson, D., Ceasar, G.P., Rohatgi, A., Eds.; IOP Publishing: Bristol, UK, 1999; p. 68, Proc. 99-11.
13. Lee, S.H.; Hwang, K.H.; Joo, S.K. Electrochromic Materials (2nd International Symposium). In Proceedings of the Electrochemical Society, Pennington, NJ, USA, 22–27 May 1994; Ho, K.-C., MacArthur, D.A., Eds.; IOP Publishing: Bristol, UK, 1994; p. 290, Proc. 94-2.

14. Chavan, G.T.; Kim, Y.; Khokhar, M.Q.; Hussain, S.Q.; Cho, E.-C.; Yi, J.; Ahmad, Z.; Rosaiah, P.; Jeon, C.-W. A Brief Review of Transparent Conducting Oxides (TCO): The Influence of Different Deposition Techniques on the Efficiency of Solar Cells. *Nanomaterials* **2023**, *13*, 1226. [CrossRef] [PubMed]
15. Ha, H.; Cheong, J.Y.; Yun, T.G.; Hwang, B. Polymeric protection for silver nanowire-based transparent conductive electrodes: Performance and applications. *Inorganics* **2023**, *11*, 409. [CrossRef]

Disclaimer/Publisher's Note: The statements, opinions and data contained in all publications are solely those of the individual author(s) and contributor(s) and not of MDPI and/or the editor(s). MDPI and/or the editor(s) disclaim responsibility for any injury to people or property resulting from any ideas, methods, instructions or products referred to in the content.

inorganics

MDPI

Article

Implementation of a Particle Swarm Optimization Algorithm with a Hooke's Potential, to Obtain Cluster Structures of Carbon Atoms, and of Tungsten and Oxygen in the Ground State

Jesús Núñez [1,2], Gustavo Liendo-Polanco [3,4], Jesús Lezama [2,3], Diego Venegas-Yazigi [5,6], José Rengel [2,7], Ulises Guevara [8], Pablo Díaz [9], Eduardo Cisternas [9], Tamara González-Vega [10], Laura M. Pérez [11] and David Laroze [12,*]

1 Plan Nacional de Formación en Procesamiento y Distribución de Alimentos (PNFPYDA), Universidad Politécnica Territorial del Oeste de Sucre "Clodosbaldo Russian", Cumaná 6101, Venezuela; jesusbnr@gmail.com
2 Instituto de Investigaciones en Biomedicina y Ciencias Aplicadas "Susan Tai", Universidad de Oriente (IIBCA-UDO), Cumaná 6101, Venezuela; jlezgar@gmail.com (J.L.); rengel66@gmail.com (J.R.)
3 Departamento de Química, Escuela de Ciencias, Núcleo de Sucre, Universidad de Oriente, Cumaná 6101, Venezuela; gustavoliendo@gmail.com
4 Grupo de Síntesis Química, Laboratorio de Fisicoquímica Orgánica, Centro de Química "Gabriel Chcuchani", Instituto Venezolano de Investigaciones Científicas (IVIC), Altos de Pipe, Miranda 1020-A, Venezuela
5 Departamento de Química de Materiales, Facultad de Química y Biología, Universidad de Santiago de Chile, USACH, Av. Libertador O'Higgins 3363, Santiago 9170022, Chile; diego.venegas@usach.cl
6 Centro para el Desarrollo de la Nanociencia y Nanotecnología, CEDENNA, Santiago 9170022, Chile
7 Departamento de Mecánica, Núcleo de Anzoátegui, Universidad de Oriente, Barcelona 8001, Venezuela
8 Vicerrectoría de Investigación y Postgrado, Universidad de La Serena, La Serena 1700000, Chile; ujguev@gmail.com
9 Departamento de Ciencias Físicas, Universidad de la Frontera, Casilla 54-D, Temuco 4780000, Chile; pablo.diaz@ufrontera.cl (P.D.); eduardo.cisternas@ufrontera.cl (E.C.)
10 Facultad de Ciencias de la Vida, Universidad Viña del Mar, Agua Santa 7055, Rodelillo, Viña del Mar 2572007, Chile; tamara.gonzalez@uvm.cl
11 Departamento de Ingeniería Industrial y de Sistemas, Universidad de Tarapacá, Casilla 7D, Arica 1000000, Chile; lperez@academicos.uta.cl
12 Instituto de Alta Investigación, Universidad de Tarapacá, Arica 1000000, Chile
* Correspondence: dlarozen@uta.cl

Abstract: Particle Swarm Optimization (PSO) is a metaheuristic optimization technique based on population behavior, inspired by the movement of a flock of birds or a school of fish. In this method, particles move in a search space to find the global minimum of an objective function. In this work, a modified PSO algorithm written in Fortran 90 is proposed. The optimized structures obtained with this algorithm are compared with those obtained using the basin-hopping (BH) method written in Python (3.10), and complemented with density functional theory (DFT) calculations using the Gaussian 09 software. Additionally, the results are compared with the structural parameters reported from single crystal X-ray diffraction data for carbon clusters C_n(n = 3–5), and tungsten–oxygen clusters, WO_n^{m-}(n = 4–6, $m = 2, 4, 6$). The PSO algorithm performs the search for the minimum energy of a harmonic potential function in a hyperdimensional space $\in \mathbb{R}^{3N}$ (where N is the number of atoms in the system), updating the global best position (g_{best}) and local best position (p_{best}), as well as the velocity and position vectors for each swarm cluster. A good approximation of the optimized structures and energies of these clusters was obtained, compared to the geometric optimization and single-point electronic energies calculated with the BH and DFT methods in the Gaussian 09 software. These results suggest that the PSO method, due to its low computational cost, could be useful for approximating a molecular structure associated with the global minimum of potential energy, accelerating the prediction of the most stable configuration or conformation, prior to *ab initio* electronic structure calculation.

 https://doi.org/10.3390/inorganics13090293

Keywords: PSO; BH; global minimum; DFT; clusters; Fortran 90; Python

1. Introduction

In science, understanding the physical and chemical properties of materials is of great significance, as these properties depend primarily on their molecular structure. Predicting the atomic structure of a molecule using computational methods, starting from an arbitrary initial position of its atoms, is a complex task, since the system may include a large number of local minima on the potential energy surface (PES) [1,2]. In recent decades, several methods have been developed to optimize molecular structures, such as minimum jumping [3,4], simulated annealing [5,6], the random sampling method [7,8], basin-hopping (BH) [9,10], and metadynamics [10,11].

As the system to be studied increases in size with the number of atoms, the problem becomes more complex, increasing the computational cost. An important aspect of the research on structural configurations of clusters is the determination of their lowest potential energy. This step is essential for the study of the properties of the clusters since the structure with the lowest potential energy corresponds to the most stable configuration [12]. This problem is known as global optimization (GO) since it involves the determination of the lowest minimum value of the energy based on the atomic coordinates of the clusters in the PES. Due to the inherent complexity of the problem, it is impossible to solve it using analytical calculation methods. In practice, the problem can only be addressed through numerical techniques. Furthermore, any algorithm that searches for a global minimum must broadly explore the energy landscape, going deep into its most relevant regions, and thus perform a more efficient exploration [13].

GO is an arduous task due to the enormous number of local equilibrium configurations (local energy minima) of the cluster structures in the PES, which increase exponentially with the number of atoms N in the system [2,14–19]. These local minima correspond to different geometric structures, which may present different properties. Thus, the further the optimized structural parameters are separated from those reported experimentally by X-ray diffraction, the less accurate the properties of the modeled structure will be.

Advances in energy landscape theory have helped narrow the search space for the diversity of more stable cluster structures in the PES [20–22]. Some studies reported in the literature implement approximations with harmonic or Hookeian potentials to estimate the structure of some clusters [23,24]. A simple harmonic oscillator is a model used to study a wide range of phenomena, such as pendulum oscillations [25], sound waves [26], and bonds between atoms [27]. The latter considers the mass of the atom as a particle oscillating around an equilibrium point under the action of a restoring force, which is proportional to the distance from the equilibrium point, like the mass-spring system in Hooke's Law, from which the oscillations or vibrations of the bonds are obtained, showing the existence of a vibrational spectrum in the infrared region [18].

One of the methods that is receiving the most attention in the optimization of molecular structures is the Particle Swarm Optimization (PSO) [28]. The algorithm is based on circumstances similar to the movement of a school of fish, a swarm of bees, or a flock of birds in motion by adjusting their positions and velocities. In general, a colony of animals, apparently without a leader, will change their speeds randomly in search of food, following the group member closest to the food source (possible solution). The group of animals optimizes its positions by following the members that have already reached a better orientation. This exploration process is repeated until the best position within the group is obtained, which is considered to be the position vector in the search space, and the velocity vector regulates the subsequent distribution or movement of the particles. The PSO method adjusts or updates the position and velocity of each particle at each step until achieving the best position experienced by the collective or swarm [29,30].

Another alternative is the basin-hopping (BH) method, which is a metaheuristic global optimization method combining local searches with stochastic jumps between different regions of the [9] space. This method involves two-step loops, a perturbation with good candidate solutions, and the application of the local search to the perturbed solution, transforming the complex energy landscape into a collection of basins that are explored by jumps. Monte Carlo random moves and the Metropolis criterion's acceptance or rejection of the solutions perform these jumps. Alternatively, this method is beneficial for solving problems with multiple local minima, such as particle cluster optimization [9,31].

In the present work, we are interested in optimizing two types of clusters: carbon and those formed by oxygen atoms and transition metals. The first group is important for organic, inorganic, and physical chemistry [32]. Furthermore, these molecules are also of great importance in astrophysics, especially regarding the chemistry of carbon stars [33], comets [34], and interstellar molecular clouds [35]. On the other hand, carbon atom clusters are vital elements in hydrocarbon flames [36], and play a crucial role in gas-phase carbon chemistry, acting as precursors in the production of fullerenes, carbon nanotubes, diamond films, and silicon carbides [37–39]. Studying and synthesizing these molecules in the laboratory is challenging due to their high reactivity [40], underscoring the interest in studying the structural information of carbon clusters in theoretical research [41–43]. On the other hand, clusters consisting of oxygen atoms and transition metals with a high oxidation state (W^{6+}, V^{5+}, or Mo^{6+}) have an electronic structure that allows them to act as oxidizing agents, making them good candidates for obtaining new materials. These clusters are precursors of polyoxometalate compounds (POM), which are formed from condensation reactions through the self-assembly of some oxometalates, such as WO_4^{2-}, MoO_4^{2-}, VO_3^-, WO_5^{4-}, or WO_6^{6-} [44,45].

The optimization was carried out using a PSO algorithm written in Fortran 90 to locate global minimum energy structures. We have used the algorithm on clusters consisting of carbon atoms C_n (n = 3–5) and tungsten–oxygen atoms WO_n^{m-} (n = 4–6, $m = 2,4,6$), where n is the number of mol of atoms of the element in the cluster, and m is the charge of the anion. In this model, the atoms have been considered rigid spheres joined by a spring (link) with a harmonic potential, where the restoring force (Hooke's Law) is proportional to the displacement from the equilibrium length. The algorithm was based on the search for the global minimum of the potential energy function of the aforementioned clusters in a multidimensional hyperspace $\in \mathbb{R}^{3N}$ (where N is the number of atoms in the system). The motions of the conformers of the clusters are guided by their own most favorable known atomic positions as well as by the best-known position of the entire swarm in the search space $\mathbb{R}^{3N}$. We have implemented a basin-hopping (BH) method and performed the DFT calculations through the Gaussian 09 software to validate our results. Furthermore, for the case of carbon clusters, we have compared with the results obtained by Jana and collaborators [46] who used a PSO algorithm written in Python, which combines an evolutionary subroutine with a variational optimization technique through an interface of the PSO algorithm with the Gaussian 09 software.

The article is divided into the following sections: Section 2, Discussion and Results, presents the results and compares them with the literature; Section 3, Methodology, presents the details of the algorithm; and Section 4, Conclusions, outlines the main results.

2. Discussion and Results

Tables S1–S4 of the supplementary material present the initial structures of the conformers of 10 clusters (swarms) of three-, four-, and five-carbon atoms, from which Jana et al. initiated their study of the PSO algorithm [46]. Using structures optimized with the commercial software Gaussian 09 as a reference, we observed that our results with PSO and BH showed good agreement with the conformations of the optimized structures reported by Jana et al. [46], for carbon clusters of 3–5 atoms (C_3–C_5; see Table 1).

Table 1. Representation of optimized geometric structures of C_3–C_5 carbon clusters using the BH and PSO algorithms, and Gaussian 09 software.

Cluster	BH	PSO		Gaussian 09
		Jana et al. [46]	In This Work	
C_3				
C_4				
	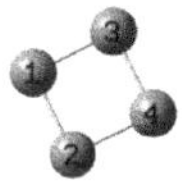			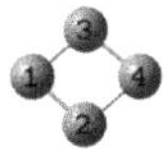
C_5				

The Python-based PSO algorithm by Jana et al. [46], implemented with a swarm of 10 clusters, operates in synergy with the Gaussian software (version 09). This commercial software optimizes cluster structures at each iteration through the gradient method; however, the article by Jana [46] does not specify whether these optimizations are sequential or parallel. At every step:

1. Gaussian software provides optimized atomic coordinates for each cluster to the PSO algorithm.
2. The PSO modifies these coordinates through displacements in the hyperspace $\mathbb{R}^{3N}$.
3. The new coordinates are fed back to Gaussian for the subsequent iteration.
4. Gaussian optimizes cluster structures in Hilbert space $\mathcal{H}$, using the Schlegel algorithm [47], which begins by minimizing the electronic energy density functional using the conjugate gradient method at each step. The optimized cluster structure is obtained by minimizing the gradient of the electronic energy concerning the nuclear coordinates, through a series of single-point calculations in the PES, applying the Hellmann–Feynman theorem [48].
5. The new atomic positions found by the Gaussian software are used by the PSO algorithm to locate the optimal nuclear coordinates, for each cluster (both p_{best} and g_{best}), based on the electronic energy system.
6. The cycle repeats during each optimization step performed by the Gaussian software until the electronic energy of the clusters converges.

The critical dependence lies in the fact that the PSO algorithm by Jana [46] lacks an intrinsic objective function. It exclusively utilizes the electronic energy computed by Gaussian software as a reference to guide atomic displacements in the hyperspace $\mathbb{R}^{3N}$. This dependence implies two fundamental limitations:

- If the electronic energy of the system converges to a local minimum on the PES through Gaussian software, the PSO algorithm is hopelessly trapped in this minimum.

- The absence of an intrinsic model to determine the energy of the system in the hyperspace $\mathbb{R}^{3N}$ restricts the PSO to functioning as an atomic configuration generator, without independent evaluation capability during the structural optimization process.

In essence, the algorithm by Jana et al. [46] acts as a conformational exploration mechanism whose efficacy is contingent upon external quantum chemical computations from Gaussian software.

However, in this work, the PSO algorithm independently optimizes the structures of 10 cluster conformers simultaneously, without executing the Gaussian software. The algorithm models atoms as rigid spheres connected by springs and minimizes the potential energy (Equation (5)) for each conformer to determine its optimal positions in the search space $\mathbb{R}^{3N}$. At each step t, the algorithm updates the best positions p_{best} and g_{best} until the potential energy of each conformer converges. This convergence is achieved when the velocity and positions vectors satisfy $\vec{v}_{i,j}^{\,t+1} \rightarrow 0$ and $\vec{x}_{i,j}^{\,t+1} \rightarrow \vec{x}_0$, respectively (Equations (3) and (4)). During PES exploration, each cluster individually stores its atomic positions from iteration $t = 1$ to $t = n - 1$; this enables each cluster to retain its exploration history of the hyperspace $\mathbb{R}^{3N}$, providing a global perspective of the energy landscape. Consequently, the possibility of cluster swarm trapping in local minima during geometric optimization is significantly reduced.

The BH algorithm is a global optimization method designed to find the global minimum energy in systems with multiple local minima. This approach combines local optimization with random jumps to explore the configuration space of molecular structures associated with local minima. The algorithm starts by assigning trial positions to each atom, and the structure is defined using the Atomic Simulation Environment (ASE) library. An energy calculator, such as the effective medium theory (EMT) model, is then used to compute the total energy of the configuration. Restrictions based on Hooke's Law were applied between specific pairs of atoms, imposing a restoring force proportional to the deviation from a target distance, to maintain physically reasonable structures during the optimization process. The algorithm initiates local optimization from the trial structure configuration using the Limited-Memory Broyden–Fletcher–Goldfarb–Shanno (L-BFGS) algorithm [49]. This step searches for the local energy minimum near the current configuration. Once a local minimum is found, the algorithm makes a random jump in the configuration space. This jump allows the exploration of new configurations that may have lower energies. The energy of the new configuration is evaluated, and if it is lower than that of the previous configuration, it is accepted as the new current configuration. The new configuration remains acceptable under specific conditions, determined by the system temperature, which enables the algorithm to escape deep local minima when the energy is higher.

Tables S5–S7 in the supplementary material present the initial structures of the 10 conformers of the tungsten–oxygen clusters WO_4^{2-}, WO_5^{4-}, and WO_6^{6-} used for geometry optimization. Table 2 displays the optimized structures obtained with the BH and PSO algorithms (this work), the commercial Gaussian 09 software, and the experimentally reported structures from single-crystal X-ray diffraction [50–52]. The structural conformations of the tungsten–oxygen clusters WO_n^{m-} (n = 4–6, $m = 2,4,6$) indicate that the PSO and BH algorithms provide a close approximation to both the commercial Gaussian 09 software results and the experimentally reported structures from single-crystal X-ray diffraction (see Table 2).

The system energy was computed using a Hookeian potential (Equation (5)), with explicit omission of the electrostatic interaction term. This approach is justified because electrostatic contributions are implicitly incorporated in the force parameters (bond constants K_r and K_a in Equation (5)), which were adjusted using experimental Differential Scanning Calorimetry (DSC) data [53–56]. Consequently, the force constants inherently account for electrostatic effects associated with interatomic bond energies.

Table 2. Representation of the optimized geometric structures of tungsten and oxygen clusters WO_n^{m-} (n = 4–6, m = 2, 4, 6), obtained using the BH and PSO (this work) algorithms, the Gaussian 09 commercial software, and those reported experimentally via single-crystal X-ray diffraction.

Cluster	BH	PSO	Gaussian 09	X-Ray [50–52]
WO_4^{2-} W O		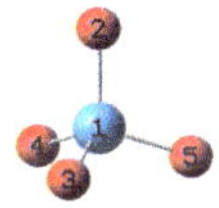	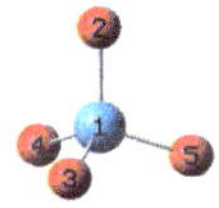	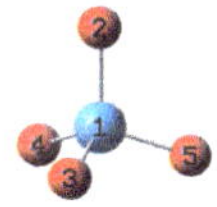
WO_5^{4-}		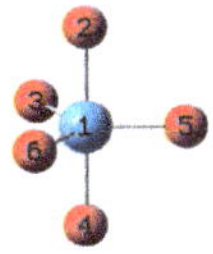	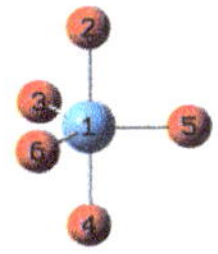	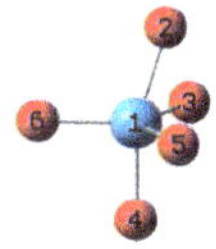
WO_6^{6-}	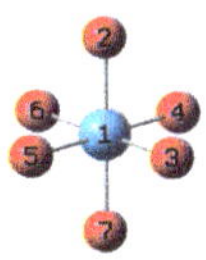	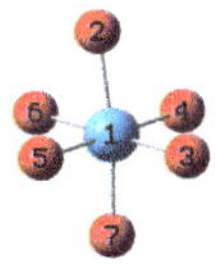	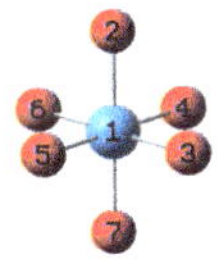	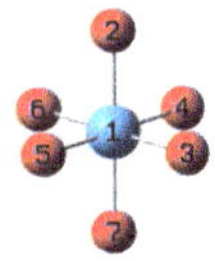

The structural parameters obtained with the PSO algorithm, shown in Table 3 for the C_3–C_5 carbon clusters, are comparable to those derived from the BH algorithm and the Gaussian 09 software. The optimization of the three-carbon cluster showed the highest accuracy in its average bond lengths and angles, except for the C–C single bond length predicted by the BH algorithm, which was slightly elongated compared to those from the PSO algorithms (Jana et al. [46] and the one in this work) and the commercial software Gaussian 09. The average bond lengths for both cyclic and acyclic four-carbon cluster structures showed good agreement, with deviations of 0.02–0.30 Å relative to the Gaussian 09 results. However, a discrepancy of 47° was observed in the angle of the acyclic C_4 cluster structures optimized using the PSO algorithm from Jana et al. [46]. For the C_5 cluster, the average bond angles (~180.00°) obtained with our PSO algorithm demonstrated higher accuracy than those from the algorithm reported by Jana et al. [46], aligning closely with the structural parameters derived from Gaussian 09 optimizations.

Table 3. Average bond lengths (*r*) and angles (*ang*) of (C_3–C_5) carbon clusters, optimized using the BH and PSO algorithms (Jana et al. [46] and this work), and the Gaussian 09 commercial software.

Reference	Cluster	*r* [a]	*ang* [b]
BH	C_3	1.58	180.00
Jana et al. [46] [c]		1.29	179.89
In this work [c]		1.29	179.98
Gaussian 09		1.29	180.00
BH	C_4 (acyclic)	1.60	179.99
Jana et al. [46] [c]		1.32	132.73
In this work [c]		1.35	178.74
Gaussian 09		1.30	179.94
BH	C_4 (cyclic)	1.58	90.00
Jana et al. [46] [c]		1.49	88.31
In this work [c]		1.37	90.00
Gaussian 09		1.45	90.00
BH	C_5	1.52	179.60
Jana et al. [46] [c]		1.29	156.56
In this work [c]		1.29	180.00
Gaussian 09		1.29	179.97

[a] r = Length (Å); [b] *ang* = Angle (°); [c] PSO.

Table 4 presents the average bond lengths and angles of tungsten–oxygen clusters WO_n^{m-} (n = 4–6, $m = 2,4,6$), optimized using BH and our PSO algorithms, Gaussian 09 software, and the values reported from single-crystal X-ray diffraction studies [50–52]. The results show that the structural parameters of the WO_n^{m-} (n = 4–6, $m = 2,4,6$) tungsten–oxygen clusters obtained via the BH and PSO methods are highly consistent with those obtained with Gaussian 09. Furthermore, the calculated average bond lengths and angles closely match the experimental XRD data [50–52]. Notably, only the optimized angle ang_3 of the WO_5^{4-} cluster exhibited a deviation 20.00° compared to the XRD result, likely due to intermolecular interactions associated with crystal lattice packing [51,52].

Table 4. Average bond lengths (r) and angles (ang_s) of tungsten–oxygen clusters WO_n^{m-} (n = 4–6, $m = 2,4,6$), obtained through structure optimizations using the BH and PSO algorithms, the commercial software Gaussian 09, and single-crystal X-ray diffraction [50–52].

Reference	Cluster	r [a]	ang_1 [b]	ang_2 [b]	ang_3 [b]
BH	WO_4^{2-}	1.96	109.47	–	–
PSO (in this work)		1.73	109.53	–	–
Gaussian 09		1.80	109.47	–	–
X-ray [50–52]		1.79	110.26	–	–
BH	WO_5^{4-}	1.97	90.00	120.00	180.00
PSO (in this work)		1.92	90.00	120.00	179.88
Gaussian 09		1.93	90.00	120.00	179.95
X-ray [50–52]		1.86	93.52	120.00	158.22
BH	WO_6^{6-}	1.97	90.00	180.00	–
PSO (in this work)		2.24	91.10	175.24	–
Gaussian 09		1.99	90.00	180.00	–
X-ray [50–52]		1.95	92.64	174.87	–

[a] r = Length (Å); [b] ang_s = Angle (°).

The evolution of the geometric structure optimization for cluster $n°$ 1 (composed of three carbon atoms, C_3) is illustrated in Figure 1, which displays snapshots of the process across iterations. The cluster bonds (r_{13} and r_{23}) are initially elongated during the first m (Figure 2b), after they gradually decrease to reach a minimum value of 1.29 Å m and Table 5). Concurrently, the angle formed by the three carbon atoms undergoes minor variations, increasing from 110.00 to 179.98° as the iterations progress. This angle stabilizes near 180.00° when the potential energy is minimized (Figure 2a,b and Table 5). Similar harmonic motion trends—though with higher frequency—are observed in the C–C bond lengths and the angle increase (denoted as Ang) for clusters $n°$ 8 and 10 (Figure 2c–f and Table 5). Notably, within the swarm of 10 clusters, each conformation exhibits distinct dynamics, geometric configurations, and energy profiles during optimization via the PSO algorithm (see Figures 2–6 and Table 5).

Table 5. Bond lengths (r), bond angles (ang), and potential energies (V) of C_3 clusters n° 1, 8, and 10, optimized using the PSO algorithm.

Cluster	r_{13} (Å)	r_{23} (Å)	ang_1 (°)	V (Ha) × 10^{-5}
1	1.29149	1.29341	179.80617	4.48600
8	1.29336	1.29080	179.71400	7.62570
10	1.30979	1.26546	178.49936	4.46674

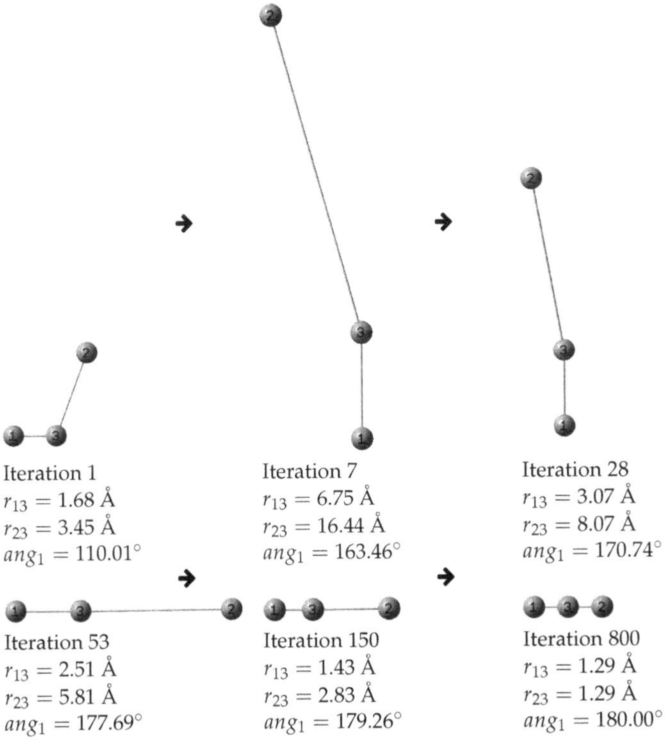

Figure 1. Conformational changes of carbon cluster $n°$ 1 of 3 atoms (C_3) as a function of iterations number (1, 7, 28, 53, 150, and 800) during its geometric optimization via the PSO algorithm.

Figures 3–5 show 3D relief maps of the angle Ang (a) and the potential energy V (b), with their respective magnifications (c,d), as functions of the lengths r_{13} and r_{23}, for the C_3 clusters n° 1, 8, and 10, over 600 iterations. Each C_3 conformer within the swarm exhibits distinct energy landscapes. However, all converge to the same minimum potential energy value, V_0, as a function of the positional changes of the clusters during the dynamics of the swarm in the 3N-dimensional hyperspace $\mathbb{R}^{3N}$. The blank spaces correspond to regions where bond lengths, angles, and energies of the C_3 cluster conformations are undefined in the energy landscape during geometry optimization. The highest Ang values are located at the peaks (red values near 180.00° in parts (a,c) of Figures 3–5), protruding from the paper plane. In contrast, the lowest values are found in the valleys (dark blue and black regions near 100.00° in the same figures), oriented towards the paper plane. These features occur when the C–C bond lengths, r_{13} and r_{23}, approach 1.29 Å (see Table 5 and Figures 3–5).

Similarly, the highest potential energy values V are observed at the peaks (red regions in parts (b) of Figures 3–5). Additional intermediate values lie on the slopes (gray regions in the magnified parts (d) of Figures 3–5), extending outward from the paper plane. The lowest energy values, however, are concentrated in the valleys of the energy landscape (dark blue regions in parts (b,d) of Figures 3–5), oriented toward the paper plane. These observations correspond to the geometry optimization process of the C_3 cluster structures.

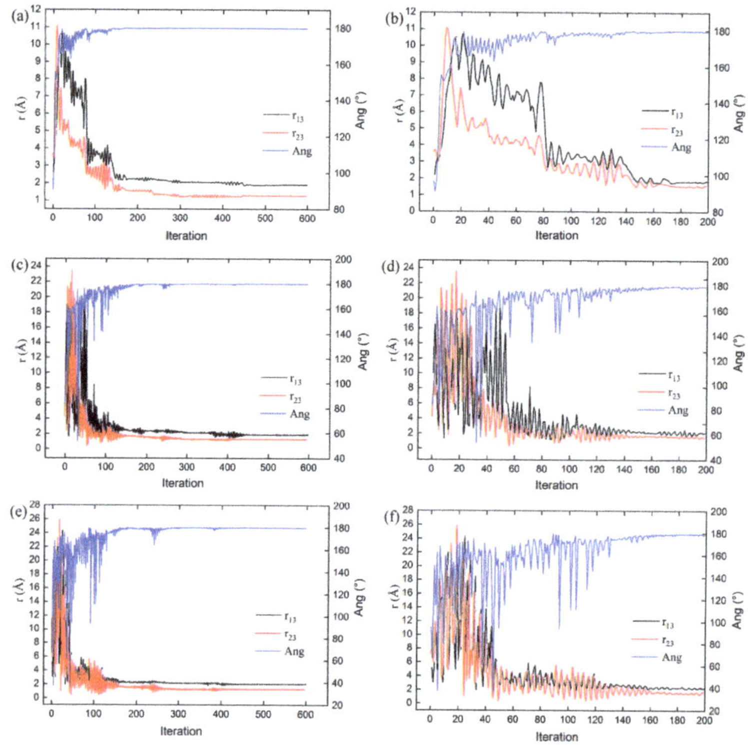

Figure 2. Evolution of bond lengths r_{13}, r_{23}, and angle ang_1 for three-atom carbon clusters (C_3) n° 1 (**a**), 8 (**c**), and 10 (**e**) over 600 iterations, along with their respective zoomed-in views near 200 iterations (**b**,**d**,**f**) during geometric optimization using the PSO algorithm.

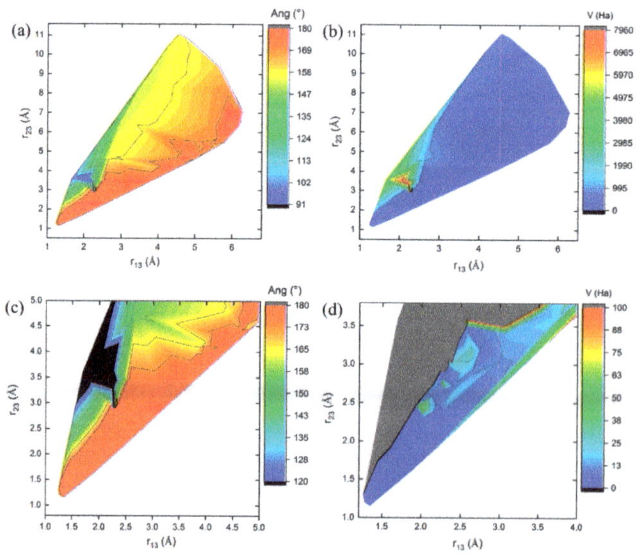

Figure 3. Maps of: (**a**) Angles Ang (°), (**b**) potential energy V (Ha), (**c**,**d**) magnifications near 5 Å of maps (**a**) and (**b**), respectively, as a function of bond lengths r_{13} and r_{23} of the cluster n° 1 of three-carbon atom (C_3), over 600 iterations.

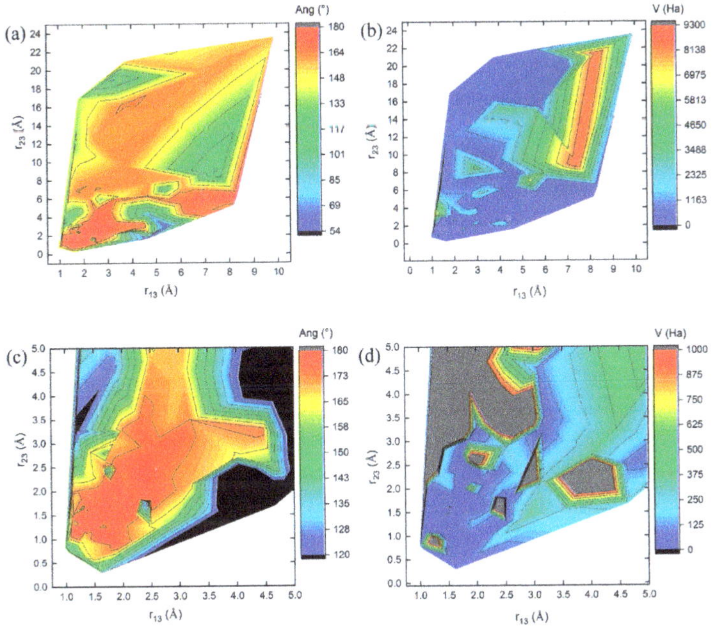

Figure 4. Maps of: (**a**) Angles Ang (°), (**b**) potential energy V (Ha), (**c**,**d**) magnifications near 5 Å of maps (**a**) and (**b**), respectively, as a function of bond lengths r_{13} and r_{23} of the cluster n° 8 of three-atom carbon (C_3), over 600 iterations.

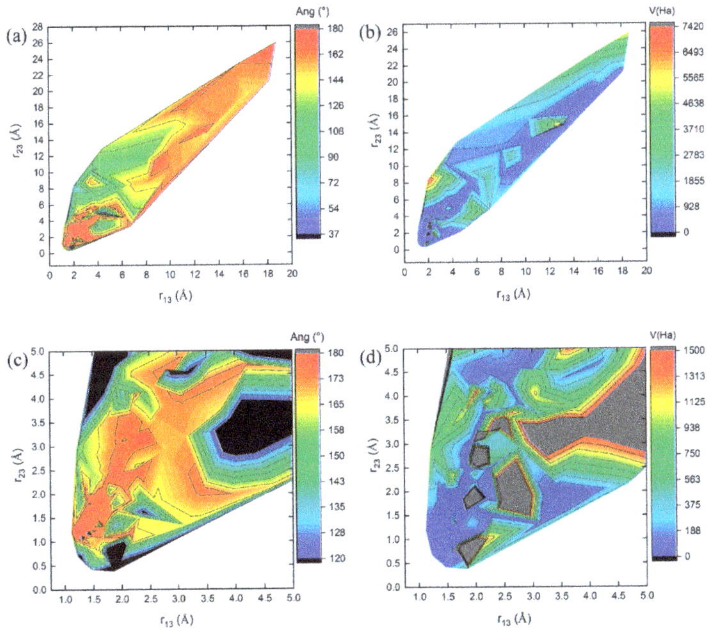

Figure 5. Maps of: (**a**) Angles Ang (°), (**b**) potential energy V (Ha), (**c**,**d**) magnifications near 5 Å of maps (**a**) and (**b**), respectively, as a function of bond lengths r_{13} and r_{23} of the cluster n° 10 of three-carbon (C_3) cluster, over 600 iterations.

Figure 6 shows the potential energy behavior of the swarm of 10 three-carbon-atom clusters over the first 120 iterations. The cluster n° 8 exhibits an intermediate potential energy during the first 60 iterations, subsequently reaching one of the lowest potential energy values in the swarm as its trajectory progresses. However, clusters 4, 5, 9, and 10 exhibit conformations with alternating high potential energy values. At the same time, the remaining conformers maintain a potential energy close to the average for the swarm during the energy minimization of each conformer.

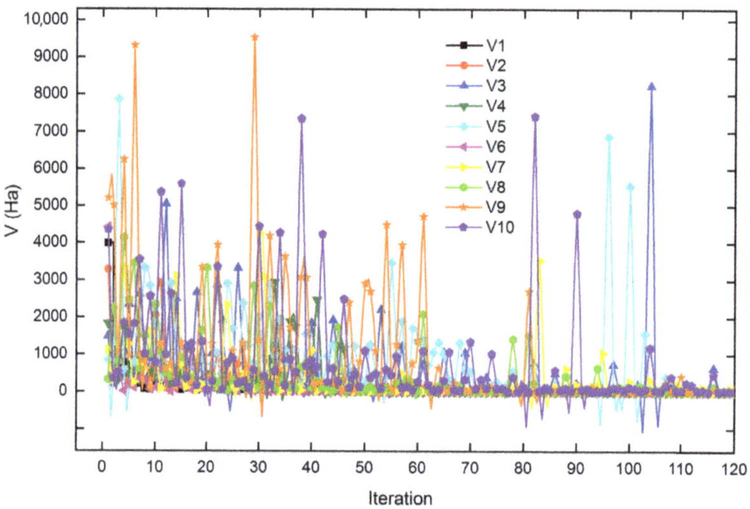

Figure 6. Behavior of the potential energy V of a swarm of 10 clusters of three-carbon atoms as a function of 120 iterations.

The potential energy behavior of the clusters, as their atoms undergo positional changes, illustrates the various conformations adopted during the search for the global potential energy minimum, a process often characterized by cooperative swarm dynamics. Conformers with higher potential energy require larger displacements within the $\mathbb{R}^{3N}$ hyperspace (relative to the average position of the swarm) to contribute to the energy minimization potential [Equation (5)]; this accounts for the bond elongation observed in the three-carbon-atom cluster $n^{\circ}1$ during early iterations. Only by iteration 22 (Figures 1 and 2) do these bond lengths begin to shorten, demonstrating the integration of the cluster $n^{\circ}1$ conformer into the swarm collective, where it assumes positions similar to the other swarm conformers, thereby reducing its potential energy as the system progresses toward convergence.

Table 6 presents the calculated relative electronic energy values for carbon, tungsten, and oxygen cluster structures studied in this work. To evaluate the precision of the electronic energies for the structures optimized with the PSO and BH methods, we used the absolute electronic energy obtained from Gaussian as a reference (Table 6). Table 6 shows the difference between the electronic energy calculated by Gaussian in a single-point calculation for a optimized structure obtained with PSO/BH, and the corresponding electronic energy obtained directly from Gaussian. In this way, values closest to zero indicate higher precision in the respective optimization method. The small relative electronic energy differences ($\sim 10^{-5}$– 10^{-1} Ha) between BH/PSO-optimized and DFT-optimized structures suggest that both BH and PSO exhibit high accuracy in predicting the electronic energies of these molecular structures, carbon, tungsten and oxygen clusters systems. However, PSO demonstrated slightly higher accuracy compared to BH, particularly for carbon clusters

C_n (n = 3–5) and the tungsten–oxygen clusters WO_n^{m-} (n = 4–6, $m = 2,4,6$), where lower relative energy electronic values were observed.

Table 6. Single-point relative electronic energies (Ha) obtained with Gaussian 09 for carbon, tungsten, and oxygen atom clusters, based on structures optimized using the BH, PSO, and DFT methods.

Cluster		BH [a]	PSO (In This Work) [b]	PSO (Jana et al.) [b] [46]	Gaussian 09 [c]
C_3		1.64×10^{-1}	1.20×10^{-5}	0.00	0.00
C_4	acyclic	3.02×10^{-1}	5.79×10^{-2}	2.66×10^{-2}	0.00
	cyclic	2.69×10^{-1}	2.26×10^{-1}	1.57×10^{-2}	0.00
C_5		3.54×10^{-1}	5.85×10^{-4}	0.00	0.00
WO_4		6.41×10^{-2}	1.05×10^{-1}	–	0.00
WO_5		1.01×10^{-2}	9.16×10^{-3}	–	0.00
WO_6		7.60×10^{-5}	1.29×10^{-2}	–	0.00

[a] Relative electronic energy calculated using Gaussian 09 at a single point, based on its BH-optimized structure.
[b] Relative electronic energy calculated using Gaussian 09 at a single point, based on its PSO-optimized structure.
[c] Relative electronic energies of the optimized structures, calculated using DFT with Gaussian 09.

The computational cost of the PSO, BH, and molecular mechanics (using the Dreiding force field in Gaussian), was assessed by tracking the energy minimization of cluster C_5 during geometric optimization as a function of optimization steps (Figure 7b,d), starting from the initial structure of cluster C_5 (Figure 7a). Figure 7 highlights that the structural optimization of the C_5 cluster required the least time with the Gaussian software (69 steps), followed by the PSO algorithm (141 steps), and the BH algorithm (1592 steps). All calculations employed the same initial structure of the C_5 cluster. Although Gaussian software optimizes the C_5 cluster structure more quickly, it predicts triple bond lengths (1.19 Å) for the carbon atoms inside the cluster. This value differs slightly from the results obtained (1.29 Å in Table 3) when the same structure is optimized at the DFT level of theory using the same software. However, our PSO algorithm predicts double bonds (1.29 Å) in the optimized structure of the C_5 cluster (see Figure 7b). Additionally, the structure of the C_5 cluster was optimized using the UFF (Unified Force Field) in Gaussian software, but an unexpected cyclic structure was obtained.

Contrarily, it is important to note that the PSO algorithm implemented by Jana et al. [46] uses Gaussian 09 software to calculate the atomic positions at each step of the geometric optimization of cluster structures via DFT. In contrast, our PSO algorithm does not depend on any external software to determine atomic positions during the simultaneous optimization of ten conformer structures. Instead, it relies solely on Equations (1)–(4) to locate each atom of each cluster within the $\mathbb{R}^{3N}$ search hyperspace at each iteration and uses the potential energy of each cluster [Equation (5)] as the objective function to be minimized.

A precise understanding of the structures of these clusters is crucial, particularly for compounds of tungsten and oxygen atoms, as they serve as fundamental building blocks for the growth of larger metal oxide arrangements. Through self-assembly, larger compounds, such as polyoxometalates, are formed, which have numerous applications due to their electronic and magnetic properties, which in turn depend on their structural arrangement [57,58].

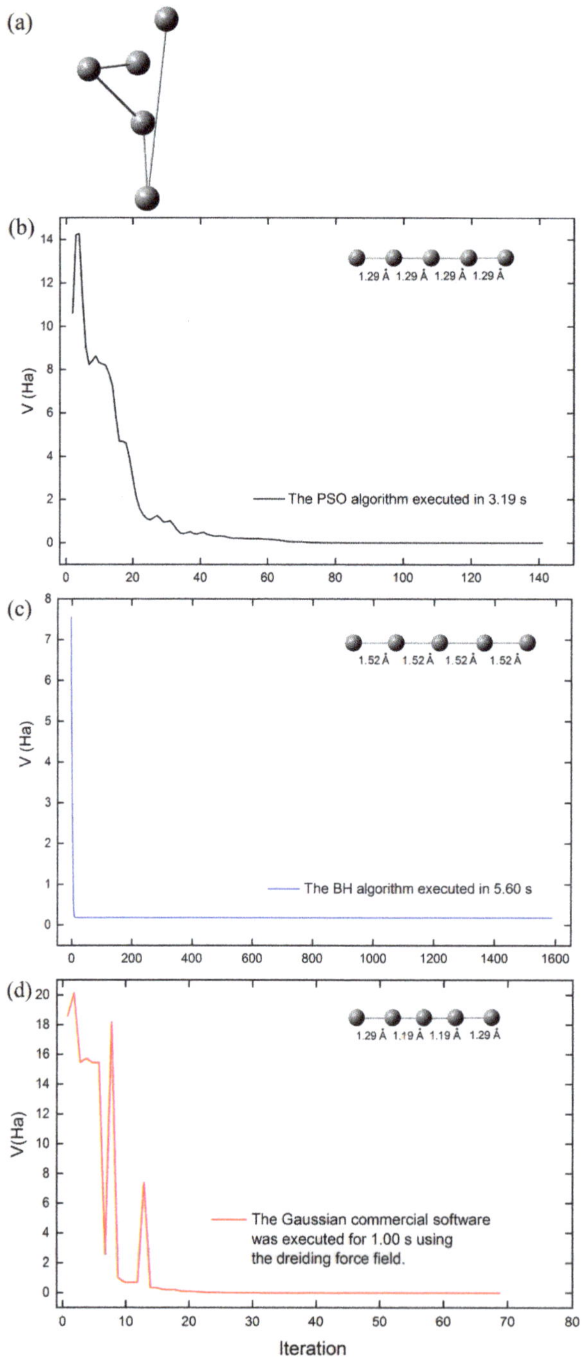

Figure 7. (**a**) Initial structure of the C_5 cluster. Energy minimization versus optimization steps for: (**b**) PSO (in this work; maximum interatomic force threshold: 1.00×10^{-7} Ha/Å), (**c**) BH (maximum interatomic force threshold: 5.00×10^{-7} Ha/Å) and (**d**) Dreiding force field via molecular mechanics (MM) in Gaussian software (maximum interatomic force threshold: 4.50×10^{-4} Ha/Å).

3. Methodology

3.1. Algorithm and Computational Details

The PSO is an algorithm that explores a set of randomly generated solutions, where each structural conformer learns from its own experience and that of other conformers through their movements during optimization [59]. All conformer structures are located within a hyperspace $\in \mathbb{R}^{3N}$ [60], where N is the number of atoms in the cluster, i is the cluster index ($i = 1, \ldots, M$), and j ($j = 1, \ldots, 3N$) is the number of dimensions of the hyperspace $\mathbb{R}^{3N}$ (Figure 8). Figure 8 shows the equivalence between the coordinates $x^t_{1,1}, \ldots, x^t_{1,9} \in \mathbb{R}^{3N}$ and $x_1, y_1, z_1, \ldots, x_3, y_3, z_3 \in \mathbb{R}^3$ of the position vector $\vec{x}^{\,t}_{1,1}$ for the conformer of the cluster $n°$ 1 (with three carbon atoms) in the swarm. From Equations (1)–(4), the velocities and positions of the conformers of each cluster in the hyperspace are defined, and Equation (5) represents the potential energy.

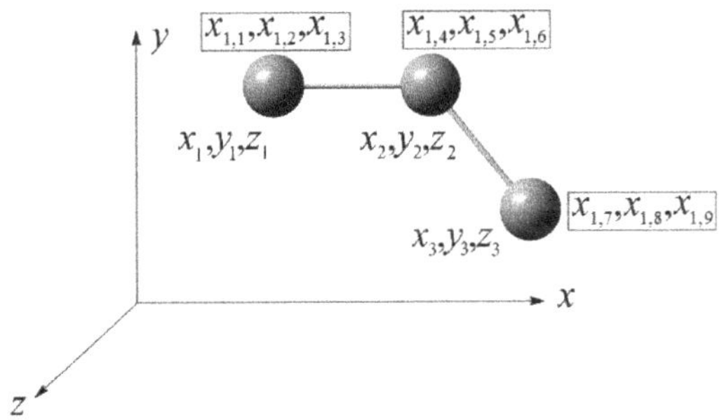

Figure 8. Schematic representation of the equivalence between the coordinates $x^t_{1,1}, \ldots, x^t_{1,9} \in \mathbb{R}^{3N}$ and $x_1, y_1, z_1, \ldots, x_3, y_3, z_3 \in \mathbb{R}^3$ of the position vector $\vec{x}^{\,t}_{1,1}$ of the conformer of cluster $n°$ 1 of three carbon atoms.

$$\vec{x}^{\,t}_{i,j} = x^t_{i,j}, \ldots, x^t_{M,3N} \tag{1}$$

$$\vec{v}^{\,t}_{i,j} = v^t_{i,j}, \ldots, v^t_{M,3N} \tag{2}$$

$$\vec{v}^{\,t+1}_{i,j} = w \cdot \vec{v}^{\,t}_{i,j} + d_1 \cdot \varepsilon_1 \left(p_{best} - \vec{x}^{\,t}_{i,j} \right) + d_2 \cdot \varepsilon_2 \left(g_{best} - \vec{x}^{\,t}_{i,j} \right) \tag{3}$$

$$\vec{x}^{\,t+1}_{i,j} = \vec{x}^{\,t}_{i,j} + \vec{v}^{\,t+1}_{i,j} \tag{4}$$

$$V^t_i \left(x^t_{i,j}, \ldots, x^t_{M,3N} \right) = \frac{1}{2} \left[\sum_{k=1}^{p} K_r (r_k - l_k)^2 + \sum_{s=1}^{q} K_a (ang_s - \theta_s)^2 \right] \tag{5}$$

where $\vec{x}^{\,t}_{i,j}$ and $\vec{x}^{\,t+1}_{i,j}$ are the positions of the cluster i in iterations t and $t+1$ (Equations (1) and (4)), and $\vec{v}^{\,t}_{i,j}$ and $\vec{v}^{\,t+1}_{i,j}$ are the velocities of the cluster i in iterations t and $t+1$ (Equations (2) and (3)), respectively. The coefficients ε_1 and ε_2 in Equation (3) are randomly selected within the interval [0, 1]. The inertia coefficient w governs the tendency of the conformer to remain in its current position, while d_1 and d_2 (adjustable if necessary) correspond to the individual and global acceleration coefficients, respectively. These coefficients guide the positions of the clusters in the hyperspace $\mathbb{R}^{3N}$ to ensure convergence, allowing all candidate solutions to reach the global minimum of the potential energy V_0 efficiently. Table 7 shows the allowed values for these parameters, which fall within the optimization ranges reported in previous studies [46,61–63]. In these studies, the parameters have demonstrated stability during the process and ensured energy convergence of the cluster structures. The variables g_{best} and p_{best} are the best positions of the swarm and each cluster, respectively, in each iteration t (Equation (3)).

Table 7. PSO parameters.

Parameters	Value
Number of particles or clusters, i	10.00
Inertia coefficient, w	0.90
Individual coefficient of acceleration, d_1	1.00–1.50
Global coefficient of acceleration, d_2	1.00–1.50
Random coefficients (ε_1 and ε_2)	0.00–1.00

Equation (5) shows that r_k represents the k bonds and ang_s are the s bond angles in the cluster, where r_k and ang_s are correlated through their components $x_{i,j}^t, \ldots, x_{M,3N}^t$. In this equation, K_r (9.53 mdyn/Å for C=C bonds and 9.73 mdyn/Å for W−O bonds) and K_a (0.372 mdyn/degree) denote the force constants of the k bonds and the angles s in the studied systems. The values l_k and θ_s correspond to the lengths and angles of the experimental bonds, respectively, for each cluster [64,65]. Table 8 shows the values of θ_s when $l_k = 1.29$ Å for clusters of three and four atoms.

Table 8. θ_s values for the clusters of three and four carbon atoms ($l_k = 1.29$ Å).

Number of Atoms	Structure	$\theta_s(^\circ)$
3		180.00
		60.00
4		180.00
	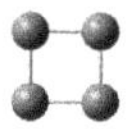	90.00

The learning acquired by the conformers is referred to as collective learning, while that acquired by an individual conformer is termed cognitive learning. Starting from the initial structures, a set of random atomic coordinates was considered, where each conformer was assigned its position $\vec{x}_i^t$ and trial velocity $\vec{v}_i^t$. During optimization, the position $\vec{x}_i^t$ and velocity $\vec{v}_i^t$ vectors are updated using the Equations (3) and (4) at each iteration t, to locate the global minimum energy configuration. As a result of collective learning, each conformer stores in its memory the best position obtained by itself or any other conformer in the swarm during iteration t, denoted as g_{best} (see Equation (3)). The g_{best} is selected by comparing the components $x_{i,j}^t$ of the vector $\vec{x}_{i,j}^t$ for each conformer in the swarm (10 conformers, where $i = 1, \ldots, 10$) at each iteration t. Throughout the process, each conformer stores in its memory the best position obtained up to iterations $n - 1$, denoted as p_{best} [30]. The p_{best} values are selected by each conformer i in the swarm by comparing its components $x_{i,j}^t$ (from vector $\vec{x}_{i,j}^t$) across iterations $t = 1, \ldots, n-1$ stored in its memory.

The change in direction of any conformer is determined by its velocity, which represents the rate of change of its position concerning iteration t. In a $3N$-dimensional search space, the position of the i-th conformer in the swarm at iteration t is represented by the vector $\vec{x}_i^t = \left(x_{i,1}^t, x_{i,2}^t, \ldots, x_{M,3N}^t\right)$, and its velocity by the vector $\vec{v}_i^t = \left(v_{i,1}^t, v_{i,2}^t, \ldots, v_{M,3N}^t\right)$ [30]. The potential energy function of the system V_i^t follows Hooke's Law [Equation (5)], where atoms are considered rigid spheres connected by a spring. The positions p_{best} and g_{best} in Equation (3) correspond to the minimum value of the objective potential function $V_0\left(x_{i,j}^t\right) =$

min $V_i^t\left(x_{i,j}^t,\ldots,x_{M,3N}^t\right)$ [Equation (5)] calculated iteratively from $t = 1$ to n. At each iteration t, the updated components ($x_{i,j}^{t+1}$ and $v_{i,j}^{t+1}$) of the atoms in the conformers are obtained utilizing Equations (3) and (4). Vectors $\vec{x}_i^{\,t}$ and $\vec{v}_i^{\,t}$ are updated in each iteration t until their components $x_{i,j}^t$ and $v_{i,j}^t$ converge ($\vec{x}_{i,j}^{\,t} \rightarrow \vec{x}_0$ and $\vec{v}_{i,j}^{\,t} \rightarrow 0$) during the search for the global minimum of the objective function (potential energy, V_0) for each conformer.

3.2. Software

This work used the initial positions of carbon atoms C_n (n = 3–5) in the cluster structures described by Jana et al. [46]. However, the initial positions of the tungsten and oxygen atoms in WO_n^{m-} (n = 4–6, $m = 2,4,6$) were randomly generated. The PSO algorithm written in Fortran 90, the commercial software Gaussian 09, and the Atomic (ASE) BH simulation environment in Python 3.10 were used to optimize the structures of the carbon, tungsten and oxygen atom clusters. In the PSO algorithm, a Hooke potential was utilized, while in the ASE-Python algorithm, the basin-hopping (BH) global optimization method was applied using an effective medium theory (EMT) potential to model the interaction between atoms. The force constants used were K = 9.53 mdyn/Å for C=C bonds and 9.73 mdyn/Å for W–O bonds. These parameters enabled comparative analysis of the optimized structures.

In the Gaussian 09 software, the 6-311+G** basis set was used for carbon (C) and oxygen (O) atoms [66,67], along with the B3LYP exchange-correlation functional. The Los Alamos Effective Core Potential (ECP) combined with the LANL2DZ double zeta basis set [68,69] and its corresponding pseudopotential was used to approximate the electron of the tungsten (W) atom. Subsequently, single-point electronic energy calculations were carried out on the optimized cluster structures (obtained via PSO, BH, and DFT methods) using DFT as implemented in Gaussian 09 software [70].

The PSO and BH algorithms do not rely on the Gaussian 09 software to perform geometric optimization of the studied clusters' structures. These algorithms were compiled and executed on a high-performance computing node equipped with two Intel Xeon E5-2660 v2 processors (Intel Corporation, Santa Clara, CA, USA), each with 10 cores operating at 2.20 GHz, and 48 GB of RAM.

3.3. PSO Algorithm

The following steps describe the cluster structure prediction algorithm using the PSO technique:

1. An initial set of cluster structures with random atomic coordinates is generated without any symmetry restrictions.
2. An improved set of atomic coordinates for the conformers is sought in hyperspace $\mathbb{R}^{3N}$.
3. At each step or iteration t, the potential energy $V_i^t\left(x_{i,j}^t,\ldots,x_{M,3N}^t\right)$ is calculated for each conformer in the swarm of M clusters.
4. The convergence criteria are verified under the condition V_i^t = constant.
5. The velocity vectors ($\vec{v}_i^{\,t}$) and position vectors ($\vec{x}_i^{\,t}$) of each conformer in the swarm are updated in hyperspace $\mathbb{R}^{3N}$.
6. The best global position (g_{best}) is obtained by comparing and selecting the conformer with the lowest potential energy V_i^t in the swarm at each iteration t.
7. Each conformer i in the swarm stores its best individual position (p_{best}) by comparing its lowest potential energy, V_i^t, obtained iteratively from $t = 1$ to $n - 1$.
8. The output structures generate a new set of initial structures for the subsequent iteration t, continuing until the lowest potential energy V_i^t for each structural conformer converges to a constant value.

Figure 9 shows the general flowchart of the PSO algorithm.

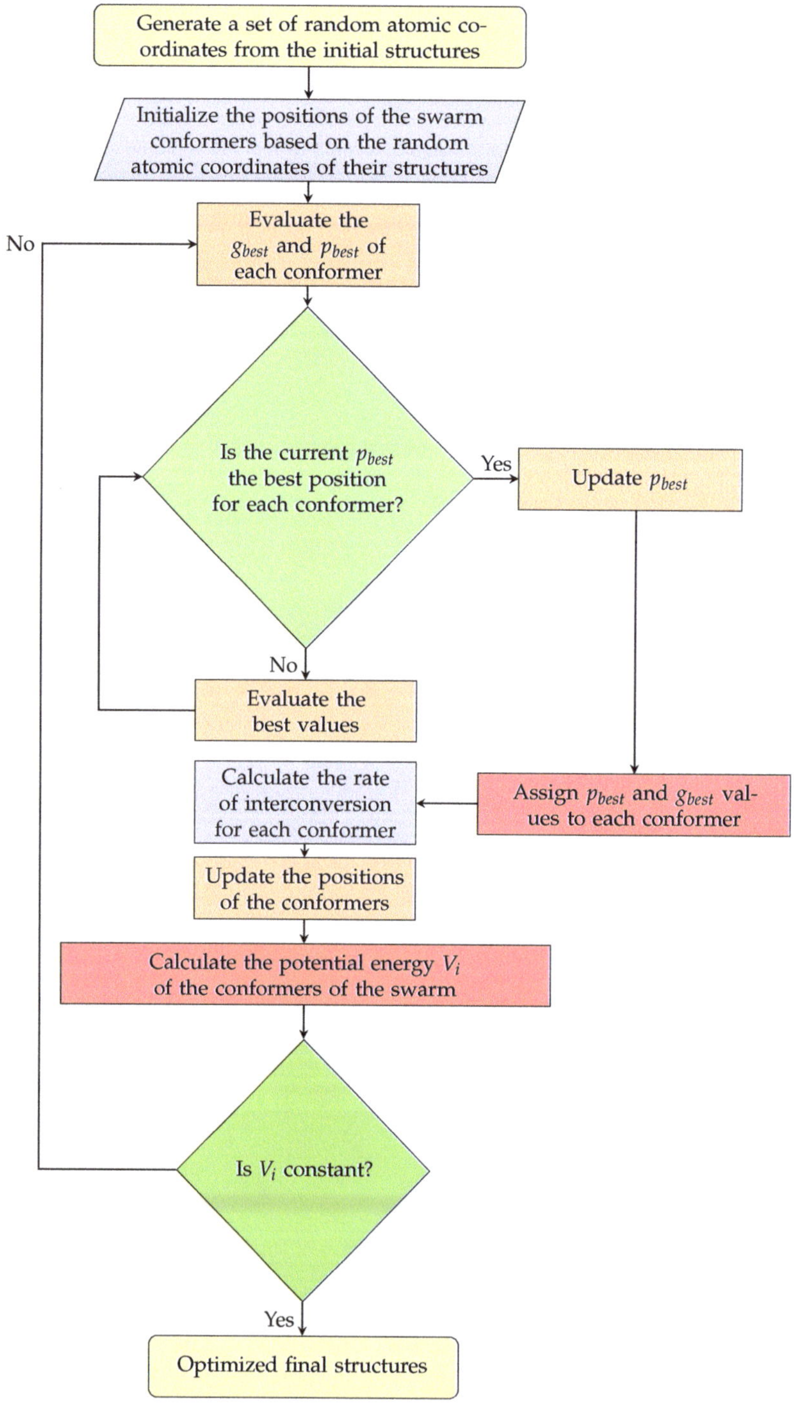

Figure 9. General flowchart of the PSO algorithm implemented in Fortran 90.

4. Conclusions

Good approximations of the molecular structures of carbon atom clusters C_n (n = 3–5) and tungsten oxide clusters, WO_n^{m-} ($n = 4$–$6, m = 2, 4, 6$), were obtained through geometric optimization using the PSO method with Hooke's potential. These structures were compared with those optimized using the BH method and DFT implemented in the Gaussian 09 software, as well as with structures reported in X-ray diffraction studies. The algorithm, written in Fortran 90, was based on a metaheuristic model designed to find the minimum potential energy of a system of ten conformers simultaneously in a hyperdimensional space $\mathbb{R}^{3N}$. This optimization was performed concerning to the positions of the atoms (connected by spring), to determine the molecular structure of the cluster in its ground state within the three-dimensional space $\mathbb{R}^3$. The implementation of the PSO algorithm represents a non-assumptions option, avoiding constraints such as cluster symmetry and external factors imposed by pressure or temperature. This method operates effectively, converging toward geometric configurations presumed to correspond to the global energy minimum or very close to it. The algorithm could serve as an alternative to gradient descent, conjugate gradient, or Newton methods, which are commonly employed in molecular structure optimization but occasionally fail to converge at specific steps or iterations. Additionally, it offers a computational low-cost option for obtaining molecular structures prior to performing *ab initio* electronic structure calculations.

Supplementary Materials: The following are available online at https://www.mdpi.com/article/10.3390/inorganics13090293/s1, Table S1: Representation of the initial geometric structures of three-carbon atom clusters (C_3), Table S2: Representation of the initial geometric structures of four-carbon atom clusters C_4 acyclic, Table S3: Representation of the initial geometric structures of four-carbon atom clusters C_4 cyclic, Table S4: Representation of the initial geometric structures of five-carbon atom clusters C_5, Table S5: Representation of the initial geometric structures of tungsten and oxygen clusters $WO_4{}^{2-}$, Table S6: Representation of the initial geometric structures of tungsten and oxygen clusters $WO_5{}^{4-}$, Table S7: Representation of the initial geometric structures of tungsten and oxygen clusters $WO_6{}^{6-}$.

Author Contributions: Conceptualization, J.N., U.G., G.L.-P., J.L., D.V.-Y., J.R., P.D., E.C. and D.L.; methodology, J.N., U.G., J.L., J.R., D.V.-Y. and D.L.; software, J.N., U.G., J.L., J.R. and L.M.P.; validation, J.N., U.G. and D.L.; formal analysis, J.N., U.G., G.L.-P., J.L., D.V.-Y., P.D., E.C., T.G.-V., L.M.P. and D.L.; investigation, J.N., U.G., G.L.-P., J.L., D.V.-Y., P.D. and E.C.; resources, D.L. and L.M.P.; data curation, J.N. and U.G.; writing—original draft preparation, J.N., U.G., G.L.-P., J.L., D.V.-Y., J.R., P.D., E.C., T.G.-V., L.M.P. and D.L.; writing—review and editing, J.N., U.G., G.L.-P., J.L., D.V.-Y., J.R., P.D., E.C., T.G.-V., L.M.P., L.M.P. and D.L.; visualization, J.N., U.G., G.L.-P., J.L., D.V.-Y., J.R., P.D., E.C., T.G.-V., L.M.P. and D.L.; supervision, J.N., U.G., G.L.-P., J.L., D.V.-Y., J.R., P.D., E.C., T.G.-V., L.M.P. and D.L.; project administration, U.G., L.M.P. and D.L. All authors have read and agreed to the published version of the manuscript.

Funding: L.M.P. and D.L. acknowledge partial financial support from FONDECYT 1240985 and from the Centers of Excellence with BASAL/ANID financing, CIA250002, CEDENNA. P.D., L.M.P. and D.L. acknowledge partial financial support from FONDECYT 1231020. U.G., E.C. and P.D. acknowledge partial financial support from the project "Implementación de una unidad interdisciplinar para el desarrollo de Tecnologías Aplicadas y Ciencias (InTec)", Code FRO2395, from the Ministry of Education of Chile.

Institutional Review Board Statement: Not applicable.

Informed Consent Statement: Not applicable.

Data Availability Statement: The original contributions presented in the study are included in the article/supplementary material, further inquiries can be directed to the corresponding author.

Conflicts of Interest: The authors declare no conflicts of interest.

References

1. Császár, A.G.; Fábri, C.; Szidarovszky, T.; Mátyus, E.; Furtenbachera, T.; Czakó, G. The fourth age of quantum chemistry: Molecules in motion. *Phys. Chem. Chem. Phys.* **2012**, *14*, 1085–1106. [CrossRef]
2. Rapetti, D.; Roncaglia, C.; Ferrando, R. Optimizing the Shape and Chemical Ordering of Nanoalloys with Specialized Walkers. *Adv. Theory Simul.* **2023**, *6*, 2300268. [CrossRef]
3. Sun, L.; Marques, M.A.L.; Botti, S. Direct insight into the structure-property relation of interfaces from constrained crystal structure prediction. *Nat. Commun.* **2021**, *12*, 811. [CrossRef]
4. Khajehpasha, E.R.; Goedecker, S.; Ghasemi, S.A. New strontium titanate polymorphs under high pressure. *J. Compt. Chem.* **2021**, *42*, 699–705. [CrossRef]
5. Delahaye, D.; Chaimatanan, S.; Mongeau, M. Gendreau, M., Potvin, J.Y., Eds.; Simulated Annealing: From Basics to Applications. In *Handbook of Metaheuristics*; International Series in Operations Research & Management Science; Springer: Cham, Switzerland, 2019; Volume 272, pp. 1–35. [CrossRef]
6. Pan, X.; Xue, L.; Lu, Y.; Sun, N. Hybrid particle swarm optimization with simulated annealing. *Multimed. Tools Appl.* **2019**, *78*, 29921–29936. [CrossRef]
7. Manzhos, S.; Carrington, T., Jr. Neural Network Potential Energy Surfaces for Small Molecules and Reactions. *Chem. Rev.* **2021**, *121*, 10187–10217. [CrossRef] [PubMed]
8. Liu, Z.; Yang, Y.; Li, D.; Lv, X.; Chen, X.; Dai, Q. Prediction of the RNA Tertiary Structure Based on a Random Sampling Strategy and Parallel Mechanism. *Front. Genet.* **2022**, *12*, 813604. [CrossRef]
9. Wales, D.J.; Doye, J.P.K. Global Optimization by Basin-Hopping and the Lowest Energy Structures of Lennard-Jones Clusters Containing up to 110 Atoms. *J. Phys. Chem. A* **1997**, *101*, 5111–5116. [CrossRef]
10. Mackie, C.J.; Lu, W.; Liang, J.; Kostko, O.; Bandyopadhyay, B.; Gupta, I.; Ahmed, M.; Head-Gordon, M. Magic Numbers and Stabilities of Photoionized Water Clusters: Computational and Experimental Characterization of the Nanosolvated Hydronium Ion. *J. Phys. Chem. A* **2023**, *127*, 5999–6011. [CrossRef] [PubMed]
11. Martoňák, R.; Laio, A.; Parrinello, M. Predicting Crystal Structures: The Parrinello-Rahman Method Revisited. *Phys. Rev. Lett.* **2003**, *90*, 075503. [CrossRef]
12. Wales, D.J. Guet, C., Hobza, P., Speigelman, F., David, F., Eds.; Energy Landscapes. In *Atomic Clusters and Nanoparticles. Agregats Atomiques et Nanoparticules*; Les Houches—Ecole d'Ete de Physique Theorique; Springer: Berlin/Heidelberg, Germany, 2001; Volume 73. [CrossRef]
13. Jørgensen, M.S.; Larsen, U.F.; Jacobsen, K.W. Exploration versus Exploitation in Global Atomistic Structure Optimization. *J. Phys. Chem. A* **2018**, *122*, 1504–1509. [CrossRef]
14. Stillinger, F.H.; Weber, T.A. Hidden structure in liquids. *Phys. Rev. A* **1982**, *25*, 978–989. [CrossRef]
15. Stillinger, F.H.; Weber, T.A. Packing Structures and Transitions in Liquids and Solids. *Science* **1984**, *225*, 983–989. [CrossRef] [PubMed]
16. Doye, J.P.K.; Wales, D.J. Calculation of thermodynamic properties of small Lennard-Jones clusters incorporating anharmonicity. *J. Chem. Phys.* **1995**, *102*, 9659–9672. [CrossRef]
17. Stillinger, F.H. Exponential multiplicity of inherent structures. *Phys. Rev. E* **1999**, *59*, 48–51. [CrossRef]
18. Gupta, V.P. *Principles and Applications of Quantum Chemistry*; Academic Press: London, UK, 2016. [CrossRef]
19. Mai, G.; Hong, Y.; Fu, S.; Lin, Y.; Hao, Z.; Huang, H.; Zhu, Y. Optimization of Lennard-Jones clusters by particle swarm optimization with quasi-physical strategy. *Swarm Evol. Comput.* **2020**, *57*, 100710. [CrossRef]
20. Oganov, A.R.; Lyakhov, A.O.; Valle, M. How Evolutionary Crystal Structure Prediction Works—And Why. *Acc. Chem. Res.* **2011**, *44*, 227–237. [CrossRef]
21. Jäger, M.; Schäfer, R.; Johnston, R.L. First principles global optimization of metal clusters and nanoalloys. *Adv. Phys. X* **2018**, *3*, 1516514. [CrossRef]
22. Lepeshkin, S.V.; Baturin, V.S.; Uspenskii, Y.A.; Artem, R.; Oganov, A.R. Method for Simultaneous Prediction of Atomic Structure and Stability of Nanoclusters in a Wide Area of Compositions. *J. Phys. Chem. Lett.* **2019**, *10*, 102–106. [CrossRef]
23. Ignatov, S.K.; Razuvaev, A.G.; Loginova, A.S.; Masunov, A.E. Global Structure Optimization of Pt Clusters Based on the Modified Empirical Potentials, Calibrated using Density Functional Theory. *J. Phys. Chem. C* **2019**, *123*, 29024–29036. [CrossRef]
24. Ünal, A.; Ayın, Ö. A Density Functional Investigation on Li_nI (n = 1–8) Clusters. *J. Clust. Sci.* **2021**, *32*, 507–516. [CrossRef]
25. Bertollucci Colherinhas, G.; Vinicius Girão de Morais, M.; Rodrigues Machado, M. Spectral Model of Offshore Wind Turbines and Vibration Control by Pendulum Tuned Mass Dampers. *Int. J. Struct. Stab. Dyn.* **2022**, *22*, 2250053. [CrossRef]
26. Osada, A.; Yamazaki, R.; Noguchi, A. Harmonic Oscillator. In *Introduction to Quantum Technologies*; Lecture Notes in Physics; Springer: Singapore, 2022; Volume 1004. [CrossRef]

27. Maddah, H.A.; Berry, V.; Behura, S.K. Simple harmonic oscillation model explaining MA torsional locking in surface passivated $MAPbI_3$ crystal. *Chem. Phys. Lett.* **2022**, *806*, 139967. [CrossRef]
28. Kennedy, R.; Eberhart, R.; Smietana, M. Particle swarm optimization. Proceedings of ICNN'95—International Conference on Neural Networks, Perth, WA, Australia, 27 November 1995–1 December 1995; Volume 4, pp. 1942–1948. [CrossRef]
29. Tong, Q.; Lv, J.; Gao, P.; Wang, Y. The CALYPSO methodology for structure prediction. *Chin. Phys. B* **2019**, *28*, 106105. [CrossRef]
30. Bansal, J.C. Bansal, J.C., Singh, P.K., Pal, N.R., Eds.; Particle Swarm Optimization. In *Evolutionary and Swarm Intelligence Algorithms*; Springer International Publishing: Berlin/Heidelberg, Germany, 2019; pp. 11–23. [CrossRef]
31. Keith, A.D.; Brichtová, E.P.; Barber, J.G.; Wales, D.J.; Jackson, S.E.; Röder, K. Energy Landscapes and Structural Ensembles of Glucagon-like Peptide-1 Monomers. *J. Phys. Chem. B* **2024**, *128*, 5601–5611. [CrossRef]
32. Jayaramulu, K.; Mukherjee, S.; Morales, D.M.; Dubal, D.P.; Nanjundan, A.K.; Schneemann, A.; Masa, J.; Kment, S.; Schuhmann, W.; Otyepka, M.; et al. Graphene-Based Metal–Organic Framework Hybrids for Applications in Catalysis, Environmental, and Energy Technologies. *Chem. Rev.* **2022**, *122*, 17241–17338. [CrossRef]
33. Hrodmarsson, H.R.; Bouwman, J.; Tielens, A.G.G.; Linnartz, H. Fragmentation of the PAH cations of Isoviolanthrene and Dicoronylene: A case made for interstellar cyclo[*N*]carbons Prod. Univers. Fragm. Processes. *Int. J. Mass Spectrom.* **2023**, *485*, 116996. [CrossRef]
34. McKay, A.J.; Roth, N.X. Organic Matter in Cometary Environments. *Life* **2021**, *11*, 37. [CrossRef]
35. McGuire, B.A.; Loomis, R.A.; Burkhardt, A.M.; Kelvin Lee, K.L.; Shingledecker, C.N.; Charnley, S.B.; Cooke, I.R.; Cordiner, M.A.; Herbst, E.; Kalenskii, S.; et al. Detection of two interstellar polycyclic aromatic hydrocarbons via spectral matched filtering. *Science* **2021**, *371*, 1265–1269. [CrossRef]
36. Hyun, K.; Saito, N. The solution plasma process for heteroatom-carbon nanosheets: The role of precursors. *Sci. Rep.* **2017**, *7*, 3825. [CrossRef]
37. Sacco, L.N.; Vollebregt, S. Overview of Engineering Carbon Nanomaterials Such As Carbon Nanotubes (CNTs), Carbon Nanofibers (CNFs), Graphene and Nanodiamonds and Other Carbon Allotropes inside Porous Anodic Alumina (PAA) Templates. *Nanomaterials* **2023**, *13*, 260. [CrossRef]
38. Niu, C.; Zhao, J.; Xia, W.; Chen, G.; Luo, W.; Zhang, J.; Ahuja, R.; Wang, X. Configuration stability and physical properties of new diamondene structure. *Mater. Today Commun.* **2023**, *36*, 106465. [CrossRef]
39. Papanasam, E.; Prashanth, K.B.; Chanthini, B.; Manikandan, E.; Agarwal, L. A Comprehensive Review of Recent Progress, Prospect and Challenges of Silicon Carbide and its Applications. *Silicon* **2022**, *14*, 12887–12900. [CrossRef]
40. Gessner, V.H. Stability and reactivity control of carbenoids: Recent advances and perspectives. *Chem. Commun.* **2016**, *52*, 12011–12023. [CrossRef] [PubMed]
41. Marlton, S.J.P.; Buntine, J.T.; Watkins, P.; Liu, C.; Jacovella, U.; Carrascosa, E.; Bull, J.N.; Bieske, E.J. Probing Colossal Carbon Rings. *J. Phys. Chem. A* **2023**, *127*, 1168–1178. [CrossRef]
42. Sharma, M.P.; Jaffe, R.L.; Panesi, M. Carbon Clusters: Thermochemistry and Electronic Structure at High Temperatures. *J. Phys. Chem. A* **2021**, *125*, 7038–7051. [CrossRef] [PubMed]
43. Kumar, S.; Dhilip Kumar, T.J. Quantum Scattering Calculations for Rotational Excitations of C_3 by Hydrogen Atom: Potential Energy Surfaces and Rate Coefficients. *J. Phys. Chem. A* **2019**, *123*, 7296–7302. [CrossRef]
44. Prabhakaran, V.; Lang, Z.; Clotet, A.; Poblet, J.M.; Johnson, G.E.; Julia Laskin, J. Controlling the Activity and Stability of Electrochemical Interfaces Using Atom-by-Atom Metal Substitution of Redox Species. *ACS Nano* **2019**, *13*, 458–466. [CrossRef]
45. Anyushin, A.V.; Kondinski, A.; Parac-Vogt, T.N. Hybrid polyoxometalates as post-functionalization platforms: From fundamentals to emerging applications. *Chem. Soc. Rev.* **2020**, *49*, 382–432. [CrossRef] [PubMed]
46. Jana, G.; Mitra, A.; Pan, S.; Sural, S.; Chattaraj, P.K. Modified Particle Swarm Optimization Algorithms for the Generation of Stable Structures of Carbon Clusters, C_n (n = 3–6, 10). *Front. Chem.* **2019**, *7*, 485. [CrossRef]
47. Schlegel, H.B. Optimization of equilibrium geometries and transition structures. *J. Comp. Chem.* **1982**, *3*, 214–218. [CrossRef]
48. Feynman, R.P. Forces in Molecules. *Phys. Rev.* **1939**, *56*, 340–343. [CrossRef]
49. Liu, D.C.; Nocedal, J. On the limited memory BFGS method for large scale optimization. *Math. Program.* **1989**, *45*, 503–528. [CrossRef]
50. Koster, A.S.; Kools, F.X.N.M.; Rieck, G.D. The crystal structure of potassium tungstate, K_2WO_4. *Acta. Cryst. B* **1969**, *25*, 1704–1708. [CrossRef]
51. Yagoubi, S.; Obbade, S.; Benseghir, M.; Abraham, F.; Saadi, M. Synthesis, crystal structure, cationic mobility, thermal evolution and spectroscopic study of $Cs_8(UO_2)_4(WO_4)_4(WO_5)_2$ containing infinite uranyl tungstate chains. *Solid State Sci.* **2007**, *9*, 933–943. [CrossRef]
52. King, G.; Abakumov, A.M.; Hadermann, J.; Alekseeva, A.M.; Rozova, M.G.; Perkisas, T.; Woodward, P.M.; Tendeloo, G.V.; Antipov, E.V. Crystal Structure and Phase Transitions in Sr_3WO_6. *Inorg. Chem.* **2010**, *49*, 6058–6065. [CrossRef]
53. Kalescky, R.; Kraka, E.; Cremer, D. Identification of the Strongest Bonds in Chemistry. *J. Phys. Chem. A* **2013**, *117*, 8981–8995. [CrossRef] [PubMed]

54. Luo, Y.-R. *Comprehensive Handbook of Chemical Bond Energies*, 1st ed.; CRC Press, Taylor and Francis: Boca Raton, FL, USA, 2007; 1688p. [CrossRef]
55. Haynes, W.M.; Lide, D.R.; Bruno, T.J. *CRC Handbook of Chemistry and Physics*, 97th ed.; CRC Press, Taylor and Francis: Boca Raton, FL, USA, 2017; 2643p.
56. Bagheri, M.; Berger, E.; Komsa, H.-P. Identification of Material Dimensionality Based on Force Constant Analysis. *J. Phys. Chem. Lett.* **2023**, *14*, 7840–7847. [CrossRef]
57. Horn, M.R.; Singh, A.; Alomari, S.; Goberna-Ferrón, S.; Benages-Vilau, R.; Chodankar, N.; Motta, N.; (Ken) Ostrikov, K.; MacLeod, J.; Sonar, P.; et al. Polyoxometalates (POMs): From electroactive clusters to energy materials. *Energy Environ. Sci.* **2021**, *14*, 1652–1700. [CrossRef]
58. Granadeiro, C.M.; Julião, D.; Ribeiro, S.O.; Cunha-Silva, L.; Salete, S.; Balula, S.S. Recent advances in lanthanide-coordinated polyoxometalates: From structural overview to functional materials. *Coord. Chem. Rev.* **2023**, *476*, 214914. [CrossRef]
59. Parsopoulos, K.E. Martí, R., Panos, P., Resende, M., Eds.; Particle Swarm Methods. In *Handbook of Heuristics*; Springer: Cham, Switzerland, 2015. [CrossRef]
60. Rhodes, C.J.; Macrae, R.M. "Vibrational Bonding": A New Type of Chemical Bond is Discovered. *Sci. Progr.* **2015**, *98*, 12–33. [CrossRef]
61. John, C.; Owais, C.; James, A.; Swathi, S. Swarm Intelligence Steers a Global Minima Search of Clusters Bound on Carbon Nanostructures. *J. Phys. Chem. C* **2021**, *125*, 2811–2823. [CrossRef]
62. Owais, C.; John, C.; Swathi, S. Swarm intelligence unravels the confinement effects for tiny noble gas clusters within carbon nanotubes. *Eur. Phys. J. D* **2021**, *75*, 16. [CrossRef]
63. Rajeevan, M.; John, C.; Swathi, R.S. On assessing the carbon capture performance of graphynes with particle swarm optimization. *Phys. Chem. Chem. Phys.* **2024**, *26*, 23152. [CrossRef] [PubMed]
64. Delgado, A.A.A.; Humason, A.; Kalescky, R.; Freindorf, M.; Kraka, E. Exceptionally Long Covalent CC Bonds—A Local Vibrational Mode Study. *Molecules* **2021**, *26*, 950. [CrossRef]
65. Zhao, L.; Zhi, M.; Frenking, G. The strength of a chemical bond. *Int. J. Quantum Chem.* **2022**, *122*, e26773. [CrossRef]
66. Becke, A.D. Density-functional thermochemistry. III. The role of exact exchange. *J. Chem. Phys.* **1993**, *98*, 5648–5652. [CrossRef]
67. Lee, C.; Yang, W.; Parr, R.G. Development of the Colle-Salvetti correlation-energy formula into a functional of the electron density. *Phys. Rev. B* **1988**, *37*, 785–789. [CrossRef]
68. Hay, P.J.; Wadt, W.R. *Ab Initio* Effective Core Potentials for Molecular Calculations. Potentials for the Transition Metal Atoms Sc to Hg. *J. Chem. Phys.* **1985**, *82*, 270–283. [CrossRef]
69. Wadt, W.R.; Hay, P.J. *Ab Initio* Effective Core Potentials for Molecular Calculations. Potentials for Main Group Elements Na to Bi. *J. Chem. Phys.* **1985**, *82*, 284–298. [CrossRef]
70. Frisch, M.J.; Trucks, G.W.; Schlegel, H.B.; Scuseria, G.E.; Robb, M.A.; Cheeseman, J.R.; Scalmani, G.; Barone, V.; Petersson, G.A.; Nakatsuji, H.; et al. *Gaussian 09, Revision D.01*; Gaussian, Inc.: Wallingford, CT, USA, 2016. Available online: https://Gaussian.com/ (accessed on 18 November 2022).

Disclaimer/Publisher's Note: The statements, opinions and data contained in all publications are solely those of the individual author(s) and contributor(s) and not of MDPI and/or the editor(s). MDPI and/or the editor(s) disclaim responsibility for any injury to people or property resulting from any ideas, methods, instructions or products referred to in the content.

inorganics MDPI

Article

A Study on the Morphology of Poly(Triaryl Amine)-Based Hole Transport Layer via Solvent Optimization for High-Performance Inverted Perovskite Solar Cells

Xiaoyin Xie [1], Xi Liu [1,2], Chufei Ding [1], Han Yang [1], Xueyi Liu [1], Guanchen Liu [1,*], Zhihai Liu [2,*] and Eun-Cheol Lee [3]

1 School of Chemistry and Chemical Technology, Hubei Polytechnic University, Huangshi 435003, China
2 School of Physics and Electronic Information, Yantai University, Yantai 264005, China
3 Department of Physics, Gachon University, Seongnam-si 13120, Republic of Korea; eclee@gachon.ac.kr
* Correspondence: gcliu@hbpu.edu.cn (G.L.); zhliu@ytu.edu.cn (Z.L.); Tel.: +86-15549716558 (G.L.); +86-18662573383 (Z.L.)

Abstract: Poly[bis(4-phenyl) (2,5,6-trimethylphenyl) amine (PTAA), as a hole transfer material, has been widely used in perovskite solar cells (PSCs). However, the optimal solvent for preparing the PTAA solution and coating the PTAA layer is still uncertain. In this work, we investigated three types of organic solvents (toluene, chlorobenzene and dichlorobenzene) for processing PTAA layers as the hole transport layer in PSCs. Based on the experimental verification and molecular dynamics simulation results, all the evidence indicated that toluene performs best among the three candidates. This is attributed to the significant polarity difference between toluene and PTAA, which leads to the formation of a uniform surface morphology characterized by granular protuberances after spin coating. The contact area of the hole transfer layer with the surface aggregation is increased in reference to the rough surface, and the hydrophilicity of the PTAA layer is also increased. The improvement of these two aspects are conducive to the effective interfacial charge transfer. This leads to the generation of more photocurrent. The PSCs employing toluene-processed PTAA exhibit an average power conversion efficiency (PCE) of 19.1%, which is higher than that of PSCs using chlorobenzene- and dichlorobenzene-processed PTAA (17.3–17.9%). This work provides a direct optimization strategy for researchers aiming to fabricate PSCs based on PTAA as a hole transport layer and lays a solid foundation for the development of high-efficiency inverted PSCs.

Keywords: perovskite; PTAA; morphology; solvent; aggregation

1. Introduction

Recently, perovskite solar cells (PSCs) have become one of the most promising photovoltaic technologies due to their low production cost and simple preparation process [1–9]. The development of PSCs exceeds people's expectation, and the power-conversion-efficiency (PCE), which increases by about 2.5% annually, keeps breaking the record [10–13]. At present, the efficiency of the highest certified single-layer PSCs has exceeded 27.0% [14]. Owing to their numerous advantages, PSCs have a very bright industrialization prospect and are therefore a potential competitor of next-generation solar cells. However, PSCs also suffer from some disadvantages, such as poor stability and lead toxicity, which limit their commercial applications [15,16].

PSCs usually have regular n-i-p and inverted p-i-n stack structures [12,13]. In such devices, perovskite materials are sandwiched between electron and hole transport layers. $MAPbI_3$ (MA is methylammonium) is widely used as the energy conversion material of the perovskite layer; TiO_2 and 2,2′,7,7′-Tetrakis[*N*,*N*-di(4-methoxyphenyl)amino]-9,9′-spirobifluorene (spiro-OMeTAD) are commonly used as electron and hole transport materials, respectively [15–20]. To further improve the performance and stability of PSCs, mix-cation perovskites have been developed, which include three cations of formamidinium (FA), MA and Cs [10]. Hole transfer materials such as poly(3,4-ethylenedioxythiophene):poly(styrenesulfonate) (PEDOT:PSS), nickel oxide, and poly[bis(4-phenyl)(2,5,6-trimethylphenyl)amine (PTAA) have been investigated in PSCs [21–24]. Although searching for new materials plays an important role in improving the performance of perovskite solar cells, controlling the morphology of each layer is also crucial to the performance of the devices [25–28]. For example, the morphology of perovskite films could directly affect the light absorption and charge generation [29–32]. The morphology of the electron and hole transport layer is of great significance for transport, separation and collection which affect the contact resistance and current density of the device [33,34]. Generally, in order to control the morphology of each functional layer, the commonly used methods are solvent optimization, thermal annealing and solvent annealing technology [22,35–38]. For example, in order to obtain a high-quality spiro-OMeTAD layer, spiro-OMeTAD is usually dissolved in chloroform, chlorobenzene or ethylacetate and then spin-coated. Similarly, controlling the morphology of the perovskite ore layer in devices by solvent treatment is frequently reported [39–42].

PTAA, as a kind of conductive polymer material, is increasingly being used to replace spiro-OMeTAD as the hole transport layer in perovskite solar cells [43–47]. This is because its lowest unoccupied molecular orbital (LUMO) levels (about −5.2 eV) can match well with the valence band of perovskite and the work function of ITO. In addition to energy level alignment, the contact between the hole transport layer and the perovskite absorber should also be optimized to more effectively extract holes from the perovskite and transport them to the anode. Jacak et al. demonstrated that using some nanometallic particles could enhance the hole capturing, which is beneficial for boosting the performance of PSCs [48]. Although in these reports, the devices using PTAA as the hole transfer layer have shown good performances. There is still no agreement on the solvents used in the treatment of PTAA in these reports [44–47]. Wu et al. [44] employed chlorobenzene as the solvent in their work, while Fu et al. and Deng et al. [46,47] used toluene. However, aside from the difference in PTAA solvents, other preparation conditions are not uniform, so it is difficult to decide which solvent is the best choice for the preparation of the PTAA coating liquid from their conclusions. This may cause some confusion for other researchers in the preparation of perovskite solar cells based on PTAA. In this work, we investigated the solvent effect for controlling the morphology of the PTAA layer, which further influences the performance of PSCs. We selected three solvents in the preparation of perovskite solar cells, i.e., toluene, chlorobenzene and dichlorobenzene, and carried out experimental and theoretical methods to obtain the data related to the device's performance. We found that using toluene-processed PTAA resulted in the best performance for the PSCs, which exhibited an average PCE of 19.1%, whereas the PSCs using chlorobenzene or dichlobenzene-processed PTAA showed lower PCEs of 17.3% and 17.9%, respectively. The optimal solvent selection is illustrated by parameters such as molecular dynamics, surface morphology, hydrophobicity and device performance. Our work may provide an effect reference for other researchers in the future when they process a device with PTAA as the hole transfer layer.

2. Results and Discussion

As shown in Figure 1a, the PSCs were fabricated with a structure of glass/ITO/PTAA/perovskite/PCBM/BCP/Ag. As seen in the SEM image in Figure 1b, the PSCs have a standard layer-by-layer structure, with a perovskite thickness of about 500 nm. As indicated in the image, the processed PTAA and PCBM show thicknesses of about 30 and 65 nm, respectively. To find the indirect relationship between the solvent characteristics and the device performance, we used toluene, chlorobenzene and 1,2-dichlorobenzene as the coating solvent of PTAA to spin-coat the PTAA layer on ITO glass, using the same preparation method and conditions. To ensure the statistical validity of the related data, we prepared 15 samples for each group. The collected scatter statistical values of the photoelectric parameters of the samples are shown in the following figure (Figure 1).

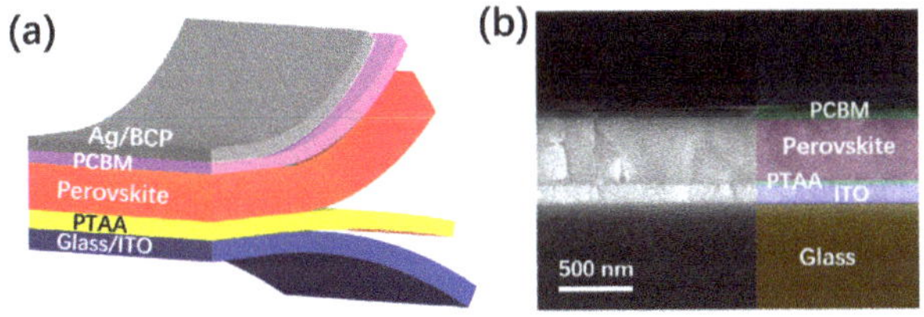

Figure 1. (**a**) Schematic structure of the PSCs in this work. (**b**) Cross-sectional SEM image of the PSC.

As shown in Figure 2a–d, the short-circuit current density (J_{sc}) and the PCE of the toluene-based group are obviously higher than those of the other two groups. In contrast, the open-circuit voltage (V_{oc}) exhibit only minor variations among the three groups, with average values of 1.06 V, 1.05 V and 1.06 V, respectively. For the fill factor, no obvious change rule can be drawn from the three groups of data (0.71, 0.69 and 0.70 in average, respectively), This phenomenon will be discussed in detail later. If the influence of FF is temporarily disregarded, it can be inferred from the figure that the variation in J_{sc} is primarily determined by the choice of solvent, and the same applies to the PCE. Table 1 shows all the parameters, which were extracted from the forward scanning J–V characteristics. The PCEs (17.3–19.1%) in this work are lower than the record efficiency (about 27%). This may be related to the device's configuration, the perovskite formula, the experimental conditions, etc. According to these data, it can be seen that for the J_{sc}, the device using toluene as solvent is superior to the other two groups. The trend of these values is in accordance with Figure 2a,b,d. For the fill factor, there are slight differences between the average values of the three groups and their best values. The results show that the PCE variation was mainly influence by the J_{sc} change, which affected the selection of the solvent.

Table 1. Device parameters for the PSCs based on PTAA processed from toluene, chlorobenzene and dichlorobenzene.

Processed Solvent	V_{oc} (V)	J_{sc} (mA cm^{-2})	FF (%)	Average PCE (%)	Best PCE (%)
Toluene	1.11 ± 0.01	22.5 ± 0.3	76.5 ± 1.2	19.1 ± 0.3	19.5
Chlorobenzene	1.08 ± 0.01	21.6 ± 0.3	74.2 ± 1.1	17.3 ± 0.3	17.7
Dichlorobenzene	1.09 ± 0.01	21.8 ± 0.3	75.2 ± 1.1	17.9 ± 0.3	18.3

To further investigate the mechanism by which solvent selection influences device performance, we measured the J-V characteristics of PSCs, with PTAA layers processed using toluene, chlorobenzene and 1,2-dichlorobenzene. As shown in Figure 3a, the use of toluene as the processing solvent significantly enhances the J_{sc} as compared with chlorobenzene and 1,2-dichlorobenzene. However, the influence of the three solvents on the V_{oc} is negligible, which is consistent with the aforementioned conclusion. Furthermore, to verify the reliability of the measured J_{sc}, we measured the IPCE spectra of the three best-performing devices and integrated the IPCE data over the entire wavelength range, as shown in Figure 3b. The integrated J_{sc} from the IPCE (on average) are 20.72, 19.48 and 19.42 mA/cm^2 for the devices processed with toluene, chlorobenzene and 1,2-dichlorobenzene, respectively. These results are in good agreement with the current densities obtained from the *J-V* measurements, further confirming the reliability of our measurements.

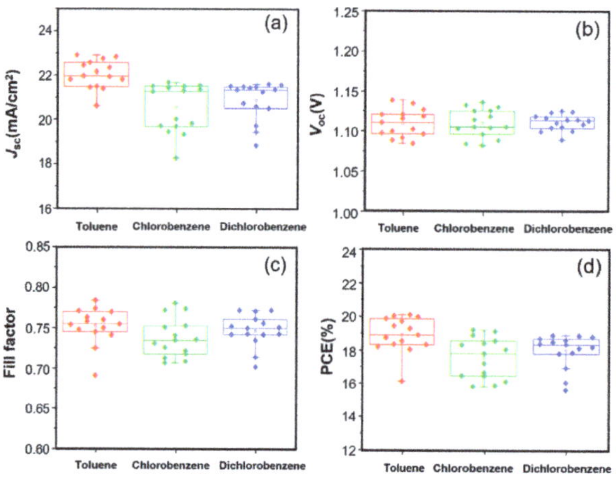

Figure 2. Scatter statistics of the photoelectric parameters. (**a**), J_{sc}, (**b**), V_{oc}, (**c**) fill factor, (**d**), PCE of the samples prepared by using the PTAA solution treated with toluene (red), chlorobenzene (green) and 1,2-dichlorobenzene (blue) as the electron transport layer of the spin-coating solution, with 15 samples for each group.

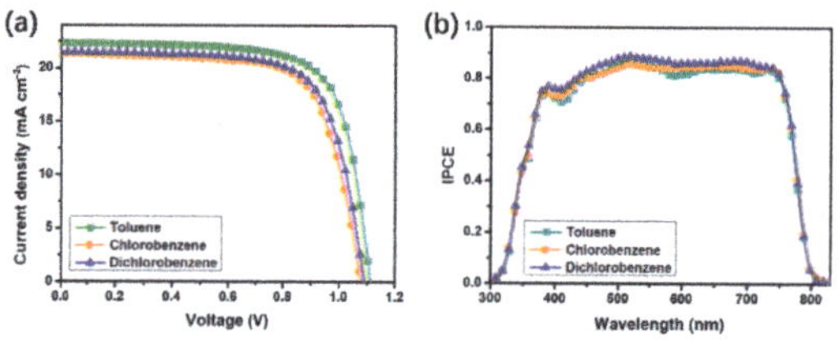

Figure 3. *J–V* characteristics (**a**) and IPCE spectra (**b**) of the PSCs, with PTAA processed from toluene, chlorobenzene and dichlorobenzene.

Because the above evidence indicates that the J_{sc} is influenced by the choice of solvent, the following section will focus on analyzing the various factors affecting the J_{sc} of the devices in order to elucidate the underlying mechanism. As is well known, the generation of photocurrent in PSCs involves the following process: first, the perovskite layer absorbs photons, exciting the electrons, then the photogenerated electron–hole pairs are separated and transported to the electron transport layer and hole transport layer, respectively; finally, the charges are collected by the corresponding electrodes. In this process, the J_{sc} is closely associated with the photon absorption capability of the perovskite layer and the interfacial charge transfer efficiency between the perovskite and hole transport layers [21,23–25].

We investigated whether the photoelectron collection process can be influenced by the choice of solvent. According to previous studies, the crystallization and morphology of the perovskite's active layer are critical factors in determining the efficiency of photoelectron collection [29,30]. We then characterized the perovskite films processed with the three solvents by X-Ray diffraction (XRD, Figure S1), scanning electron microscopy (SEM, Figure S2) and ultraviolet–visible absorption spectroscopy (UV–vis, Figure S3). The SEM images show that the morphology of the perovskite layer remains largely unaffected by the different solvent treatments. This can be explained as follows: during the preparation process, the thickness of the PTAA layer is much thinner than that of the perovskite layer. Moreover, as indicated in Figure 4, all the PTAA films processed from the different solvents showed a hydrophobic surface property. Therefore, variations in the PTAA layer do not significantly impact the surface morphology of the perovskite. However, the XRD and UV–vis absorption results indicate that the PSCs using toluene-processed PTAA exhibit the best performance. This aligns well with the FF results, strongly suggesting that the choice of organic solvent slightly affects the perovskite quality, although its impact on this is less significant than its impact on the charge transfer process.

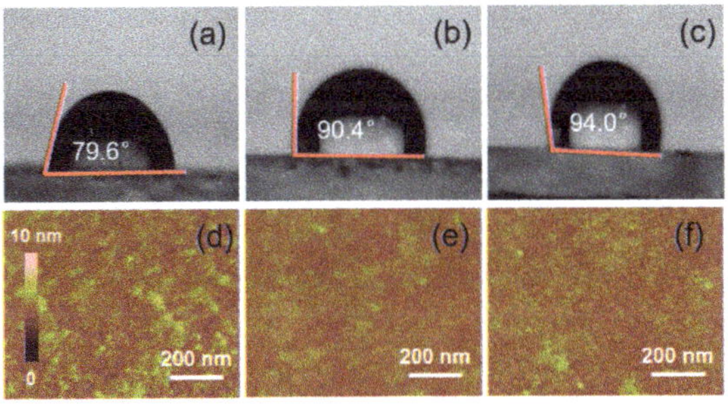

Figure 4. Contact angle of the PTAA layer prepared by using the PTAA solution treated with toluene (**a**), chlorobenzene (**b**) and 1,2-dichlorobenzene (**c**) and AFM images of the PTAA layer prepared by using the PTAA solution treated with toluene (**d**), chlorobenzene (**e**) and 1,2-dichlorobenzene (**f**), respectively.

To understand the influence of solvent selection on the interface charge transfer, we tested the surface morphology of the PTAA layer. First, the contact angles and atomic force microscopy (AFM) patterns of the PTAA layers, with toluene, chlorobenzene and 1,2-dichlorobenzene as solvents, were measured. Compared with Figure 4a–c, it can be seen that the contact angle of the PTAA layer with toluene is the smallest at 79.6°, while the contact angles of the PTAA surfaces with chlorobenzene and 1,2-dichlorobenzene as the solvents are more than 90° (90.4° and 94.0°, respectively), which indicates that the hydrophilicity of the PTAA layer can be obviously improved with toluene as the solvent. Then, atomic force microscopy was used to test their surface morphologies. AFM imaging (Figure 4d–f) shows the reason for the different hydrophobicity levels of the PTAA layers. Compared with the PTAA layers with chlorobenzene and 1,2-dichlorobenzene as solvents, the surface of the PTAA layer with toluene as the solvent shows obvious and dense protrusions, and the diameters of these protrusions are about 15 nm each. However, the surfaces of the PTAA layers with the other two solvents are much smoother, and almost no obvious protuberances can be seen. The RMS roughness values of the PTAA films processed from toluene, chlorobenzene and dichlorobenzene are 2.3, 1.8 and 1.6 nm, respectively. This phenomenon also corresponds to the contact angle obtained above. On the other hand, the existence of these protrusions is equivalent to increasing the interface area between the PTAA layer and the perovskite layer, which also increases the effective transfer probability of electrons between the interfaces, and this improvement can indeed improve the interfacial charge transfer efficiency. From the calculation mode of the dynamic simulations, the calculated interface area of the toluene-processed sample is 10.4%, which is 9.2% higher than that of the chlorobenzene- and 1,2-dichlorobenzene-processed samples. This conclusion is also consistent with the EIS (Figure S4) results, in which we tested the EIS to study the variation coursed by the different organic solution-treated PTAAs. Figure S4 shows the Nyquist plots with two characteristic arcs and equivalent circuits. The plots show that the recombination resistance (R_{rec}), the series resistance (R_s) and capacitor C can be observed. The toluene-treated sample exhibited an R_s of 26.3 Ω and an R_{rec} of 362.7 Ω, while the chlorobenzene and 1,2-dichlorobenzene-treated samples presented an R_s of 30.6 Ω with an R_{rec} of 477.5 Ω and an R_s of 29.3 Ω with an R_{rec} of 402.3 Ω, respectively. The decreased R_s and increased R_{rec} indicate the improved electron-transporting performance and suppressed charge recombination [40], which are conducive to an enhanced interface area and an enlarged J_{sc}.

To investigate the origin of the surface protrusions observed in the AFM images, molecular dynamics simulations were conducted to analyze the distribution of PTAA molecular chains in different solvents. As shown in Figure 5, in toluene, the PTAA molecules tend to aggregate into large clusters due to their poor solubility, leading to an inhomogeneous distribution in the solution. This aggregation is likely due to the polarity difference between PTAA and toluene molecules (Table 2). Upon solvent evaporation, the aggregated state of PTAA is retained, leading to the formation of surface protrusions, as observed in the AFM images. Compared with toluene, PTAA molecules are more uniformly dispersed in the other two solvents, which accounts for the relatively smooth morphology of the PTAA layers formed using these solvents.

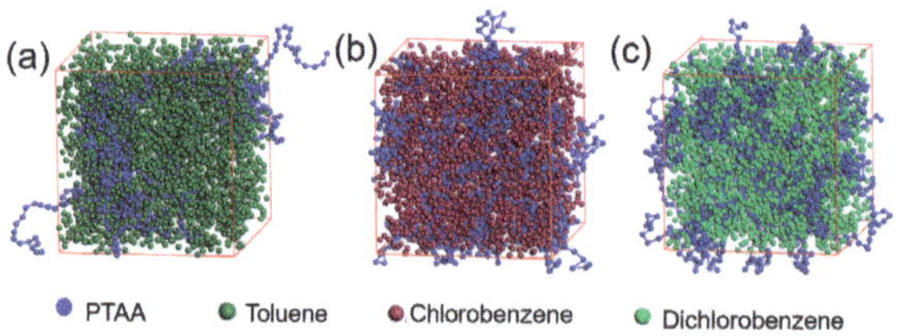

Figure 5. Molecular dynamic simulation of the solvent and PTAA's interaction mechanisms: toluene (**a**), chlorobenzene (**b**) and 1,2-dichlorobenzene (**c**).

Table 2. The polarities of the solvents used in this work.

Solvent Parameter	Toluene	Chlorobenzene	Dichlorobenzene
Polarity	2.4	2.7	2.7

3. Materials and Methods

3.1. Materials

The patterned indium tin oxide (ITO) glass substrate and [6,6]-phenyl-C_{61} butyric acid methyl ester (PCBM) were purchased from Kaivo Co., Ltd. (Zhuhai, China) and Nano-C (Westwood, MA, USA), respectively. Methylammonium bromide (MABr), PTAA, FAI, cesium iodide (CsI) and lead bromide ($PbBr_2$) were purchased from Xi'an Polymer Light Technology Corp. (Xi'an, China). Lead iodide (PbI_2), bathocuproine (BCP) and all solvents [*N*,*N*-dimethylmethanamide (DMF), dimethylsulfoxide (DMSO), isopropanol, chlorobenzene, 1,2-dichlorobenzene and toluene] were supplied by Sigma-Aldrich (St. Louis, MO, USA).

3.2. Device Preparation

We employed $Cs_{0.05}FA_{0.81}MA_{0.14}PbI_{2.55}Br0.45$ perovskite as the absorber for photoelectric conversion. First, the etched ITO glass was washed in deionized water, acetone and isopropanol for 10 min and dried with a nitrogen gun. Next, clean ITO glass panels were treated in an ultraviolet ozone cleaner for 15 min. PTAA solutions (10 mg/mL in chlorobenzene, toluene or dichlorobenzene) were spin-coated onto the ITO glass substrates at 4000 r/min for 30 s. The spin-coated glass substrates were then annealed in air (140 °C,

15 min) to thoroughly evaporate the residual solvents. The perovskite precursor was prepared in a mix solvent (DMF:DMSO = 85:15) by dissolving FAI, MABr, CsI, PbI_2 and $PbBr_2$ into DMF [with molar ratios of $PbI_2/PbBr_2$ and FAI/MABr, both fixed at 0.85:0.15; CsI/(FAI + MABr) fixed at 0.05:0.95; and (FAI + MABr + CsI)/(PbI_2 + $PbBr_2$) fixed at 1:1)]. The final mass fraction was 40%. Before spin-coating, the perovskite precursor was heated and stirred for 12 h at 60 °C. The spin-coating of the perovskite and PCBM layers was accomplished in glove box filled with nitrogen: 150 μL perovskite solution was dripped on the spin-coated PTAA film and then spin-coated at 4500 r/min for 30 s. After spin-coating, the samples were heated for 10 min at 100 °C until the perovskite film completely changed from yellow to reddish brown.

Subsequently, PCBM was dissolved in chlorobenzene to prepare a 20 mg/mL solution. This PCBM solution was then dropped onto the formed perovskite film, spin-coated at 1000 rpm for 30 s and annealed at 40 °C for 10 min to remove the residual solvent. Finally, the spin-coated samples were transferred to a vacuum evaporator, where 0.6 nm of BCP was deposited at a rate of 0.2–0.3 Å/s under a vacuum pressure below 1×10^{-5} Torr. Subsequently, 100 nm thick Ag electrodes were deposited at a rate of 1–2 Å/s. Subsequently, the device was encapsulated using ultraviolet-cured epoxy resin and a glass cover slide, after which its performance was evaluated. The effective active area of the PSCs was 0.04 cm^2.

3.3. Characterizations

The *J–V* characteristic curve of the solar cells were tested under 1 sunlight intensity (100 MW/cm), and the Keithley 2400 was used to collect data. The incident photon-to-current efficiency (IPCE) spectra of the PSCs were measured by a SolarCellScan100 (Zolix, Beijing, China). The cross-sectional image of the PSCs and the morphology of the perovskite layer were measured by using a SEM (SU8020, Tokyo, Japan) with an acceleration voltage of 8000 V. The contact angles were measured by dropping water onto the surface of the PTAA films. The morphologies of the PTAA films were measured by an AFM (Veeco, San Jose, CA, USA) in the tapping mode.

3.4. Simulation Background

In the Mesocite module of Quantum Espresso (QE) 7.4, we used the dissipative particle dynamics method to study the effect of the solvent. A cubic hypermonomer with a transverse dimension of 100 Å × 100 Å × 100 Å was used, which contained 7750 PTAA monomers and 2461 solvent molecules. The average density of PTAA was 1.09 g cm^{-3}. Newton's equations of motion were integrated by using the Verlet algorithm with a time step of 1 *fs*. The simulated temperature and pressure were set to 413.15 K (140 °C) and 101.325 kPa (1atm), respectively.

4. Conclusions

In conclusion, we systematically investigated three different organic solvents (toluene, chlorobenzene and dicholobenzene) for processing PTAA layers as the hole transport layer in perovskite solar cells. Both the experimental results and molecular dynamics simulations reveal that PTAA exhibits significant molecular aggregation in toluene, whereas such aggregation is not observed in the cases of chlorobenzene and dichlorobenzene. The evaporation of different solvents resulted in distinct surface morphologies of the PTAA layers. On one hand, the formation of surface protrusions increases the interfacial area for excited-state charge transfer, thereby enhancing the probability of efficient hole transport and improving the current density of the device. On the other hand, the roughened surface enhances the hydrophilicity of the PTAA layer, which is beneficial for achieving higher device efficiency. Although different solvents influence the morphology of the PTAA layer, this morphological variation plays a key role in determining device performance. Our study demonstrates that the choice of PTAA solvent has a significant influence on the properties of the perovskite layer. The above evidence indicates that toluene is the most suitable solvent for processing PTAA in perovskite solar cells.

Supplementary Materials: The following supporting information can be downloaded at https://www.mdpi.com/article/10.3390/inorganics13070232/s1, Figure S1: XRD patterns of the perovskite layers; Figure S2: Top-view SEM images of the perovskite layers; Figure S3: Absorption spectra of the peroskite layers; Figure S4: EIS spectra of the devices.

Author Contributions: Conceptualization, X.X. and Z.L.; methodology, X.L. (Xi Liu); investigation, C.D.; resources, X.X.; data curation, H.Y. and E.-C.L.; writing—original draft preparation, X.X.; writing—review and editing, Z.L.; visualization, C.D.; supervision, G.L. and Z.L.; project administration, G.L.; funding acquisition, X.L. (Xueyi Liu). All authors have read and agreed to the published version of the manuscript.

Funding: This research was funded by Hubei Provincial Key R&D Program Projects, NO.: 2023DJC187 and the Hubei Provincial Natural Science Foundation Project, NO.: 2024AFD003; Hubei Polytechnic University Research Foundation, No.: ky2025-119; Hubei Provincial Natural Science Foundation of China, No. 2025AFB357.

Data Availability Statement: All relevant data are contained within the manuscript and its Supplementary Materials files.

Conflicts of Interest: The authors declare no conflicts of interest, and the funders had no role in the design of the study; in the collection, analyses, or interpretation of data; in the writing of the manuscript; or in the decision to publish the results.

Abbreviations

The following abbreviations are used in this manuscript:

PTAA	Poly[bis(4-phenyl) (2,5,6-trimethylphenyl) amine
PSCs	perovskite solar cells
HOMO	highest occupied molecular orbital
LUMO	lowest unoccupied molecular orbital
PCE	power conversion efficiency
IPCE	incident photon to current efficiency

References

1. Zhou, H.; Chen, Q.; Li, G.; Luo, S.; Song, T.; Duan, H.S.; Hong, Z.; You, J.; Liu, Y.; Yang, Y. Interface engineering of highly efficient perovskite solar cells. *Science* **2014**, *345*, 542–546. [CrossRef] [PubMed]
2. Lee, M.M.; Teuscher, J.; Miyasaka, T.; Murakami, T.N.; Snaith, H.J. Efficient Hybrid Solar Cells Based on Meso-Superstructured Organometal Halide Perovskites. *Science* **2012**, *338*, 643–647. [CrossRef] [PubMed]
3. Yang, W.S.; Park, B.W.; Jung, E.H.; Jeon, N.J.; Kim, Y.C.; Lee, D.U.; Shin, S.S.; Seo, J.; Kim, E.K.; Noh, J.H.; et al. Iodide management in formamidinium-lead-halide–based perovskite layers for efficient solar cells. *Science* **2017**, *356*, 1376–1379. [CrossRef]
4. Yang, W.S.; Noh, J.H.; Jeon, N.J.; Kim, Y.C.; Ryu, S.; Seo, J.; Seok, S.I. High-performance photovoltaic perovskite layers fabricated through intramolecular exchange. *Science* **2015**, *348*, 1234–1237. [CrossRef]
5. Zhang, X.; Ren, X.; Liu, B.; Munir, R.; Zhu, X.; Yang, D.; Li, J.; Liu, Y.; Smilgies, D.M.; Li, R.; et al. Stable high efficiency two-dimensional perovskite solar cells via cesium doping. *Energy Environ. Sci.* **2017**, *10*, 2095–2102. [CrossRef]
6. Hu, X.; Wang, H.; Wang, M.; Zang, Z. Interfacial defects passivation using fullerene-polymer mixing layer for planar-structure perovskite solar cells with negligible hysteresis. *Sol. Energy* **2020**, *206*, 816–825. [CrossRef]
7. Grancini, G.; Roldán-Carmona, C.; Zimmermann, I.; Mosconi, E.; Lee, X.; Martineau, D.; Narbey, S.; Oswald, F.; De Angelis, F.; Graetzel, M.; et al. One-Year stable perovskite solar cells by 2D/3D interface engineering. *Nat. Commun.* **2017**, *8*, 15684. [CrossRef]

8. El-Ballouli, A.O.; Bakr, O.M.; Mohammed, O.F. Structurally Tunable Two-Dimensional Layered Perovskites: From Confinement and Enhanced Charge Transport to Prolonged Hot Carrier Cooling Dynamics. *J. Phys. Chem. Lett.* **2020**, *11*, 5705–5718. [CrossRef] [PubMed]
9. Zhong, Y.; Hufnagel, M.; Thelakkat, M.; Li, C.; Huettner, S. Role of PCBM in the Suppression of Hysteresis in Perovskite Solar Cells. *Adv. Funct. Mater.* **2020**, *30*, 1908920. [CrossRef]
10. Xie, Y.; Yu, H.; Duan, J.; Xu, L.; Hu, B. Enhancing Device Performance in Quasi-2D Perovskite ($(BA)_2(MA)_3Pb_4I1_3$) Solar Cells Using $PbCl_2$ Additives. ACS Appl. *Mater. Interfaces* **2020**, *12*, 11190–11196. [CrossRef]
11. Gong, X.; Wang, Y.; Kuang, T. ZIF-8-Based Membranes for Carbon Dioxide Capture and Separation. *ACS Sustain. Chem. Eng.* **2017**, *5*, 11204–11214. [CrossRef]
12. Li, Y.; Miao, P.; Zhou, W.; Gong, X.; Zhao, X. N-doped carbon-dots for luminescent solar concentrators. *J. Mater. Chem. A* **2017**, *5*, 21452–21459. [CrossRef]
13. Shen, C.; Courté, M.; Krishna, A.; Tang, S.; Fichou, D. Quinoidal 2,2′,6,6′-Tetraphenyl-Dipyranylidene as a Dopant-Free Hole-Transport Material for Stable and Cost-Effective Perovskite Solar Cells. *Energy Technol.* **2017**, *5*, 1852–1858. [CrossRef]
14. Green, M.A.; Dunlop, E.D.; Yoshita, M.; Kopidakis, N.; Bothe, K.; Siefer, G.; Hao, X.; Jiang, J.Y. Solar Cell Efficiency Tables (Version 66). *Prog. Photovoltaics* **2025**, *33*, 795–810. [CrossRef]
15. Tsai, H.; Nie, W.; Blancon, J.C.; Stoumpos, C.C.; Asadpour, R.; Harutyunyan, B.; Neukirch, A.J.; Verduzco, R.; Crochet, J.J.; Tretiak, S.; et al. High-efficiency two-dimensional Ruddlesden–Popper perovskite solar cells. *Nature* **2016**, *536*, 312–316. [CrossRef] [PubMed]
16. Zhang, X.; Munir, R.; Xu, Z.; Liu, Y.; Tsai, H.; Nie, W.; Li, J.; Niu, T.; Smilgies, D.; Kanatzidis, M.G.; et al. Phase Transition Control for High Performance Ruddlesden–Popper Perovskite Solar Cells. *Adv. Mater.* **2018**, *30*, 1707166. [CrossRef]
17. Cao, D.H.; Stoumpos, C.C.; Farha, O.K.; Hupp, J.T.; Kanatzidis, M.G. 2D Homologous Perovskites as Light-Absorbing Materials for Solar Cell Applications. *J. Am. Chem. Soc.* **2015**, *137*, 7843–7850. [CrossRef]
18. Zhang, X.; Wu, G.; Fu, W.; Qin, M.; Yang, W.; Yan, J.; Zhang, Z.; Lu, X.; Chen, H. Orientation Regulation of Phenylethylammonium Cation Based 2D Perovskite Solar Cell with Efficiency Higher Than 11%. *Adv. Energy Mater.* **2018**, *8*, 1702498. [CrossRef]
19. Chen, Y.; Sun, Y.; Peng, J.; Zhang, W.; Su, X.; Zheng, K.; Pullerits, T.; Liang, Z. Tailoring Organic Cation of 2D Air-Stable Organometal Halide Perovskites for Highly Efficient Planar Solar Cells. *Adv. Energy Mater.* **2017**, *7*, 1700162. [CrossRef]
20. Ma, C.; Shen, D.; Ng, T.W.; Lo, M.F.; Lee, C.S. 2D Perovskites with Short Interlayer Distance for High-Performance Solar Cell Application. *Adv. Mater.* **2018**, *30*, 1800710. [CrossRef]
21. Stoumpos, C.C.; Cao, D.H.; Clark, D.J.; Young, J.; Roudinelli, J.M.; Jang, J.I.; Hupp, J.T.; Kanatzidis, M.G. Ruddlesden–Popper Hybrid Lead Iodide Perovskite 2D Homologous Semiconductors. *Chem. Mater.* **2024**, *20*, 2405953. [CrossRef]
22. Mann, D.S.; Thakur, S.; Sangale, S.S.; Jeong, K.-U.; Kwon, S.-N.; Na, S.-I. Interfacial engineering of nickel oxide-perovskite interface with amino acid complexed NiO to improve perovskite solar cell performance. *Small* **2015**, *8*, 1602–1608. [CrossRef] [PubMed]
23. Wang, Y.; Duan, L.; Hameiri, Z.; Liu, X.; Bai, Y.; Hao, X. PTAA as efficient hole transport materials in perovskite solar cells: A review. *Sol. RRL* **2022**, *6*, 2200234. [CrossRef]
24. Lim, K.G.; Ahn, S.; Kim, Y.H.; Qi, Y.; Lee, T.W. Universal energy level tailoring of self-organized hole extraction layers in organic solar cells and organic–inorganic hybrid perovskite solar cells. *Energy Environ. Sci.* **2016**, *9*, 932–939. [CrossRef]
25. Kim, H.; Lim, K.G.; Lee, T.W. Planar heterojunction organometal halide perovskite solar cells: Roles of interfacial layers. *Energy Environ. Sci.* **2016**, *9*, 12–30. [CrossRef]
26. Xie, X.; Liu, G.; Xu, C.; Li, S.; Liu, Z.; Lee, E.C. Tuning the work function of indium-tin-oxide electrodes for low-temperature-processed, titanium-oxide-free perovskite solar cells. *Org. Electron.* **2017**, *44*, 120–125. [CrossRef]
27. Liu, Y.; Yang, Z.; Cui, D.; Ren, X.; Sun, J.; Liu, X.; Zhang, J.; Wei, Q.; Fan, H.; Yu, F.; et al. Two-Inch-Sized Perovskite $CH_3NH_3PbX_3$ (X = Cl, Br, I) Crystals: Growth and Characterization. *Adv. Mater.* **2015**, *27*, 5176–5183. [CrossRef] [PubMed]
28. Luo, D.; Zhao, L.; Wu, J.; Hu, Q.; Zhang, Y.; Xu, Z.; Liu, Y.; Liu, T.; Chen, K.; Yang, W.; et al. Dual-Source Precursor Approach for Highly Efficient Inverted Planar Heterojunction Perovskite Solar Cells. *Adv. Mater.* **2017**, *29*, 1604758. [CrossRef]
29. Dong, Q.; Fang, Y.; Shao, Y.; Mulligan, P.; Qiu, J.; Cao, L.; Huang, J. Electron-hole diffusion lengths > 175 μm in solution-grown CH3NH3PbI3 single crystals. *Science* **2015**, *347*, 967–970. [CrossRef]
30. Wang, Z.B.; Helander, M.G.; Greiner, M.T.; Qiu, J.; Lu, Z.H. Analysis of charge-injection characteristics at electrode-organic interfaces: Case study of transition-metal oxides. *Phys. Rev. B* **2009**, *80*, 235325. [CrossRef]
31. Liu, Z.; Xie, X.; Lee, E.C. Effects of organic solvents for the phenyl-C61-butyric acid methyl ester layer on the performance of inverted perovskite solar cells. *Org. Electron.* **2018**, *56*, 247–253. [CrossRef]
32. Leijtens, T.; Eperon, G.E.; Noel, N.K.; Habisreutinger, S.N.; Petrozza, A.; Snaith, H.J. Stability of Metal Halide Perovskite Solar Cells. *Adv. Energy Mater.* **2015**, *5*, 1500963. [CrossRef]
33. Liu, G.; Xie, X.; Zeng, F.; Liu, Z. Improving the Performance of Perovskite Solar Cells Through Solvent Vapor Annealing-based Morphology Control of the Hole-Transport Layer. *Energy Technol.* **2018**, *6*, 1283–1289. [CrossRef]

34. Kim, B.J.; Kim, D.H.; Lee, Y.Y.; Shin, H.W.; Han, G.S.; Hong, J.S.; Mahmood, K.; Ahn, T.K.; Joo, Y.C.; Hong, K.S.; et al. Highly efficient and bending durable perovskite solar cells: Toward a wearable power source. *Energy Environ. Sci.* **2015**, *8*, 916–921. [CrossRef]
35. Chen, L.; Xie, X.; Liu, Z.; Lee, E.C. A transparent poly(3,4-ethylenedioxylenethiophene):poly(styrene sulfonate) cathode for low temperature processed, metal-oxide free perovskite solar cells. *J. Mater. Chem. A* **2017**, *5*, 6974–6980. [CrossRef]
36. Chen, W.; Zhang, Q. Recent progress in non-fullerene small molecule acceptors in organic solar cells (OSCs). *J. Mater. Chem. C* **2017**, *5*, 1275–1302. [CrossRef]
37. Lin, Y.; Wang, J.; Zhang, Z.G.; Bai, H.; Li, Y.; Zhu, D.; Zhan, X. An Electron Acceptor Challenging Fullerenes for Efficient Polymer Solar Cells. *Adv. Mater.* **2015**, *27*, 1170–1174. [CrossRef] [PubMed]
38. Bin, H.; Zhang, Z.G.; Gao, L.; Chen, S.; Zhong, L.; Xue, L.; Yang, C.; Li, Y. Non-Fullerene Polymer Solar Cells Based on Alkylthio and Fluorine Substituted 2D-Conjugated Polymers Reach 9.5% Efficiency. *J. Am. Chem. Soc.* **2016**, *138*, 4657–4664. [CrossRef]
39. Lin, Y.; Zhao, F.; He, Q.; Huo, L.; Wu, Y.; Parker, T.C.; Ma, W.; Sun, Y.; Wang, C.; Zhu, D.; et al. High-Performance Electron Acceptor with Thienyl Side Chains for Organic Photovoltaics. *J. Am. Chem. Soc.* **2016**, *138*, 4955–4961. [CrossRef]
40. Xie, X.; Liu, G.; Cheng, G.; Liu, Z.; Lee, E.C. Improving performance of organic solar cells by supplying additional acceptors to surface of bulk-heterojunction layers. *J. Mater. Chem. C* **2018**, *6*, 2793–2800. [CrossRef]
41. Liu, Z.; Lee, E.C. Solvent engineering of the electron transport layer using 1,8-diiodooctane for improving the performance of perovskite solar cells. *Org. Electron.* **2015**, *24*, 101–105. [CrossRef]
42. Li, S.; Ye, L.; Zhao, W.; Zhang, S.; Mukherjee, S.; Ade, H.; Hou, J. Energy-Level Modulation of Small-Molecule Electron Acceptors to Achieve over 12% Efficiency in Polymer Solar Cells. *Adv. Mater.* **2016**, *28*, 9423–9429. [CrossRef] [PubMed]
43. Qin, P.; Tetreault, N.; Dar, M.; Gao, P.; Call, K.L.M.; Rutter, S.R.; Ogier, S.D.; Forrest, N.D.; Bissett, J.S.; Simms, M.J.; et al. A Novel Oligomer as a Hole Transporting Material for Efficient Perovskite Solar Cells. *Adv. Energy Mater.* **2015**, *5*, 1400980. [CrossRef]
44. Wu, Y.; Wang, P.; Wang, S.; Wang, Z.; Cai, B.; Zheng, X.; Chen, Y.; Yuan, N.; Ding, J.; Zhang, W.H. Heterojunction Engineering for High Efficiency Cesium Formamidinium Double-Cation Lead Halide Perovskite Solar Cells. *Chemsuschem* **2017**, *5*, 11.
45. Song, J.; Li, J.; Xu, L.; Li, J.; Zhang, F.; Han, B.; Shan, Q.; Zeng, H. Room-Temperature Triple-Ligand Surface Engineering Synergistically Boosts Ink Stability, Recombination Dynamics, and Charge Injection toward EQE-11.6% Perovskite QLEDs. *Adv. Mater.* **2018**, *30*, 1800764. [CrossRef]
46. Deng, Y.; Zheng, X.; Bai, Y.; Wang, Q.; Zhao, J.; Huang, J. Surfactant-controlled ink drying enables high-speed deposition of perovskite films for efficient photovoltaic modules. *Nat. Energy* **2018**, *3*, 560–566. [CrossRef]
47. Fu, F.; Feurer, T.; Weiss, T.P.; Pisoni, S.; Avancini, E.; Andres, C.; Buecheler, S.; Tiwari, A.N. High-efficiency inverted semi-transparent planar perovskite solar cells in substrate configuration. *Nat. Energy* **2016**, *2*, 16190. [CrossRef]
48. Jacak, J.E.; Jacak, W.A. Routes for Metallization of Perovskite Solar Cells. *Materials* **2022**, *15*, 2254. [CrossRef]

Disclaimer/Publisher's Note: The statements, opinions and data contained in all publications are solely those of the individual author(s) and contributor(s) and not of MDPI and/or the editor(s). MDPI and/or the editor(s) disclaim responsibility for any injury to people or property resulting from any ideas, methods, instructions or products referred to in the content.

inorganics

MDPI

Article

Synergistic Enhancement of Near-Infrared Electrochromic Performance in $W_{18}O_{49}$ Nanowire Thin Films via Copper Doping and Langmuir–Blodgett Assembly

Yueyang Wu , Honglong Ning *, Ruiqi Luo, Muyun Li, Zijian Zhang , Rouqian Huang, Junjie Wang, Mingyue Peng, Runjie Zhuo, Rihui Yao * and Junbiao Peng

Guangdong Basic Research Center of Excellence for Energy & Information Polymer Materials, State Key Laboratory of Luminescent Materials and Devices, School of Materials Sciences and Engineering, South China University of Technology, Guangzhou 510640, China
* Correspondence: ninghl@scut.edu.cn (H.N.); yaorihui@scut.edu.cn (R.Y.)

Abstract: The development of high-performance electrochromic materials demands innovative approaches to simultaneously control the nanoscale architecture and the electronic structure. We present a dual-modification strategy that synergistically combines copper doping with the Langmuir–Blodgett (LB) assembly to overcome the traditional performance trade-offs in tungsten oxide-based electrochromic systems. Cu-doped $W_{18}O_{49}$ nanowires with varying Cu concentrations (0–12 mol%) were synthesized hydrothermally and assembled into thin films via the LB technique, with LB precursors characterized by contact angle, surface tension, viscosity, and thermogravimetric-differential scanning calorimetry (TG-DSC) analyses. The films were systematically evaluated using scanning electron microscopy, X-ray photoelectron spectroscopy, chronoamperometry, and transmittance spectroscopy. Experimental results reveal an optimal Cu-doping concentration of 8 mol%, achieving a near-infrared optical modulation amplitude of 76.24% at 1066 nm, rapid switching kinetics (coloring/bleaching: 5.0/3.0 s), and a coloration efficiency of 133.00 cm^2/C. This performance is speculated to be a balance between Cu-induced improvements in ion intercalation kinetics and LB-ordering degradation caused by lattice strain and interfacial charge redistribution, while mitigating excessive doping effects such as structural deterioration and thermodynamic instability. The work establishes a dual-modification framework for designing high-performance electrochromic interfaces, emphasizing the critical role of surface chemistry and nanoscale assembly in advancing adaptive optoelectronic devices like smart windows.

Keywords: Langmuir–Blodgett assembly; copper doping; $W_{18}O_{49}$; nanowires; thin films; near-infrared region; electrochromic materials

1. Introduction

The escalating global energy crisis and environmental challenges have intensified demands for energy-efficient technologies and sustainable functional materials, particularly in the field of optoelectronic devices. Electrochromic materials, capable of reversibly modulating optical properties under external electric fields, have emerged as pivotal components for smart windows, adaptive displays, and energy-saving coatings, offering transformative potential for reducing building energy consumption [1–10]. Among these materials, tungsten oxide (WO_3) has distinguished itself as a frontrunner for next-generation electrochromic devices due to its high coloration efficiency, strong optical contrast, cost-effectiveness, and long-term operational stability [11,12].

Recent advancements have focused on doping strategies and optimized thin-film fabrication processes to enhance electrochromic performance. For instance, Yong et al. reported molybdenum-doped WO_3 films achieving 55% optical modulation at 550 nm [13], while Nguyen et al. demonstrated 59% modulation via spin-coated WO_3 [14]. However, persistent limitations in film uniformity, slow switching kinetics, and insufficient spectral coverage continue to impede their practical implementation. These challenges underscore an urgent need for innovative manufacturing strategies.

In this work, we address these limitations through the synergistic integration of two modification approaches: (1) precise engineering of $W_{18}O_{49}$ nanowires via copper (Cu) doping, and (2) the novel application of Langmuir–Blodgett (LB) assembly in thin-film fabrication. The LB technique [15–17], leveraging amphiphilic molecular self-organization at air-water interfaces, enables the deposition of ordered, homogeneous monolayers with controlled thickness—a critical advancement over conventional methods prone to inhomogeneity and impurity incorporation. By systematically regulating Cu-doping concentrations (0%, 4%, 8%, 12%), we achieved exceptional electrochromic performance, including a 76.24% modulation amplitude at 1066 nm (near-infrared, NIR), rapid switching kinetics (5.0 s coloration and 3.0 s bleaching), and a coloration efficiency of 133.00 cm^2/C.

Furthermore, extending electrochromic activity to the NIR spectrum [18–22] represents a strategic imperative, given that nearly 50% of solar energy resides in this region. For instance, Zhou et al. reported WO_3/Ag NW films achieving 68.3% of optical modulation at 1100 nm [23], while Huang et al. demonstrated 72% modulation at 1200 nm via hydrothermal WO_3 nanosheets [24]. Our Cu-doped $W_{18}O_{49}$ films, fabricated via LB technology, also effectively address this requirement by demonstrating robust NIR modulation capabilities that synergize energy-saving benefits with radiative thermal control. This study not only advances the fundamental understanding of doping-engineered electrochromism but also establishes a scalable paradigm for multifunctional smart coatings in sustainable architecture and adaptive optics, bridging material innovation with practical energy management solutions.

2. Materials and Methods

2.1. Material and Reagent Information

Tungsten hexachloride (WCl_6, 99.9%), polyvinylpyrrolidone (PVP, average molecular weight 2500, K12), copper(II) chloride dihydrate ($CuCl_2 \cdot 2H_2O$, 99%), n-hexane (99%), anhydrous ethanol (99.5%), and propylene carbonate (PC, 99%) were procured from Macklin Biochemical Co., Ltd. (Shanghai, China). Lithium perchlorate ($LiClO_4$, 99%) was supplied by Tongyuan Chemical Co., Ltd. (Guangzhou, China). Indium tin oxide (ITO)-coated glass substrates (1×1 cm^2, 2×2 cm^2; sheet resistance: 10 Ω/cm^2) were purchased from New Vision Optoelectronics Technology Co., Ltd. (Guangzhou, China). All chemicals were used as received without further purification.

2.2. Fabrication of Thin Films and Devices

2.2.1. Synthesis of $W_{18}O_{49}$ Nanowires

Four batches of precursors were prepared by dissolving 0.125 g WCl_6, 0.003 g PVP, and varying masses of $CuCl_2 \cdot 2H_2O$ (0 g, 0.0021 g, 0.0043 g, and 0.0064 g, corresponding to Cu molar ratios of 0%, 4%, 8%, and 12%, respectively) in 50 mL anhydrous ethanol under magnetic stirring until complete dissolution. Each solution was diluted to 150 mL with ethanol and transferred into a Teflon-lined autoclave for hydrothermal treatment at 180 °C for 10 h. The resulting suspensions were centrifuged at 8000 rpm for 5 min, after which the supernatant was discarded, and the blue gel-like precipitates were collected for subsequent use.

2.2.2. Preparation of LB Solutions

The nanowire gels from each doping group were redispersed in 2 mL ethanol and 4 mL n-hexane, followed by homogenization via mechanical stirring. The mixtures were sonicated for 2 min in a 200 mL water-filled beaker and aged for 24 h to stabilize the colloidal suspensions.

2.2.3. LB Film Deposition

Figure 1 shows the experimental steps and the schematic diagram of the LB method for film formation. ITO substrates were ultrasonically cleaned sequentially in deionized water, ethanol, and water (10 min each) and dried under nitrogen. Then, 0.6 mL of the aged LB dispersion was spread onto a 30 mL deionized water subphase in a 50 mL beaker. After 5 min of solvent evaporation, ITO substrates were vertically immersed into the subphase at a 30° angle and withdrawn at a controlled speed (1 min immersion and 1 min withdrawal) to deposit monolayers. Substrates were air-dried to evaporate residual solvent and annealed at 150 °C for 1 min (single-layer films for SEM) or 30 min (7-layer films for device integration). Film deposition was performed on the glass side of 1 × 1 cm^2 substrates and the ITO side of 2 × 2 cm^2 substrates.

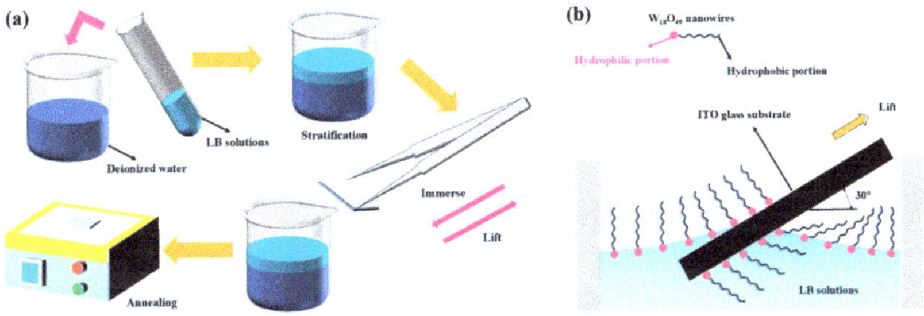

Figure 1. LB film formation: (**a**) experimental process, (**b**) mechanism.

2.2.4. Device Assembly

Electrochromic devices were fabricated by sandwiching the LB-coated ITO substrate (2 × 2 cm^2) with a bare ITO substrate using conductive adhesive, creating an active area of ~1.5 cm^2. The interlayer was filled with 1.5 mol/L $LiClO_4$-PC electrolyte to complete the "sandwich" structure.

2.3. Characterization and Analytical Methods

2.3.1. LB Precursor Solution Analysis

Interfacial Properties: Surface tension and contact angles were measured using an Attension Theta Lite optical tensiometer (TL200, Biolin Scientific, Gothenburg, Sweden).

Viscosity: Determined with a HAAKE MARS 40 rheometer (Thermo Scientific, Waltham, MA, USA) at 25 °C (Contact angle, surface tension and viscosity were tested three times).

Thermogravimetric-Differential Scanning Calorimetry (TG-DSC): Conducted on a STA 449 F5 instrument (NETZSCH, Selb, Germany) from 25 °C to 400 °C at 10 °C/min under nitrogen flow.

2.3.2. Thin Film and Device Characterization

Morphology: Surface topography was imaged by scanning electron microscopy (SEM, EVO18, Carl Zeiss AG, Oberkochen, Germany) at 20,000× magnification.

Chemical Composition: X-ray photoelectron spectroscopy (XPS, Thermo Fisher Scientific, Waltham, MA, USA) with Al Kα radiation (1486.6 eV).

Electrochemical Performance: Cyclic chronoamperometry (CA) (±2.5 V, 30 s per cycle, 5 cycles) using a CHI660E electrochemical workstation (CHI660E, CH Instruments, Shanghai, China).

Optical Modulation: Transmittance spectra at 550, 600, 700, 800, 900, 1000, 1066 nm were recorded with a microspectrometer (Morpho PG2000, Shanghai, China). Optical modulation amplitude, switching times, and coloration efficiency were calculated using synchronized electrochemical-optical data.

3. Results and Discussion

3.1. *Interfacial Properties*

The contact angle measurements (Figure 2a–d) demonstrate favorable wettability of the LB solutions on ITO substrates (contact angles < 90°), facilitating uniform thin-film spreading. Low-concentration Cu-doping (4%) increased the contact angle, whereas higher doping levels (>4%) further reduced the contact angle and enhanced substrate wettability. This behavior is corroborated by the surface tension and viscosity trends (Figure 2e–f): 4% Cu-doping marginally elevated both parameters, while higher doping concentrations monotonically decreased them.

The directional patterns in the experimental data suggest a competitive interplay between the two modification mechanisms—Cu-doping and LB-derived film formation—which necessitates the existence of a synergistic concentration threshold for optimal electrochromic performance. Subsequently, this critical concentration value will be determined through comprehensive characterization.

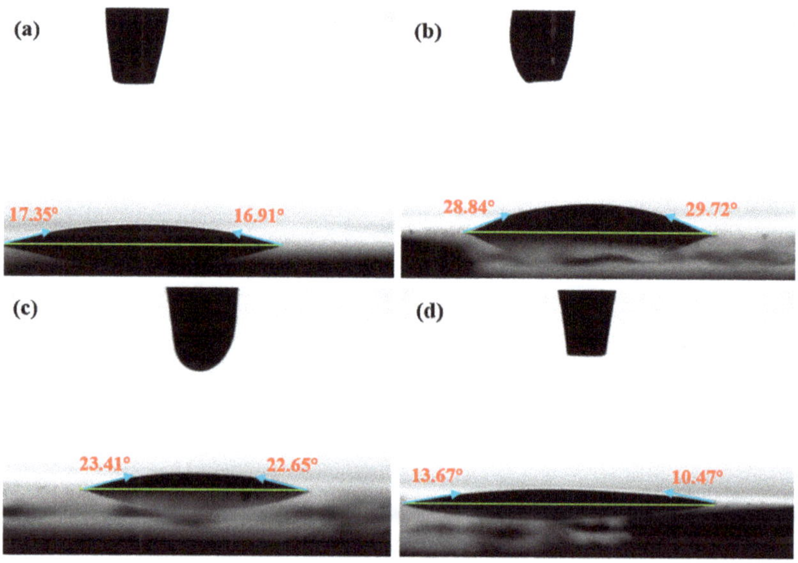

Figure 2. *Cont.*

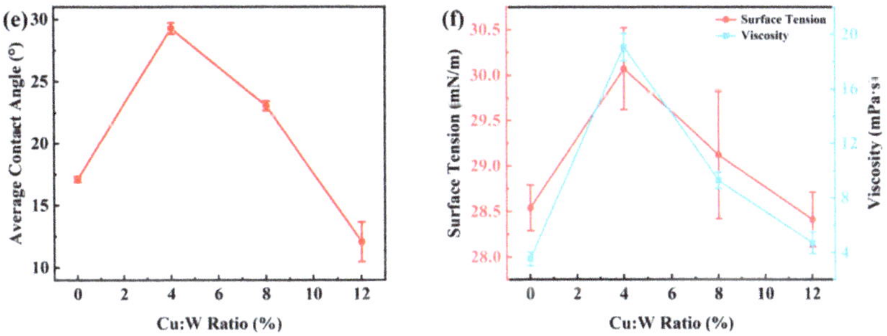

Figure 2. Interfacial properties of $W_{18}O_{49}$ nanowire LB precursor solutions with different copper doping concentrations—contact angle: (**a**) 0%, (**b**) 4%, (**c**) 8%, (**d**) 12%, (**e**) contact angle line chart, (**f**) line chart of surface tension and viscosity. In figures (**a**–**d**), the horizontal green baseline represents the surface of the ITO substrate, and the angle between the blue arrow and the green baseline is the contact angle, which is marked in red.

3.2. TG-DSC Analysis

As illustrated in Figure 3, the thermogravimetric-differential scanning calorimetry (TG-DSC) profiles of copper-doped $W_{18}O_{49}$ nanowire LB precursor solutions exhibit similar trends across varying doping concentrations. The DSC curves reveal an endothermic peak near 100 °C, while the TG curves display a monotonic mass loss from room temperature to ~110 °C, attributable to the evaporation of ethanol and n-hexane (boiling points: 78 °C and 69 °C, respectively). Above 110 °C, the TG curves plateau, indicating thermal stability of the residual material. The absence of additional endothermic or exothermic transitions in the DSC profiles confirms no phase transformations occurred during the heating process [25].

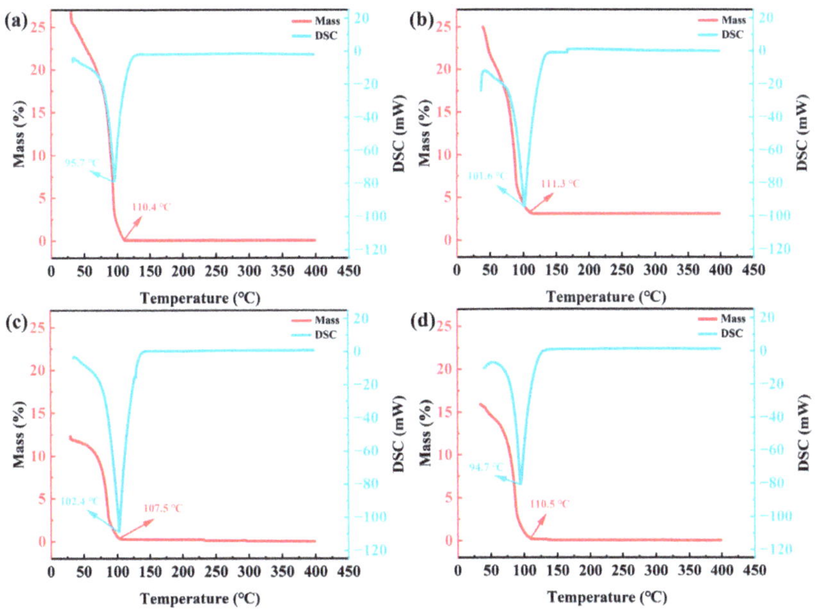

Figure 3. TG-DSC curves of $W_{18}O_{49}$ nanowire LB precursor solutions with different copper doping concentrations: (**a**) 0%, (**b**) 4%, (**c**) 8%, (**d**) 12%.

Notably, the 8% Cu-doped sample demonstrates the highest endothermic peak temperature (102.4 °C) in the DSC analysis, suggesting that solvent evaporation requires greater energy input, which reflects enhanced thermodynamic stability. This phenomenon is speculated to arise from optimized $W_{18}O_{49}$ crystal structures at moderate Cu-doping levels, characterized by reduced porosity and strengthened solvent molecule adsorption on the material surface. Such thermodynamic stability is critical for LB film fabrication, as it ensures a controlled solvent evaporation process, promoting the formation of uniform and dense thin films. The resulting compact nanostructure facilitates continuous electron transport pathways while optimizing ion insertion/extraction kinetics, thereby collectively enhancing electrochromic performance. This result indicates that 8% may be the optimal synergistic concentration to be determined, but other characterization methods are still needed for comprehensive judgment.

3.3. SEM Morphology

Scanning electron microscopy (SEM) analysis revealed distinct structural evolution in Cu-doped $W_{18}O_{49}$ thin films (Figure 4). All films exhibited interconnected nanowire networks, a morphology favorable for electrochromic applications due to their high surface area, which facilitates abundant Li^+ intercalation sites and shortened ion diffusion pathways. Compared to spin-coated films, LB-assembled films demonstrated significantly improved nanowire alignment overall, indicative of optimized assembly kinetics.

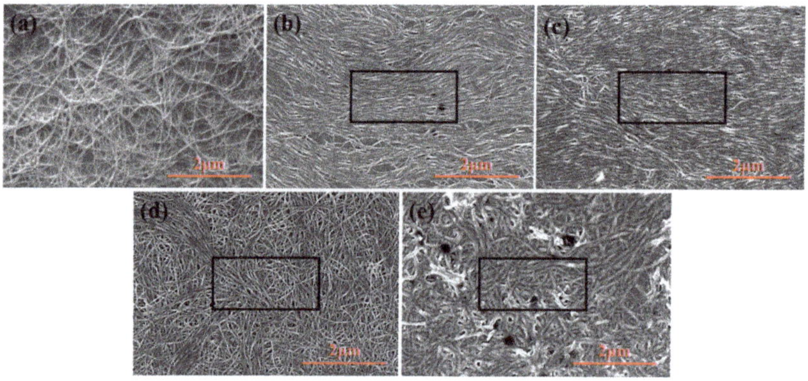

Figure 4. SEM images of $W_{18}O_{49}$ thin films with different copper doping concentrations: (**a**) spin coating method (0%); LB method: (**b**) 0%, (**c**) 4%, (**d**) 8%, (**e**) 12%. The black rectangular box is the selection area for evaluating the orderliness of orientation.

The structural ordering of Cu-doped $W_{18}O_{49}$ nanowires was quantitatively assessed by analyzing SEM images using Image Pro Plus software (v6.0, Media Cybernetics, 10 Eunos Road 8, Singapore). A rectangular region ($2.5 \times 1.2\ \mu m^2$) was selected to calculate the alignment ratio, defined as the percentage of nanowires oriented within $\pm 10°$ of the dominant alignment axis relative to the total nanowire count in the region. The alignment ratio systematically decreased with increasing Cu-doping concentration: 65.24% for 0% Cu (pristine), 50.65% for 4% Cu, 40.92% for 8% Cu, and 20.54% for 12% Cu. This structural degradation is attributed to two primary mechanisms:

(1) Lattice strain effects: Cu ions occupying interstitial sites or substituting W atoms within the $W_{18}O_{49}$ lattice induce local lattice distortions. Such strain perturbations disrupt the anisotropic growth dynamics of one-dimensional nanowires, destabilizing their ordered assembly.
(2) Interfacial charge redistribution: Doping alters the surface charge characteristics of nanowires [26], perturbing the critical electrostatic equilibrium required for LB orientation at the air-liquid interface and introducing defects in monolayer stacking.

The interplay between atomic-scale lattice strain and mesoscale charge imbalance underscores the sensitivity of LB assembly to dopant-induced perturbations, emphasizing the necessity for precise doping control in nanofilm engineering.

Significantly, Figure 4e shows localized dark spots in the 12% Cu-doped film, suggesting that excessive doping may compromise surface quality through heterogeneous phase segregation or impurity accumulation. Cross-referencing SEM observations with DSC data (Figure 3), we deduce that the optimal synergistic Cu-doping concentration lies near 8%.

3.4. XPS Analysis

X-ray photoelectron spectroscopy (XPS) surveys (Figure 5) confirm the elemental composition of W, O, Cu, and trace C/Si in all films. The Si signal originates from the ITO substrate, while residual carbon (C 1s at 284.8 eV) stems from incomplete solvent evaporation during annealing, as evidenced by localized film blackening at excessive temperatures (>150 °C). Notably, Cu 2p 3/2 peaks emerge at 931.5–932.8 eV (Figure 5a–d insets), with peak intensity scaling linearly with nominal doping concentrations, verifying controlled Cu incorporation into the $W_{18}O_{49}$ matrix.

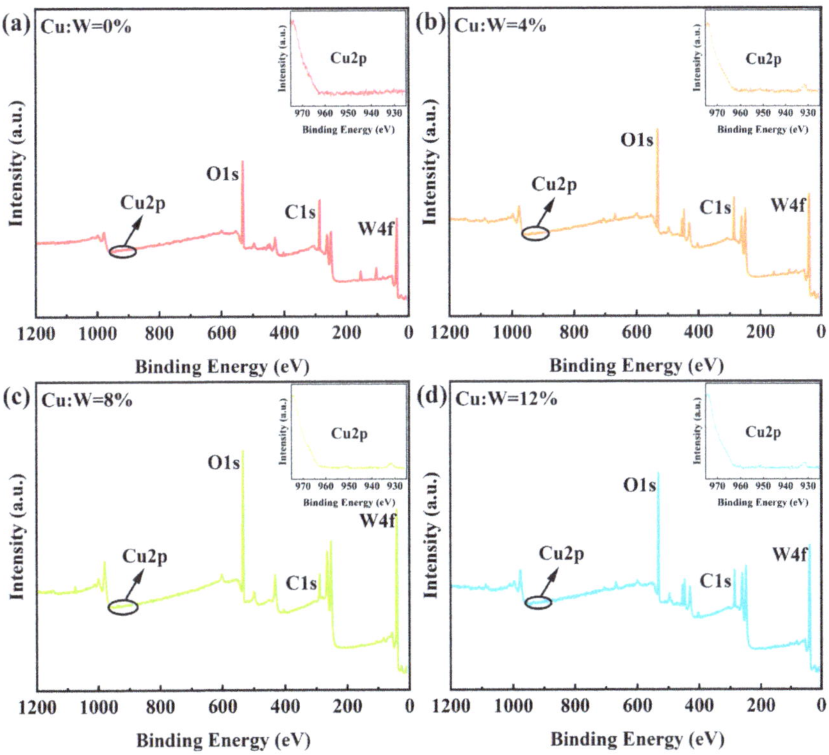

Figure 5. XPS spectra of $W_{18}O_{49}$ thin films with different copper doping concentrations: full spectra and Cu 2p spectra. (**a**) 0%, (**b**) 4%, (**c**) 8%, (**d**) 12%.

The calculated O/W atomic ratio (~2.81) slightly exceeded the stoichiometric $W_{18}O_{49}$ value (2.72), attributed to the oxygen contribution of hydroxyl groups adsorbed on the substrate and surface, but also largely confirmed the elemental composition of $W_{18}O_{49}$. O 1s deconvolution (Figure 6) quantifies lattice oxygen (529.43 ± 0.1 eV) and oxygen vacancies (530.51 ± 0.1 eV) [27,28]. As the concentration of copper doping increases, the concentration of oxygen vacancies first decreases and then increases, reaching a minimum relative content of 8%, indicating that copper ions occupy the positions of oxygen vacancies to the maximum extent at this copper doping concentration, providing a higher density of coloring sites per unit area. This also proves that the 8% is likely to be the optimal synergistic doping concentration.

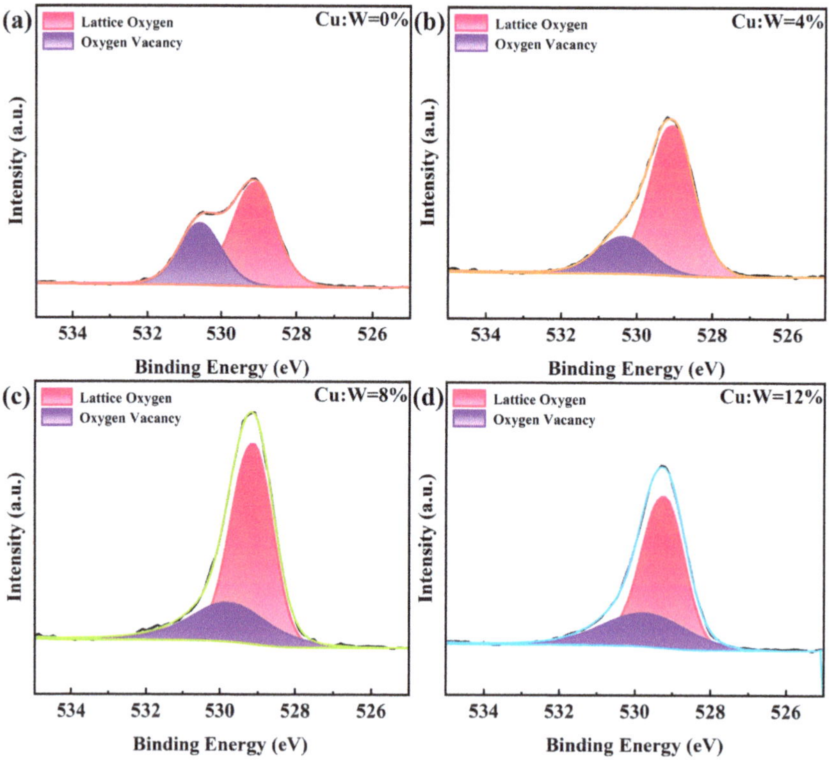

Figure 6. XPS spectra of $W_{18}O_{49}$ thin films with different copper doping concentrations: O 1s spectra. (**a**) 0%, (**b**) 4%, (**c**) 8%, (**d**) 12%. The four different colors "red, orange, green, and blue" represent the fitting peak shapes of the experimental groups with copper doping concentrations of "0%, 4%, 8%, and 12%", respectively.

3.5. Electrochemical Performance

The electrochemical dynamics of Cu-doped $W_{18}O_{49}$ electrochromic devices were investigated via chronoamperometry (CA) under $\pm$ 2.5 V cycling (Figure 7). All devices exhibit rapid current stabilization within 2–3 s during both coloration (negative bias) and bleaching (positive bias) phases, indicative of efficient double-layer charging and fast Faradaic charge transfer kinetics. When a step voltage is applied, the larger the sudden change value of the current, the faster the insertion/extraction rate of ions in the thin film, and the smaller the charge transfer resistance. The decrease in the sudden change value of current may reflect an increase in interface resistance or obstruction of ion diffusion, indicating the deterioration of material structure. Based on the above comparative analysis, the film doped with 8% copper exhibits a comprehensive combination of high-ion transport speed and strong stability, demonstrating excellent electrochemical performance. At other doping concentrations, the instability of current mutation values may be due to the inability of LB assembly process to maintain a defect minimized single-layer structure due to dopant interference, resulting in locally dense/sparse regions. This morphological heterogeneity will increase the charge transfer resistance.

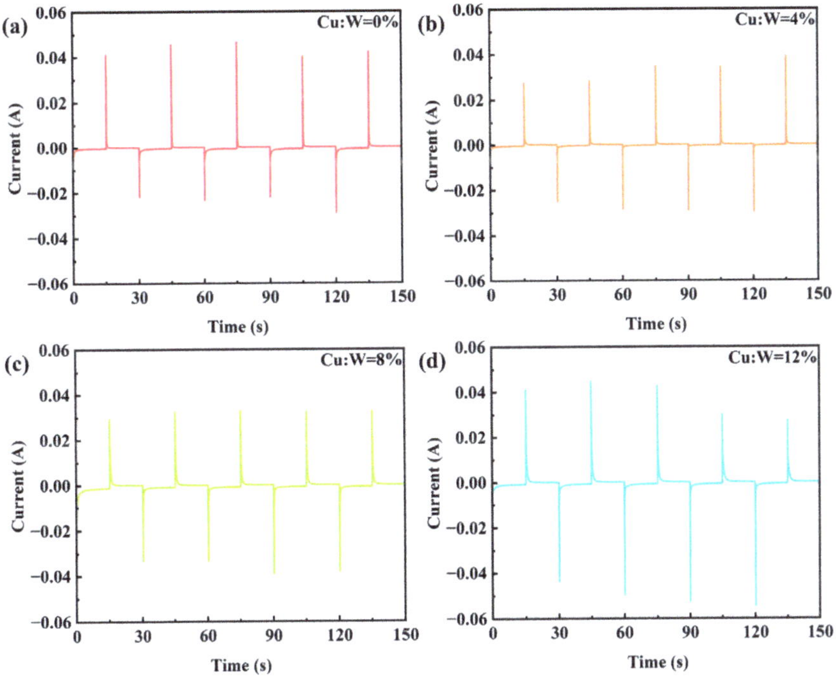

Figure 7. Current–time curves of $W_{18}O_{49}$ thin film devices with different copper doping concentrations: (**a**) 0%, (**b**) 4%, (**c**) 8%, (**d**) 12%.

3.6. Optical Modulation and Device Metrics

Figure 8 quantifies the electrochromic response of devices through synchronized transmittance–time profiling. Key performance metrics were calculated as follows:

- Optical modulation amplitude:

$$\Delta T = T_b - T_c \tag{1}$$

where T_b (bleaching state) and T_c (coloring state) represent steady-state transmittance.

- Switching time:

$$t_c/t_b = t_{0.9\Delta T} \tag{2}$$

$t_{0.9\Delta T}$: time to reach 90% of ΔT.

- Coloration efficiency:

$$\eta = \frac{\Delta A}{Q} \tag{3}$$

$$\Delta A = lg\frac{T_b}{T_c} \tag{4}$$

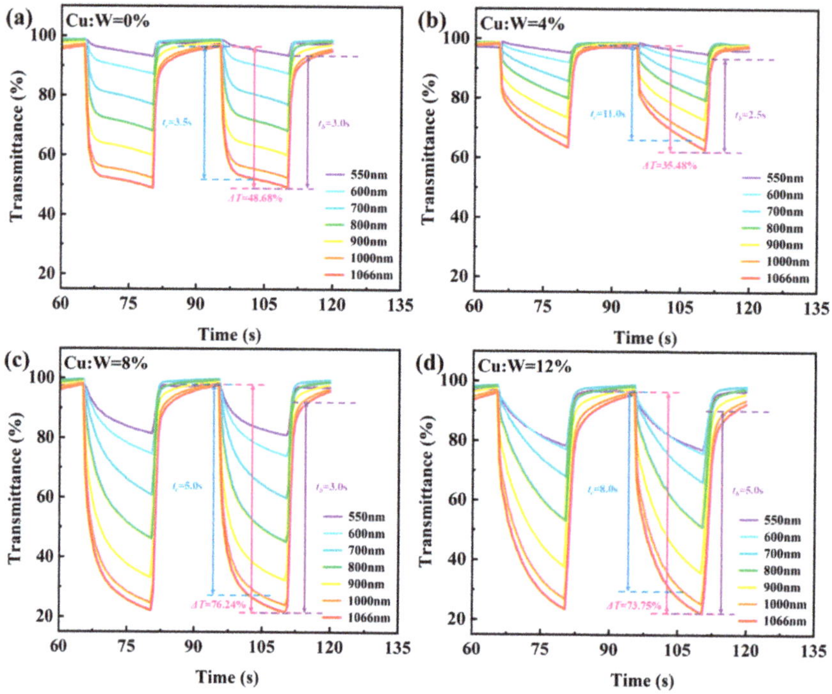

Figure 8. Transmittance–time curves of $W_{18}O_{49}$ thin film devices with different copper doping concentrations: (**a**) 0%, (**b**) 4%, (**c**) 8%, (**d**) 12%.

As shown in Figure 9 and Table 1, device performance exhibits a volcano-type dependence on Cu concentration. The longer the testing band of the device, the greater the modulation amplitude. At the testing wavelength of 1066 nm, the optimal comprehensive optical performance index was achieved when doped with 8% copper: (ΔT = 76.24%, t_c = 5.0 s, t_b = 3.0 s, η = 133.00 cm^2/C). By combining the electrochemical and optical performance tests of the device, it can be determined that the optimal synergistic concentration for achieving the best electrochromic performance through LB film formation and copper doping is 8%.

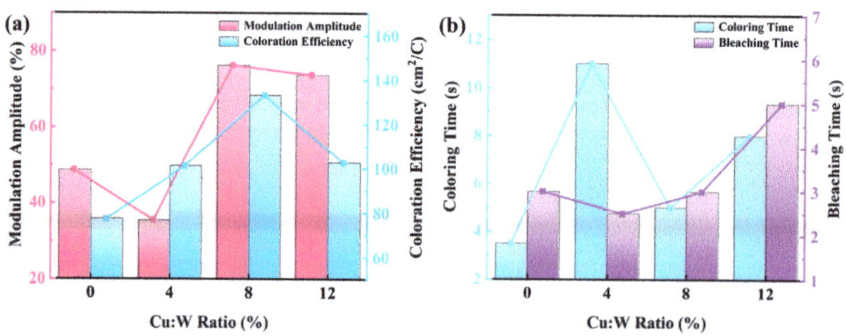

Figure 9. Line chart: (**a**) modulation amplitude and coloration efficiency, (**b**) response time (coloring/bleaching).

Table 1. Summary of device performance test data.

Copper Doping Ratio (%)	Modulation Amplitude ΔT (%)	Response Time t_c/t_b (s)	Coloration Efficiency η (cm^2/C)
0	48.68	3.5/3.0	77.17
4	35.48	11.0/2.5	101.06
8	76.24	5.0/3.0	133.00
12	73.75	8.0/5.0	102.63

In summary, we further analyze and rationalize the existence of an optimal synergistic Cu-doping concentration in LB-assembled $W_{18}O_{49}$ films through the interplay of two competing mechanisms:

(1) Copper doping increases coloring sites: When the copper doping concentration is $\leq$8%, increasing the copper doping concentration will introduce additional redox active centers, thereby increasing the number of available color development sites and improving the charge storage capacity, enhancing the electrochromic performance. When the copper doping exceeds 8%, the performance degradation originates from a critical point where structural degradation (SEM), defect site competition (XPS), and thermodynamic instability (TG-DSC) outweigh the benefits of increased redox active sites.

(2) LB-order degradation: According to the quantitative analysis of the orderliness of SEM images, higher doping levels disrupt the LB-driven structural alignment of nanowires, leading to reduced ordering and a corresponding decline in coloration site density per unit area, which detrimentally impacts ion diffusion kinetics and optical modulation efficiency.

When the doping concentration is low ($\leq$4%), the degradation of the orderliness of LB films plays a dominant role in affecting their performance, resulting in a decrease in modulation amplitude when doped with 4% copper compared to when not doped with copper. When the doping concentration is high (4–8%), copper doping increases the dominant role of coloring sites and significantly increases the modulation amplitude. When the doping concentration is too high (>8%), the damage to performance caused by excessive doping exceeds the gain of the increase in coloring sites, resulting in a slight decrease in modulation amplitude.

The intersection of these competing curves defines the optimal Cu-doping threshold (around 8%) (Figure 10). At this critical concentration, the improvement of ion intercalation kinetics caused by doping is balanced with the degradation of LB ordering caused by lattice strain and interface charge redistribution, while avoiding the adverse effects of excessive doping.

This speculative mechanistic framework aligns with experimental observations and underscores the necessity of balancing atomic-scale doping effects with nanoscale assembly control for designing high-performance electrochromic systems. The identified 8% Cu-doping level represents a "sweet spot" where competing electronic, ionic, and structural factors achieve mutual optimization, offering a universal guideline for defect engineering in nanostructured functional coatings.

In future improvements, conductivity testing, electrochemical impedance testing and Raman spectroscopy testing can be combined to further analyze the mechanism of the synergistic effect. In addition, due to the use of manual pulling in the LB film formation process in this study, the repeatability is not ideal, and it is difficult to prepare on a large scale. Machine pulling can be used for future improvements.

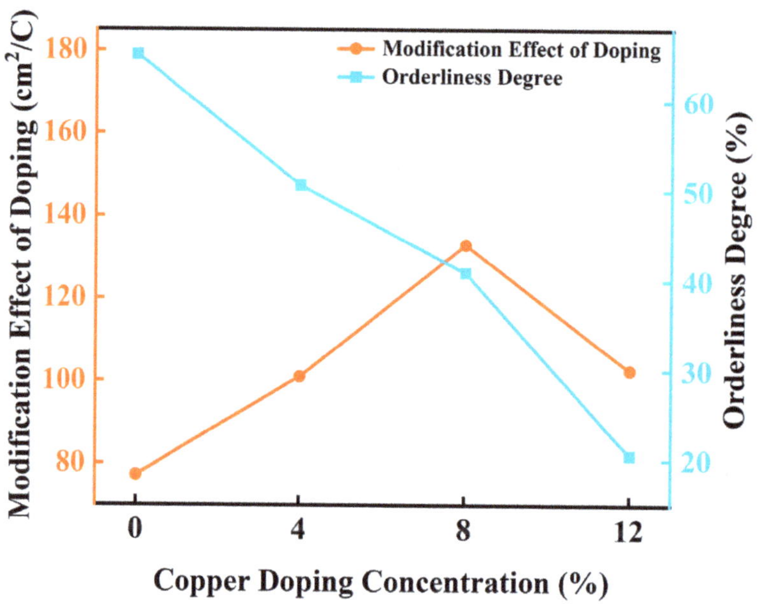

Figure 10. Speculation on the mechanism of synergistic copper doping concentration.

Table 2 summarizes some recent related studies and compares our research results.

Table 2. Comparison between this study and other studies.

Work	Film Forming Process	Material	Coloration Efficiency η (cm^2/C)	Modulation Amplitude ΔT (%)
[23]	Electrochemical deposition	WO_3/Ag NW	86.9	68.3 (1100 nm)
[24]	Hydrothermal method	h-WO_3	70.0	72.0 (1200 nm)
[29]	Magnetron sputtering	WO_3	80.5	72.5 (1000 nm)
[30]	Magnetron sputtering	WO_3	102.9	78.8 (1000 nm)
Ours	Langmuir–Blodgett method	Cu-doped $W_{18}O_{49}$	133.0	76.2 (1066 nm)

4. Conclusions

This study demonstrates a new manufacturing strategy for high-performance near-infrared (NIR) electrochromic devices by comprehensively using LB film formation and copper doping to modify $W_{18}O_{49}$ nanowire thin films. The system optimization of copper doping concentration showed that an 8% copper doped device achieved an excellent near-infrared modulation index at 1066 nm: 76.24% light modulation amplitude, fast switching kinetics (coloring: 5.0 s, bleaching: 3.0 s), and excellent coloration efficiency (133.00 cm^2/C), making it the optimal synergistic concentration for achieving excellent comprehensive electrochromic performance. At this critical concentration, it is speculated that the improvement in ion intercalation kinetics caused by doping is balanced with the degradation of LB ordering caused by lattice strain and interface charge redistribution, while avoiding the adverse effects of excessive doping. The observed near-infrared selective electrochromic response makes $W_{18}O_{49}$ a promising candidate for spectral selective smart windows, meeting the key requirements of thermal management without compromising visible transparency. This work provides a material design blueprint that goes beyond traditional visible light with electrochromic properties and has a direct impact on energy-efficient buildings and adaptive optics systems.

Author Contributions: Conceptualization, Y.W. and H.N.; data curation, Y.W.; formal analysis, Y.W.; funding acquisition, H.N. and R.Y.; investigation, M.L.; methodology, Y.W.; project administration, J.P.; resources, H.N.; software, R.L.; supervision, H.N. and R.Y.; validation, Y.W., R.L., and Z.Z.; visualization, Y.W.; writing—original draft, Y.W., R.H., J.W., M.P., and R.Z.; writing—review and editing, H.N. and R.Y. All authors have read and agreed to the published version of the manuscript.

Funding: This work was funded by National Key R&D Program of China (No.2021YFB3600604), National Natural Science Foundation of China (Grant No.62174057 and 22090024), Guangdong Natural Science Foundation (No.2024A1515012216 and 2023A1515011026), Educational Commission of Guangdong Province (Grant Nos.2022ZDZX1002), Key R&D Plan of Guangdong Province (2022B0303010001), State Key Lab of Luminescent Materials and Devices (Skllmd-2024-05), Southwest Institute of Technology and Engineering Cooperation Fund (HDHDW59A020301), and the 2025 Hundred Step Ladder Climbing Program (j2tw202502026).

Institutional Review Board Statement: Not applicable.

Informed Consent Statement: Not applicable.

Data Availability Statement: Data are contained within the article.

Conflicts of Interest: The authors declare no conflicts of interest.

References

1. Wang, Y.; Zhang, Z.; Wang, Z.; Zuo, Y.; Zhou, H.; Sun, D.; Li, Y.; Yan, Y.; Feng, T.; Xie, A. Self-seeded growth of hexagonal-phase WO_3 film by a one-step hydrothermal method for high-performance electrochromic energy storage devices. *J. Power Sources* **2025**, *633*, 236350. [CrossRef]
2. Huang, S.; Guo, H.; Xia, P.; Sun, H.; Lu, C.; Feng, Y.; Zhu, J.; Liang, C.; Xu, S.; Wang, C. Integrated device of luminescent solar concentrators and electrochromic supercapacitors for self-powered smart window and display. *Nat. Commun.* **2025**, *16*, 2085. [CrossRef] [PubMed]
3. Jiao, K.; Zhang, Z.; Liu, Y.; Li, J.; Xue, C.; Zhang, J.; Chen, Y.; Chen, Z.; Hou, X.; Zhang, L.; et al. Carbon nanotube-based solid-state mid-infrared electrochromic devices for smart outdoor thermal management. *Chem. Eng. J.* **2025**, *507*, 160802. [CrossRef]
4. Ding, Y.; Qiu, Q.; Jiang, M.; Li, Y.; Shi, J.; Du, L.; Li, C.; Wang, J.; Li, Z. $Ti_3C_2T_x$ MXene-mediated synthesis of a prussian blue nanocomposite film for a flexible large-area electrochromic device. *ACS Appl. Mater. Interfaces* **2025**, *17*, 15657–15665. [CrossRef] [PubMed]
5. Gillissen, F.; Lobet, M.; Dewalque, J.; Colson, P.; Spronck, G.; Gouttebaron, R.; Duttine, M.; Faceira, B.; Rougier, A.; Henrard, L.; et al. Mixed molybdenum-tungsten oxide as dual-band, VIS-NIR selective electrochromic material. *Adv. Opt. Mater.* **2025**, *13*, 2401995. [CrossRef]
6. Huang, Z.; Peng, Y.; Zhao, J.; Zhang, S.; Qi, P.; Qu, X.; Yan, F.; Ding, B.; Xuan, Y.; Zhang, X. An efficient and flexible bifunctional dual-band electrochromic device integrating with energy storage. *Nano Micro Lett.* **2025**, *17*, 98. [CrossRef]
7. Xu, H.; Haider, I.; Zheng, Y.; Li, W.; Zhuiykov, S.; Cui, Y. Towards the solid-state electrochromic devices: Platform based on transparent and flexible solid polymer electrolyte. *Chem. Eng. J.* **2025**, *508*, 161116. [CrossRef]
8. Wang, Z.; Liu, G.; Li, C.; Qiao, M.; Tian, M.; Lin, X.; Cui, W.; Wang, X.; Liu, J.; Xu, S. $W/WO_3/TiO_2$ multilayer film with elevated electrochromic and capacitive properties. *Materials* **2025**, *18*, 161. [CrossRef]
9. Yaseen, M.; Khattak, M.A.; Khan, A.; Bibi, S.; Bououdina, M.; Usman, M.; Khan, N.A.; Pirzado, A.A.; Abumousa, R.A.; Humayun, M. State-of-the-art electrochromic thin films devices, fabrication techniques and applications: A review. *Nanocomposites* **2024**, *10*, 1–40. [CrossRef]
10. Chen, D.; Chua, M.; He, Q.; Zhu, Q.; Wang, X.; Meng, H.; Xu, J.; Huang, W. Multifunctional electrochromic materials and devices recent advances and future potential. *Chem. Eng. J.* **2025**, *503*, 157820. [CrossRef]
11. Li, M.; Hou, M.; Wu, Y.; Li, X.; Guo, C.; Ning, H.; Li, Z.; Lu, K.; Yao, R.; Peng, J. Preparation and performance optimization of Ag-ITO composite electrode in electrochromic devices. *SID Symp. Dig. Tech. Pap.* **2024**, *55*, 1119. [CrossRef]
12. Wu, H.; Ning, H.; Li, M.; Guo, C.; Li, X.; Qiu, T.; Xu, Z.; Luo, C.; Yao, R.; Peng, J. Cu-doped $W_{18}O_{49}$ nanowire reticular films for electrochromic supercapacitors. *APL Mater.* **2023**, *11*, 111120. [CrossRef]
13. Yong, W.; Chen, N.; Xiong, T.; Fu, G. Development of high-performance Mo-doped WO_3 photo-electrochromic devices. *Mater. Today Chem.* **2024**, *38*, 102095. [CrossRef]

14. Nguyen, T.V.; Huynh, K.A.; Le Quyet, V.; Kim, H.; Ahn, S.H.; Kim, S.Y. Highly stable electrochromic cells based on amorphous tungsten oxides prepared using a solution-annealing process. *Int. J. Energy Res.* **2021**, *45*, 8061–8072. [CrossRef]
15. Kondalkar, V.V.; Yang, S.S.; Patil, P.S.; Choudhury, S.; Bhosale, P.N.; Lee, K. Langmuir-Blodgett assembly of nanometric WO_3 thin film for electrochromic performance: A new way. *Mater. Lett.* **2017**, *194*, 102–106. [CrossRef]
16. Orman, E.B.; Koca, A.; Özkaya, A.R.; Gürol, I.; Durmus, M.; Ahsen, V. Electrochemical, spectroelectrochemical, and electrochromic properties of lanthanide bis-phthalocyanines. *J. Electrochem. Soc.* **2014**, *161*, H422–H429. [CrossRef]
17. Gu, W.; Li, Q.; Wang, R.; Zhang, L.; Liu, Z.; Jiao, T. Recent progress in the applications of Langmuir–Blodgett film technology. *Nanomaterials* **2024**, *14*, 1039. [CrossRef]
18. Zhai, Y.; Li, J.; Shen, S.; Zhu, Z.; Mao, S.; Xiao, X.; Zhu, C.; Tang, J.; Lu, X.; Chen, J. Recent advances on dual-band electrochromic materials and devices. *Adv. Funct. Mater.* **2022**, *32*, 2109848. [CrossRef]
19. Bai, T.; Li, W.; Fu, G.; Zhang, Q.; Zhou, K.; Wang, H. Dual-band electrochromic smart windows towards building energy conservation. *Sol. Energy Mater. Sol. Cells* **2023**, *256*, 112320. [CrossRef]
20. Cao, S.; Zhang, S.; Zhang, T.; Yao, Q.; Lee, J. A visible light-near-infrared dual-band smart window with internal energy storage. *Joule* **2019**, *3*, 1152–1162. [CrossRef]
21. Li, R.; Lan, F.; Tang, L.; Huang, Y.; Zhao, S.; Wang, B.; Han, Y.; Gao, D.; Jiang, Q.; Zhao, Y.; et al. A local-dissociation solid-state polymer electrolyte with enhanced Li^+ transport for high-performance dual-band electrochromic smart windows. *Adv. Funct. Mater.* **2025**, *35*, 2419357. [CrossRef]
22. Wang, J.; Wang, Z.; Cui, L.; Zhang, M.; Huo, X.; Guo, M. Visible-near infrared independent modulation of hexagonal WO_3 induced by ionic insertion sequence and cavity characteristics. *Adv. Mater.* **2024**, *36*, 2406939. [CrossRef] [PubMed]
23. Zhou, K.; Wang, H.; Zhang, S.; Jiu, J.; Liu, J.; Zhang, Y.; Yan, H. Electrochromic modulation of near-infrared light by WO_3 films deposited on silver nanowire substrates. *J. Mater. Sci.* **2017**, *52*, 12783–12794. [CrossRef]
24. Huang, Q.; Liang, Y.; Cao, S.; Yang, Z. Morphology-dependent near-infrared electrochromic properties of tungsten oxide. *Coatings* **2023**, *13*, 344. [CrossRef]
25. Thummavichai, K.; Wang, N.; Xu, F.; Rance, G.; Xia, Y.; Zhu, Y. In situ investigations of the phase change behaviour of tungsten oxide nanostructures. *R. Soc. Open Sci.* **2018**, *5*, 171932. [CrossRef]
26. Ahmed, I.; Dastider, S.G.; Roy, A.; Mondal, K.; Haldar, K.K. Nitrogen doping in NiS/Ni_3S_4 nanowire-based electrocatalysts for promoting the second-order hydrogen evolution reaction. *ACS Appl. Nano Mater.* **2024**, *7*, 661–671. [CrossRef]
27. Hai, G.; Huang, J.; Cao, L.; Jie, Y.; Li, J.; Wang, X.; Zhang, G. Influence of oxygen deficiency on the synthesis of tungsten oxide and the photocatalytic activity for the removal of organic dye. *J. Alloys Compd.* **2017**, *690*, 239–248. [CrossRef]
28. Lu, D.; Chen, J.; Deng, S.; Xu, N.; Zhang, W. The most powerful tool for the structural analysis of tungsten suboxide nanowires: Raman spectroscopy. *J. Mater. Res.* **2008**, *23*, 402–408. [CrossRef]
29. Zhao, Y.; Zhang, X.; Chen, X.; Li, W.; Wang, L.; Ren, F.; Zhao, J.; Endres, F.; Li, Y. Preparation of WO_3 films with controllable crystallinity for improved near-infrared electrochromic performances. *ACS Sustain. Chem. Eng.* **2020**, *8*, 11658–11666. [CrossRef]
30. Zhang, X.; Dou, S.; Li, W.; Wang, L.; Qu, H.; Chen, X.; Zhang, L.; Zhao, Y.; Zhao, J.; Li, Y. Preparation of monolayer hollow spherical tungsten oxide films with enhanced near infrared electrochromic performances. *Electrochim. Acta* **2019**, *297*, 223–229. [CrossRef]

Disclaimer/Publisher's Note: The statements, opinions and data contained in all publications are solely those of the individual author(s) and contributor(s) and not of MDPI and/or the editor(s). MDPI and/or the editor(s) disclaim responsibility for any injury to people or property resulting from any ideas, methods, instructions or products referred to in the content.

inorganics

MDPI

Review

Comparative B3PW and B3LYP Calculations of ABO_3 (A = Ba, Sr, Pb, Ca; B = Sn, Ti, Zr) Neutral (001) and Polar (111) Surfaces

Roberts I. Eglitis [1,*], Juris Purans [1], Ran Jia [1,2], Sergei P. Kruchinin [3] and Steffen Wirth [4]

1 Institute of Solid State Physics, University of Latvia, 8 Kengaraga Str., LV1063 Riga, Latvia; juris.purans@cfi.lu.lv (J.P.); jiaran@jlu.edu.cn (R.J.)

2 Laboratory of Theoretical and Computational Chemistry, Institute of Theoretical Chemistry, Jilin University, Changchun 130023, China

3 Bogolyubov Institute for Theoretical Physics, National Ukrainian Academy of Science, 03143 Kyiv, Ukraine; sergeikruchinin@yahoo.com

4 Max-Planck-Institute for Chemical Physics of Solids, D-01187 Dresden, Germany; wirth@cpfs.mpg.de

* Correspondence: rieglitis@gmail.com; Tel.: +371-26426703

Abstract: We completed B3LYP and B3PW computations for AO- and BO_2-terminated (001) as well as AO_3- and B-terminated (111) surfaces of BSO, BTO, STO, PTO, CTO, BZO, SZO, and CZO perovskites. In particular, we performed the first B3LYP computations for polar BSO (111) surfaces. We observed that most of the upper-layer atoms for AO- and BO_2-terminated ABO perovskite (001) surfaces relax inward. In contrast, practically all second-layer atoms relax upward. Lastly, almost all third-layer atoms relax inward. This tendency is less pronounced for atomic relaxation of first, second, and third layer atoms for AO_3- and B-terminated ABO perovskite (111) surfaces. For almost all ABO perovskites, their (001) surface rumplings s are considerably larger for AO-terminated compared to BO_2-terminated surfaces. On the contrary, the ABO perovskite (001) surface energies, for both AO and BO_2-terminations, are essentially equivalent. The ABO perovskite polar (111) surface energies are always substantially larger than their neutral (001) surface energies. In most cases, the surface energies of AO_3-terminated ABO perovskite polar (111) surfaces are considerably larger than their B-terminated surface energies. Our computations illustrate a noticeable boost in the B-O bond covalency near the BO_2-terminated (001) surface related to the bulk. Our computed ABO perovskite bulk Γ-Γ band gaps are almost always reduced near the AO- and BO_2-terminated neutral (001) surfaces as well as in most cases also near the AO_3- and B-terminated polar (111) surfaces.

Keywords: ABO_3 perovskites; (001) surfaces; (111) surfaces; surface energies; B3LYP; B3PW

1. Introduction

Crucial role of surface as well as interface phenomena taking place in the ABO_3 (in the following called ABO for simplicity) perovskites are very serious issues in today's condensed matter physics [1–25]. All $BaSnO_3$ (BSO), $BaTiO_3$ (BTO), $SrTiO_3$ (STO), $PbTiO_3$ (PTO), $CaTiO_3$ (CTO), $BaZrO_3$ (BZO), $SrZrO_3$ (SZO), and $CaZrO_3$ (CZO) perovskites belong to the group of ABO perovskite oxides [26–29]. In our case, A = Ba, Sr, Pb, or Ca, whereas B = Sn, Ti, or Zr. ABO perovskites have plenty of commercially essential functions. Applications include actuators, capacitors, charge storage apparatus, and countless others [30,31]. For example, BSO may be used as the protonic conductor [32], with application potential for fuel cells [32]. BTO is examined as one of the inexpensive preparation price substitutes for wide band gap semiconductors with application prospects in numerous optoelectric devices [33]. Doped STO is an excellent anode material for solid oxide fuel cells [34] as well

https://doi.org/10.3390/inorganics13040100

as a perfect choice for photocatalytic applications [35]. PTO ferroelectric perovskite is an interesting material for numerous high-temperature as well as high- frequency piezoelectric applications [36]. CTO perovskite oxide has received recognition in the last years due to strong catalytic activity [37]. BZO perovskite is attractive as a thermal barrier coating material, used in gas turbine engines, which work at elevated temperatures [38]. SZO perovskite has been widely used as a catalyst [39], luminescent material [40], and proton conductor [41]. CZO perovskite has numerous technologically important applications, including capacitors and resonators, and also as humidity sensors [42]. For that reason, it is evident that in the last 25 years, ABO perovskite (001) surfaces were massively explored experimentally as well as theoretically [43–82].

In contrast to the ABO perovskite neutral (001) surfaces, their charged, polar, and thereby very complex (111) surfaces are considerably less studied. For example, to the best of our knowledge, BSO (111) surfaces have never been investigated before, neither experimentally nor theoretically. Accordingly, in this paper, we performed the first in the world ab initio calculations, dealing with polar and charged BSO (111) surfaces. BTO perovskite (111) surfaces were analyzed experimentally by Hagendorf et al. [83,84] by means of STM, XPS, and LEED methods. Recently, Chun et al. [8] examined the (111) surface termination of a BTO single crystal employing the combined density functional theory (DFT) as well as X-ray photoelectron spectroscopy (XPS) methods. They computed the (111) surface stability of the BTO stoichiometric Ti and BaO_3-terminations applying the DFT + U formalism [8]. Finally, Eglitis [85] performed ab initio hybrid DFT computations for BTO (111) surfaces and demonstrated that the surface energy of the BaO_3-terminated (111) surface is considerably larger than for the Ti-terminated (111) surface. Pojani et al. [86] performed semi-empirical Hartree–Fock (HF) calculations and discussed the polarity effects on the STO (111) surfaces. Biswas et al. [87] detected the upper atomic layer of STO (001), (011), and (111) surfaces using the time of flight mass spectroscopy. These authors found [87] that all (001), (011), and (111) surface orientations exhibit the Ti-rich surface [87]. Sekiguchi et al. [88] investigated the changes in STO (111) polar surface structures by AFM, AES, and XPS experimental techniques as a function of the atmosphere [88]. Tanaka et al. [89] observed the clean (111) surfaces of reduced STO crystals by means of STM and RHEED experiments [89]. They observed two different STO (111) surface structures [89]. Namely, the first possible STO (111) surface structure has the SrO_{3-x} outermost layer, whereas the second possible STO (111) surface structure has the Ti outermost layer [89]. Lastly, Eglitis [90] performed ab initio computations of SrO_3- and Ti-terminated STO (111) polar surface atomic relaxations, surface rumplings, and energies, as well as charge redistributions and Γ-Γ band gaps [90]. Using the GGA exchange-correlation functional, Pang et al. [91] calculated the structural and electronic properties as well as the stabilities of four different terminations of cubic PTO (111) surfaces. For instance, they computed the directly cleaved Ti- and PbO_3-terminated as well as constructed TiO- and PbO_2-terminated PTO (111) polar surfaces [91]. Eglitis [85] performed ab initio B3LYP calculations for PbO_3 and Ti-terminated PTO (111) surfaces and found that for both terminations, the PTO (111) surface energies are considerably larger than the PTO (001) surface energies [3]. Liu et al. [92] constructed the stoichiometric as well as nonstoichiometric terminations for the CTO (111) surfaces. The LDA computations for the CTO (111) surface and cleavage energies, as well as surface electronic and atomic structure, and surface grand potential, were performed [92]. Eglitis [90,93] performed ab initio computations for the polar CTO (111) surfaces using the B3LYP hybrid exchange-correlation functional. The atomic and electronic structure as well as surface energies and Ti-O chemical bond populations of the Ti and CaO_3-terminated polar CTO (111) surfaces were computed [90,93]. The only existing ab initio computations dealing with BaO_3- and Zr-terminated polar $BaZrO_3$ (111)

surfaces were performed by Eglitis [94,95]. Namely, these authors performed ab initio B3LYP computations for BZO polar (111) surface atomic relaxations and electronic structure as well as energetics for two possible BaO_3 and Zr polar (111) surface terminations [94,95]. The only available ab initio computations dealing with polar $SrZrO_3$ (111) surfaces were performed by Eglitis et al. [85,96]. Specifically, Eglitis et al. [85,96] performed ab initio hybrid B3LYP computations for surface relaxations and electronic structure as well as surface energies of Zr- and SrO_3-terminated $SrZrO_3$ perovskite polar (111) surfaces. Lastly, the only recent existing and preliminary ab initio B3LYP computations dealing with CZO perovskite polar (111) CaO_3- and Zr-terminated surface atomic relaxations and surface energies were performed by Eglitis and Jia [11].

According to the XRD measurement results obtained by Janifer et al. [97], barium stannate (BSO) is a single-phase cubic perovskite [97,98]. Moreover, BSO has a wide optical band gap equal to 3.1 eV [97,98], and the cubic lattice parameter a is identical to 4.119 Å [97,98]. BTO perovskite exhibits three phase transitions [99,100]. At high temperatures, BTO perovskite has a cubic structure with a symmetrical group $(Pm\overline{3}m)$ [99]. The BTO perovskite structure changes from this cubic $(Pm\overline{3}m)$ to a tetragonal structure ($P4mm$) at 403 K temperature [99]. Then, at 278 K temperature, the BTO perovskite structure changes to orthorhombic ($Amm2$) [99]. Lastly, the BTO perovskite structure changes to rhombohedral ($R3m$) at a temperature of 183 K [99,100]. According to experiments performed by Wemple [101], the BTO room temperature band gaps are 3.38 eV and 3.27 eV, accordingly, for light polarized parallel as well as perpendicular to the ferroelectric c axis [101]. STO perovskite has only one structural phase transition at a temperature of 110 K [102]. This phase transition reduces the STO perovskite symmetry from high symmetry cubic to tetragonal [102]. The experimentally detected STO perovskite direct (Γ-Γ) band gap energy in the room temperature cubic phase is equal to 3.75 eV [103]. PTO perovskite displays the single-phase transition at 763 K temperature from the high-temperature cubic phase to the tetragonal ferroelectric ground state [104]. The PTO perovskite Γ-Γ band gap, measured at room temperature in its tetragonal ferroelectric ground state, is equal to 3.4 eV [105]. Ali et al. [106], using the careful Rietveld analysis of the neutron as well as X-ray powder diffraction data, discovered that the CTO perovskite displays two structural phase transitions [106]. Namely, from a high-temperature cubic CTO structure $(Pm\overline{3}m)$ at 1634 ± 13 K temperature to a tetragonal structure ($I4/mcm$) [106]. Finally, at 1498 ± 25 K temperature, CTO perovskite displays a phase transition to the ground state orthorhombic structure ($Pbnm$) [106]. The CTO perovskite Γ-Γ band gap, according to the experimental results obtained by Ueda et al. in the orthorhombic phase [107], is approximately 3.5 eV [107]. According to Knight [108], the BZO perovskite is cubic at all measured temperatures within the temperature range from 4.2 K to 450 K [108]. Namely, BZO always has the cubic perovskite structure with the symmetry group $(Pm\overline{3}m)$ [108]. Therefore, the experimental BZO Γ-Γ band gap is equal to 5.3 eV [105]. SZO perovskite goes through three structural phase transitions [109]. Initially, below 970 K temperature, SZO perovskite is orthorhombic with the symmetry group $Pnma$ [109]. In the temperature range between 970 K and 1020 K, the SZO perovskite belongs to the other orthorhombic symmetry group Cmcm [109]. At 1020 K temperature, SZO perovskite transfers into the tetragonal structure with the symmetry group $I4/mcm$ [109]. Finally, at a temperature above 1360 K, the SZO perovskite turns into a cubic structure with the symmetry group $(Pm\overline{3}m)$ [109]. According to the optical conductivity measurements performed by Lee et al. [110], the SZO Γ-Γ gap within the orthorhombic phase, which is stable at room temperature, is equal to 5.6 eV [110]. CZO perovskite, at low temperatures, has an orthorhombic structure ($Pbnm$), which is stable up to temperatures of 2173 K $\pm$ 100 [111,112]. At higher temperatures, CZO perovskite has a cubic structure with a symmetrical group $(Pm\overline{3}m)$ [111,112]. According to Rosa et al. [113], the experimental CZO band gap of the orthorhombic structure at room temperature is equal to 5.7 eV [113].

The intention of our review paper was to execute essential additional computations dealing mostly with BSO neutral (001) and especially polar (111) surfaces that have never been studied before. After completing necessary supplementary first principles computations for ABO perovskite (001) as well as (111) surfaces, we carefully analyzed our results and identified the systematic tendencies frequent for all eight BSO, BTO, STO, PTO, CTO, BZO, SZO, and CZO perovskite neutral (001) [43–82,114–120] and polar (111) [83–96] surfaces. This paper research primarily investigates the surface properties. Concretely, the atomic relaxation, surface energy, chemical bond covalency, as well as electronic band gaps of ABO perovskites (e.g., BSO, BTO, STO, etc.), are established at both neutral (001) and polar (111) surfaces. The study aims to enhance the understanding of how surface characteristics influence the properties of these materials. We want to emphasize that the topic dealing with the theoretical investigation of ABO perovskite surfaces is both original and highly relevant to the field of condensed matter physics and materials science. While ABO perovskites have been extensively studied, the specific examination of polar (111) surfaces in the context of BSO, particularly with B3LYP computations, addresses a notable gap in the literature. Previous research primarily focused on (001) surfaces [1–5,7–11,22,43,45,48,50,56–61], making this investigation significant for understanding surface phenomena that impact applications in electronics and catalysis. This study adds considerable depth to the existing literature by providing the first computational analysis of polar (111) surfaces for BSO, and it includes a comprehensive study among various perovskites. The detailed examination of atomic relaxations, surface energies, and bond covalency enhances understanding of how surface properties differ for these ABO perovskite materials compared to bulk behavior, offering insights that can guide future experimental and theoretical investigations.

2. Simulation Approach and Surface Models

In order to carry out the first principles of DFT-B3PW or DFT-B3LYP computations, we employed the CRYSTAL computer program [121]. The CRYSTAL code makes use of Gaussian-type functions (GTFs) localized on all perovskite atoms. They are the basis for the expansion of the crystalline orbitals [122]. The superiority of the CRYSTAL ab initio code [121,122] is its capability to compute confined 2D (001) and (111) ABO perovskite slabs without forced periodicity along the z-axis [121]. In order to make use of the linear combination of atomic orbitals (LCAO-GTF) method [121,122], it is advantageous to utilize the optimized basis sets (BS) [121]. The optimized BS for BTO, STO, and PTO perovskites was developed in Ref. [123]. The BS for Sn was taken from the CRYSTAL computer code [121]. The relevant BS, used for CTO, BZO, SZO, and CZO perovskites and their surfaces, are described in Refs. [2,124–126]. All our ABO perovskite bulk as well as (001) and (111) surface computations were performed by means of the B3PW [127,128] or B3LYP [129] hybrid exchange-correlation functionals. Both B3PW and B3LYP functionals give comparable results for ABO perovskites [2–4,9,56]. However, it is important to note that the hybrid exchange-correlation functionals, like B3PW or B3LYP, allow us to achieve an excellent agreement with the experiment for the Γ-Γ band gaps of the related complex oxide materials, whereas the GGA-PBE and LDA exchange-correlation functionals, as a rule, underestimate the relevant Γ-Γ band gaps [2–4,9,56]. On the other hand, it is well known that the HF approach considerably overestimates the Γ-Γ band gaps [2–4,9,56]. We performed the reciprocal space integration by checking out the Brillouin zone for the 5 atom ABO perovskite cubic unit cell, applying the $8 \times 8 \times 8$ times increased Pack Monkhorst mesh [130] for the ABO perovskite bulk [2–4,11,124–126,131] as well as the 8 × 8 × 1 times increased mesh for their (001) and (111) surfaces [2–4,11,85,90,124–126,131].

The ABO perovskite (001) surfaces were described using 2D slabs (Figures 1 and 2). These slabs subsist from nine planes located perpendicular to the [001] crystal direction.

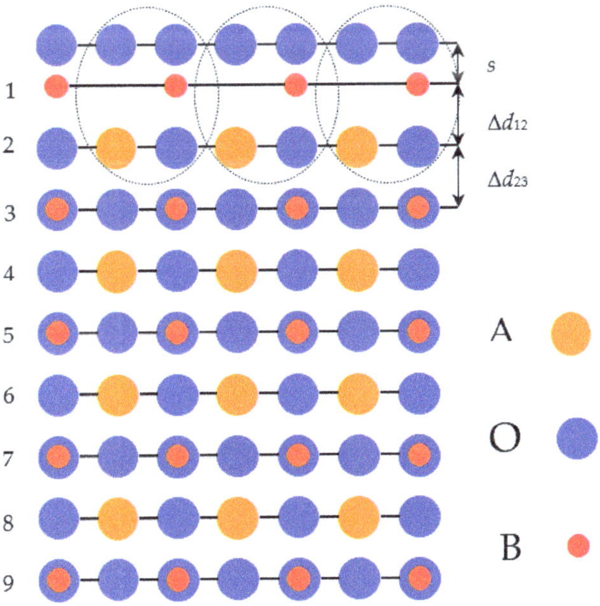

Figure 1. Outline of the BO_2-terminated (001) surface of ABO–type perovskite enclosing nine atomic layers.

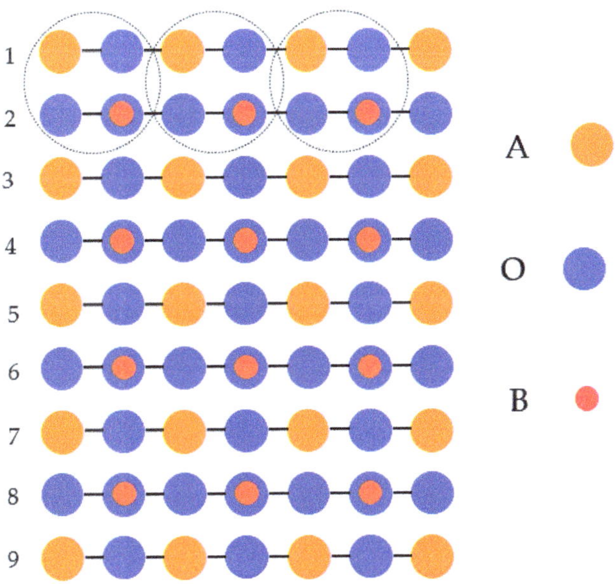

Figure 2. Outline of the AO-terminated (001) surface of ABO–type perovskite enclosing nine atomic layers.

Namely, to compute ABO perovskite (001) surfaces, we considered slabs consisting of nine alternating BO_2 and AO layers (Figures 1 and 2). The mirror symmetries of the slabs were retained regarding their central layers (Figures 1 and 2). Our computed 23-atom-containing slab with BO_2-terminated surfaces as well as the 22-atom slab with AO-terminated surfaces are depicted in Figures 1 and 2, respectively. Both these slabs are non-stoichiometric. They have unit-cell formulas $A_4B_5O_{14}$ and $A_5B_4O_{13}$, respectively (Figures 1 and 2). The succession of the ABO-perovskite (001) surface layers as well as the definitions of the interplane separations Δd_{12}, Δd_{23}, and the surface rumpling s are pictured in Figure 1.

In contrast to the ABO perovskite neutral (001) surfaces, the polar (111) surfaces subsist of charged AO_3 and B planes, as depicted in Figure 3. The ABO perovskite polar (111) surfaces have been characterized with 2D slabs, subsisting of nine planes perpendicular to the [111] direction in the crystal. In order to compute the ABO perovskite polar (111) surfaces, we employed slabs subsisting of nine alternating B and AO_3 layers (Figures 3 and 4). The first slab is terminated by B planes and subsists of a supercell consisting of 21 atoms (B-AO_3-B-AO_3-B-AO_3-B-AO_3-B) (Figure 4a). Another slab is terminated by AO_3 planes (Figure 4b). The corresponding supercell contains 24 atoms (AO_3-B-AO_3-B-AO_3-B-AO_3-B-AO_3) (Figure 4b). Both these slabs are non-stoichiometric (Figures 3 and 4). They have unit cell formulas $A_4B_5O_{12}$ and $A_5B_4O_{15}$, respectively (Figure 4). As we know from earlier ab initio computations, devoted to the polar ABO perovskite (111) surfaces [11,86,92,132–134], the powerful electron reallocation takes place for these terminations in order to eliminate the polarity [11,86,92,132–134]. Consequently, the AO_3- as well as B-terminated ABO-perovskite (111) surfaces maintain their insulating character, and such computations can thereby be realized [11,86,92,132–134].

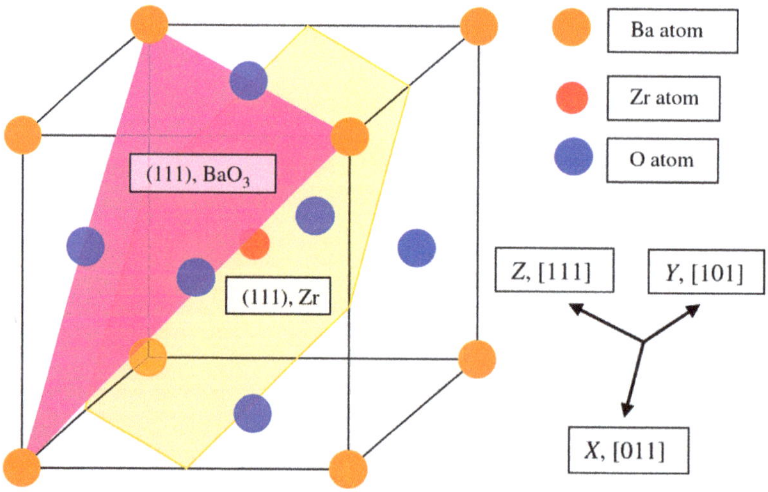

Figure 3. ABO perovskite structure displaying two (111) surface terminations: AO_3 and B.

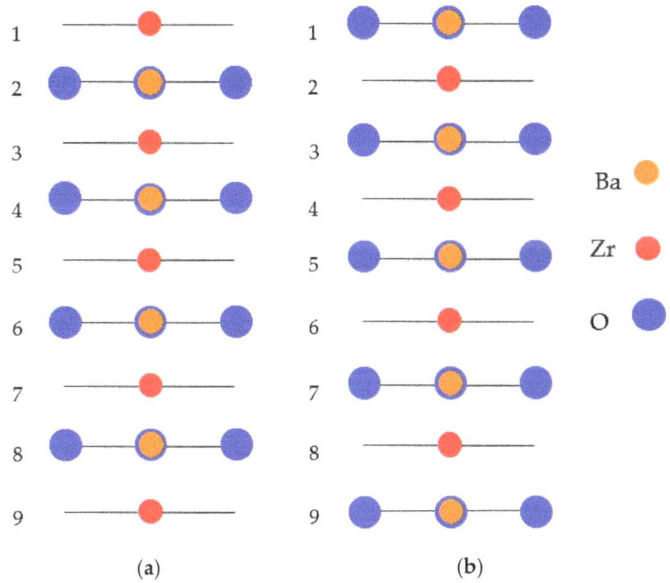

Figure 4. Outline of slab constructions used to explore ABO perovskite (111) surfaces. (**a**) Non-stoichiometric ABO slab consisting of 9 layers with B-terminated (111) surfaces. (**b**) Non-stoichiometric ABO slab consisting of 9 layers with AO_3-terminated (111) surfaces.

The first step for the ABO perovskite (001) and (111) surface energy computations is to calculate the relevant cleavage energies. Our computed cleavage energies are equally shared amid the created surfaces [2–4,11,85,94,135]. Namely, the ABO perovskite (001) and (111) surfaces arise as a result of simultaneous (001) as well as (111) cleaved perovskite crystal [2–4,11,85,94,135]. In our executed ABO perovskite (001) surface cleavage energy computations, the nine-layer AO- and BO_2-terminated slabs together embody 45 atoms corresponding to nine ABO perovskite unit cells:

$$E_{surf}^{unr}(AO + BO_2) = ¼\,[E_{slab}^{unr}(AO) + E_{slab}^{unr}(BO_2) - 9E_{bulk}], \quad (1)$$

where $E_{slab}^{unr}(AO)$ and $E_{slab}^{unr}(BO_2)$ are the unrelaxed AO- and BO_2-terminated ABO perovskite nine layer (001) slab total energies. E_{bulk} denotes the total energy of the ABO perovskite bulk unit cell containing five atoms. The factor equal to ¼ implies that four surfaces were generated due to the ABO perovskite (001) cleavage. In the second step, the AO- and BO_2-terminated nine-layer (001) slab relaxation energies, taking into account the relaxation of slabs from both sides, were computed, as follows:

$$E_{rel}(\lambda) = ½\,[E_{slab}^{rel}(\lambda) - E_{slab}^{unr}(\lambda)], \quad (2)$$

where λ is AO or BO_2. $E_{slab}^{rel}(\lambda)$ is relaxed from both sides with regard to the AO- or BO_2-terminated (001) slab total energy. $E_{slab}^{unr}(\lambda)$ is the total energy for unrelaxed AO or BO_2-terminated ABO perovskite (001) slab. Lastly, the AO- or BO_2-terminated ABO perovskite (001) surface energy should be calculated using the following equation:

$$E_{surf}(\lambda) = E_{surf}^{unr}(AO + BO_2) + E_{rel}(\lambda). \quad (3)$$

With this, we computed the ABO perovskite (111) surface as well as cleavage energies. Again, B- and AO_3-terminated ABO perovskite (111) surfaces are complementary. For that reason, the cleavage energy is the same for both AO_3 and B-terminated ABO perovskite (111) surfaces. Thereby, the cleavage energy for the complementary surface $E_{cl}(AO_3 + B)$ may be computed from the total energies of nine-layer unrelaxed AO_3- and B-terminated ABO perovskite (111) slabs, as follows:

$$E_{cl}(AO_3 + B) = ¼ [E_{slab}^{unr}(B) + E_{slab}^{unr}(AO_3) - 9E_{bulk}], \quad (4)$$

where $E_{slab}^{unr}(B)$ is our computed total energy of unrelaxed 21 atoms containing B-terminated ABO perovskite nine-layer (111) slab. $E_{slab}^{unr}(AO_3)$ is our computed total energy for 24 atoms containing a nine-layer unrelaxed AO_3-terminated ABO perovskite (111) slab. The relaxation energies for AO_3- and B-terminated ABO perovskite (111) surfaces can be obtained via the following equation:

$$E_{rel}(\beta) = ½ [E_{slab}^{rel}(\beta) - E_{slab}^{unr}(\beta)], \quad (5)$$

where $\beta = AO_3$ or B describes the ABO perovskite (111) surface termination. In the end, the ABO perovskite AO_3- or B-terminated (111) surface energy is equal to the sum of the cleavage (4) and relaxation (5) energies, as follows:

$$E_{surf}(\beta) = E_{cl}(AO_3 + B) + E_{rel}(\beta) \quad (6)$$

3. Computation Results for ABO Perovskite Bulk as Well as (001) and (111) Surfaces

3.1. ABO Perovskite Bulk Atomic and Electronic Structure

As the starting point of our B3LYP and B3PW computations, we computed the theoretical bulk lattice constants for BSO, BTO, STO, PTO, CTO, BZO, SZO, and CZO perovskites [2–4,90,94,123–126,131] (Table 1). We matched our computation results with the accessible experimental data [106,136–140] (Table 1). For some ABO perovskites, we also performed the theoretical lattice constant computations by means of the ab initio Hartree–Fock (HF) method [121,122,141–143]. For our ab initio DFT [144,145] bulk lattice constant computations, we employed the Generalized Gradient approximation [146] suggested by Perdew and Wang (PWGGA) [127,128]. As we can see from Table 1, according to our B3PW computation results, the SZO and BTO perovskite bulk lattice constants (4.155 Å [131] and 4.008 Å [3], respectively) are almost in ultimate agreement with the relevant experimental data (4.154 Å [140] and 4.00 Å [137], respectively). It must be pointed out that for BSO, BTO, STO, PTO, and SZO perovskites, our B3LYP computed bulk lattice constants (4.107 Å, 4.04 Å, 3.94 Å, 3.96 Å, and 4.195 Å, respectively) are larger than the relevant B3PW computed bulk lattice constants (4.087 Å, 4.008 Å, 3.904 Å, 3.936, Å, and 4.155 Å, respectively) (Table 1).

In order to describe the covalency effects, effective atomic charges, and chemical bondings for the ABO perovskites bulk and their (001) as well as (111) surfaces, we employed a classical Mulliken population analysis [147–150].

Our B3LYP computed effective atomic charges and bond populations for BSO as well as for seven other perovskites are collected in Table 2. As we can see from Table 2, our B3LYP computed effective atomic charges for the BSO bulk are (+1.825e) for the Ba atom, (+2.122e) for the Sn atom, and (−1.316e) for the O atom (Table 2). It is worth noting that our computed Sn atom charge in the BSO perovskite (+2.122e) is the smallest B atom charge among all eight computed ABO perovskites (Table 2). Our B3LYP-computed Sn-O chemical bond population (+0.284e) in BSO perovskite (Table 2) is considerably larger than the respective B-O chemical bond populations for the other seven computed ABO perovskites, ranging from only (+0.084e) for CTO to (+0.108e) for the BZO perovskite.

Table 1. B3LYP-, B3PW-, PWGGA-, and HF-computed bulk lattice constants (in Å) for the BSO, BTO, STO, PTO, CTO, BZO, SZO, and CZO perovskite bulk. The experimental data (in Å) are listed for comparison.

Perovskite	Functional	Calculated	Experiment
BSO	B3LYP	4.107	4.119 [136]
	B3PW	4.087	
	PWGGA	4.107	
	HF	4.078	
BTO	B3LYP	4.04 [123]	4.00 [137]
	B3PW	4.008 [3]	
STO	B3LYP	3.94 [123]	3.89 [137]
	B3PW	3.904 [4]	
PTO	B3LYP	3.96 [123]	3.97 [138]
	B3PW	3.936 [3]	
CTO	B3LYP	3.851 [90]	3.8967 [106]
	B3PW	3.851 [2]	
	PWGGA	3.884 [131]	
	HF	3.863 [131]	
BZO	B3LYP	4.234 [94]	4.199 [139]
	B3PW	4.234 [124]	
	PWGGA	4.24 [131]	
	HF	4.25 [131]	
SZO	B3LYP	4.195 [125]	4.154 [140]
	B3PW	4.155 [131]	
	PWGGA	4.176	
	HF	4.182	
CZO	B3LYP	4.157 [126]	No data for the cubic phase

Table 2. Our B3LYP or B3PW computed effective atomic charges Q (e) as well as bond populations P (e) in BSO, BTO, STO, PTO, CTO, BZO, SZO, and CZO perovskites.

Bulk Material		BSO	BTO	STO	PTO	CTO	BZO	SZO	CZO
Ion	**Property**	**B3LYP**	**B3PW**	**B3PW**	**B3PW**	**B3PW**	**B3PW**	**B3LYP**	**B3LYP**
A	*Q*	+1.825	+1.797	+1.871	+1.354	+1.782	+1.815	+1.880	+1.787
	P	−0.030	−0.034	−0.010	+0.016	+0.006	−0.012	+0.002	+0.014
O	*Q*	−1.316	−1.388	−1.407	−1.232	−1.371	−1.316	−1.351	−1.310
	P	+0.284	+0.098	+0.088	+0.098	+0.084	+0.108	+0.092	+0.086
B	*Q*	+2.122	+2.367	+2.351	+2.341	+2.330	+2.134	+2.174	+2.144

Our B3LYP- (3.65 eV) or B3PW-computed (3.68 eV) bulk Γ-Γ band gaps for BSO perovskite are in a more acceptable agreement with the experimental result of (3.1 eV) [97,98] than our HF (12.11 eV) or PWGGA (1.71 eV) computation results (Table 3).

Table 3. Our B3LYP-, B3PW-, PWGGA-, and HF-computed Γ-Γ band gaps (in eV) for ABO perovskite bulk. Experimental ABO perovskite bulk Γ-Γ band gaps are listed for comparison (in eV).

ABO Perovskite	Method	Γ-Γ Band Gap, Bulk, Theory	Experiment
BSO	B3LYP	3.65	3.1 [97,98]
	B3PW	3.68	
	PWGGA	1.71	
	HF	12.11	
BTO	B3LYP	3.49 [123]	No data for cubic phase
	B3PW	3.55 [123]	
	PWGGA	1.97 [123]	
	HF	11.73 [123]	
PTO	B3LYP	4.15 [123]	No data for cubic phase
	B3PW	4.32 [123]	
	PWGGA	2.61 [123]	
	HF	12.74 [123]	
STO	B3LYP	3.89 [123]	3.75 [103]
	B3PW	3.96 [123]	
	PWGGA	2.31 [123]	
	HF	12.33 [123]	
CTO	B3LYP	4.20 [131]	No data for cubic phase
	B3PW	4.18 [131]	
	PWGGA	2.34 [131]	
	HF	12.63 [131]	
BZO	B3LYP	4.79 [94]	5.3 [105]
	B3PW	4.93 [131]	
	PWGGA	3.24 [131]	
	HF	12.96 [131]	
SZO	B3LYP	5.31 [125]	No data for cubic phase
	B3PW	5.30 [125]	
	PWGGA	3.53 [125]	
	HF	13.54 [125]	

As we can see from Figure 5 and Table 3, our B3PW-computed BTO bulk Γ-Γ band gap is equal to 3.55 eV [123]. It is comparable with our B3LYP-computed BTO bulk Γ-Γ band gap equal to 3.49 eV (Table 3). It is worth noting that our HF-computed BTO bulk Γ-Γ band gap is really huge (11.73 eV), whereas the PWGGA computed band gap is comparatively small, only 1.97 eV (Table 3). There are no experimental data available for the high-temperature cubic BTO bulk Γ-Γ band gap. Nevertheless, Wemple suggested that the BTO room temperature Γ-Γ band gaps are equal to 3.38 and 3.27 eV for parallel and perpendicular, respectively, polarized light to the ferroelectric axis **c** [101].

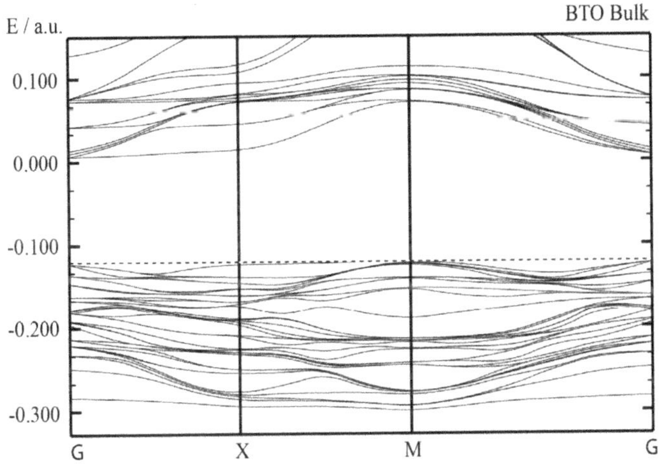

Figure 5. Our B3PW computed electronic BTO bulk band structure (Table 3). The dashed line correspond to the bulk valence band maximum.

As it is possible to see from Figure 6 and Table 3, our B3PW-computed PTO bulk Γ-Γ band gap is equal to 4.32 eV [123]. It is in the same range as our B3LYP-computed PTO bulk Γ-Γ band gap equal to 4.15 eV [123] (Table 3). It is interesting to notice that our HF-computed PTO bulk Γ-Γ band gap is very large and equal to (12.74 eV) [123]. Just the opposite, the PWGGA computed PTO bulk Γ-Γ band gap is very small, only 2.61 eV (Table 3). To the best of our knowledge, there are no experimental data available for the high-temperature cubic PTO bulk Γ-Γ band gap. As it is possible to see from Table 3, the PWGGA computed STO bulk Γ-Γ band gap is equal to 2.31 eV [123], whereas the HF computed respective band gap is 5.34 times larger and equal to 12.33 eV. It is important to note that STO bulk Γ-Γ band gaps, computed by means of B3LYP (3.89 eV) and B3PW (3.96 eV) hybrid exchange-correlation functionals, almost coincide [123]. They are both in good agreement with the available experimental data for the STO bulk Γ-Γ band gap (3.75 eV) [103].

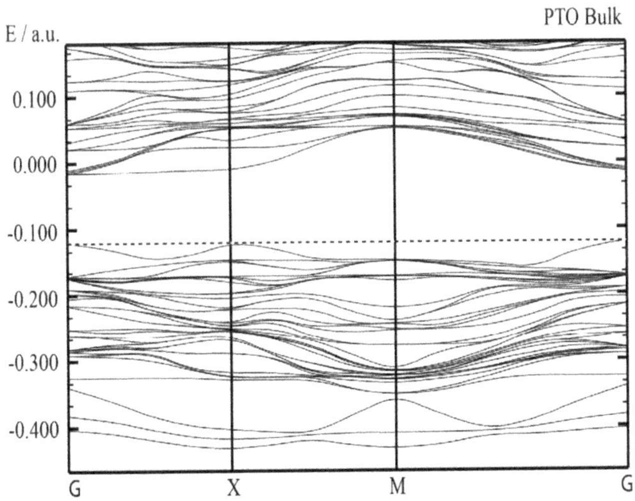

Figure 6. Our B3PW computed electronic PTO bulk band structure (Table 3). The dashed line correspond to the bulk valence band maximum.

Our PWGGA-computed BZO bulk Γ-Γ band gap is very small, equal to only 3.24 eV [131] (Table 3). In contrast, our HF computed BZO bulk Γ-Γ band gap is exactly four times larger (12.96 eV) (Table 3) [131]. Again, almost equal and in fair agreement with the experiment (5.3 eV) [105] are our B3LYP- (4.79 eV) [94] as well as B3PW-computed (4.93 eV) (Figure 7) bulk Γ-Γ band gaps.

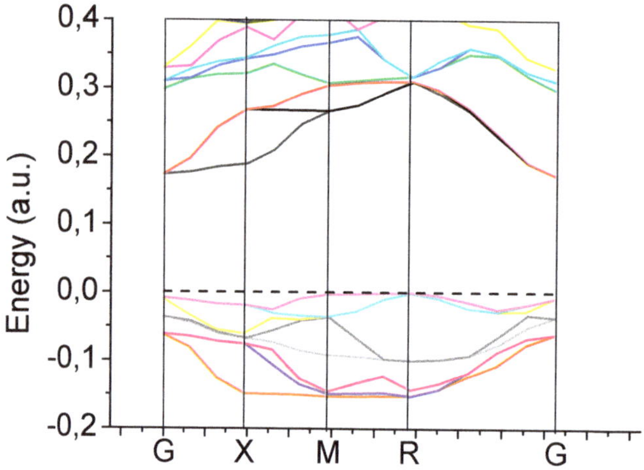

Figure 7. Our B3PW-computed electronic band structure for BZO bulk. The dashed line correspond to the bulk valence band maximum.

As we can see from Figure 8 and Table 3, our PWGGA-computed ABO perovskite bulk Γ-Γ band gaps are in the energy range from 1.71 eV (BSO) to 3.53 eV (SZO). The PWGGA-computed ABO perovskite bulk Γ-Γ band gaps (Figure 8) are always considerably smaller than relevant experimental values as well as B3LYP-, B3PW-, and HF-computed respective band gaps (Table 3). As we can see from Figure 8, our B3PW- and B3LYP-computed ABO perovskite bulk Γ-Γ band gaps always are very close and in fair agreement with the available experimental data (Table 3). In contrast, our HF-computed ABO perovskite Γ-Γ bulk band gaps are in the energy range from 11.73 eV (BTO) to 13.54 eV (SZO) and always, to a very great extent, overestimate the relevant experimental data (Table 3 and Figure 8).

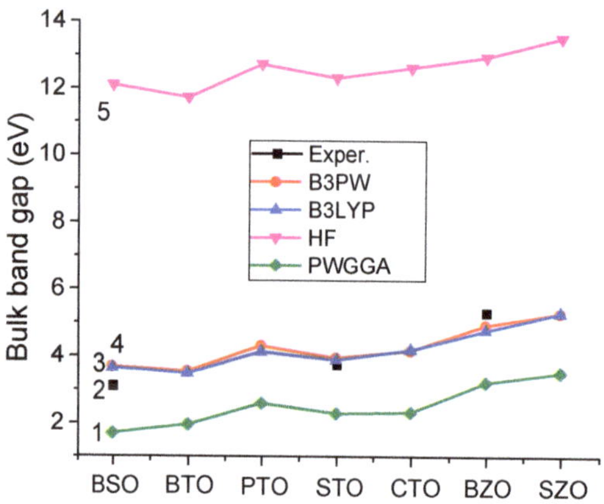

Figure 8. Our computed and experimental bulk Γ-Γ band gaps for seven ABO perovskites derived by different theoretical and experimental methods: (1) PWGGA; (2) Experiment; (3) B3LYP; (4) B3PW; (5) HF.

3.2. *ABO Perovskite (001) Surface Atomic and Electronic Structure*

Our B3LYP or B3PW computation results for the upper two- or three-layer atomic relaxations of BO_2- or AO-terminated ABO perovskite (001) surfaces (Tables 4 and 5). As we can see from Tables 4 and 5, in most cases, the ABO perovskite BO_2- and AO-terminated (001) surface upper layer atoms relax inward, toward the bulk, the second surface layer atoms relax outward, and third surface layer atoms, again, relax inward. For the AO-terminated ABO perovskite (001) surface upper two layers (Table 5), the metal atom relaxation magnitudes are always considerably larger than the respective oxygen atom relaxation magnitudes. The largest relaxation magnitude between all upper layer atoms exhibits the CaO-terminated CTO (001) surface upper layer Ca atom (−10.01% of a_0) (Table 5). Between all second-layer atoms, the largest relaxation magnitude exhibits the TiO_2-terminated PTO (001) surface Pb atom (+5.32% of a_0) (Table 4).

Table 4. Our B3LYP- or B3PW-computed atomic relaxation (in % of a_0) for BO_2-terminated BSO, BTO, STO, PTO, CTO, BZO, SZO, and CZO (001) surfaces.

Bulk Material		**BSO**	**BTO**	**STO**	**PTO**	**CTO**	**BZO**	**SZO**	**CZO**
Terminat., (001)		SnO_2	TiO_2	TiO_2	TiO_2	TiO_2	ZrO_2	ZrO_2	ZrO_2
Layer	**Ion**	**B3LYP**	**B3PW**	**B3PW**	**B3PW**	**B3PW**	**B3PW**	**B3LYP**	**B3LYP**
1	B	−0.97	−3.08	−2.25	−2.81	−1.71	−1.79	−1.38	−1.30
	O	−0.27	−0.35	−0.13	+0.31	−0.10	−1.70	−2.10	−2.31
2	A	+0.93	+2.51	+3.55	+5.32	+2.75	+1.94	+2.81	+4.23
	O	−0.04	+0.38	+0.57	+1.28	+1.05	+0.85	+0.91	+1.25
3	B	−0.11	-	-	-	-	−0.03	−0.04	−0.05
	O	+0.03	-	-	-	-	0.00	−0.05	−0.09

Table 5. Our B3LYP- or B3PW-computed atomic relaxation (in % of a_0) for AO-terminated BSO, BTO, STO, PTO, CTO, BZO, SZO, and CZO (001) surfaces.

Bulk Material		**BSO**	**BTO**	**STO**	**PTO**	**CTO**	**BZO**	**SZO**	**CZO**
Terminat., (001)		BaO	BaO	SrO	PbO	CaO	BaO	SrO	CaO
Layer	**Ion**	**B3LYP**	**B3PW**	**B3PW**	**B3PW**	**B3PW**	**B3PW**	**B3LYP**	**B3LYP**
1	A	−1.75	−1.99	−4.84	−3.82	−8.31	−4.30	−7.63	−10.01
	O	−0.39	−0.63	+0.84	−0.31	−0.42	−1.23	−0.86	−0.79
2	B	+0.39	+1.74	+1.75	+3.07	+1.12	+0.47	+0.86	+1.11
	O	−0.07	+1.40	+0.77	+2.30	+0.01	+0.18	−0.05	+0.01
3	A	−0.20	-	-	-	-	−0.01	−1.53	−2.60
	O	+0.02	-	-	-	-	−0.14	−0.45	−0.48

In order to obtain a comparison between experimental results as well as our B3LYP- or B3PW-computed ABO perovskite (001) surface atomic structures, we computed and listed the surface rumpling and the changes in interlayer distances Δd_{12} and Δd_{23} in Table 6. We computed the surface rumplings and interlayer distances Δd_{12} and Δd_{23} for completely relaxed ABO perovskite (001) surfaces (Tables 4 and 5). For B3LYP and B3PW computations of interlayer distances, we employed the positions of fully relaxed metal atoms (Tables 4 and 5).

Table 6. Our B3LYP- or B3PW-computed and experimentally detected surface rumpling s and relative displacements Δd_{ij} (in % of a_0) for the three near-surface planes [151–154].

Mater.	Method	AO-Terminated (001) Surface			BO_2-Terminated (001) Surface		
		s	Δd_{12}	Δd_{23}	s	Δd_{12}	Δd_{23}
STO	B3PW	5.66	−6.58	+1.75	2.12	−5.79	3.55
	LDA [151]		−3.4	+1.2		−3.5	+1.6
	LEED [152]	4.1 ± 2	−5 ± 1	2 ± 1	2.1 ± 2	1 ± 1	−1 ± 1
	RHEED [153]	4.1	2.6	1.3	2.6	1.8	1.3
BSO	B3LYP	+1.36	−2.14	+0.59	+0.70	−1.90	+1.04
BTO	B3PW	+1.37	−3.73	+1.74	+2.73	−5.59	+2.51
	LDA [151]		−2.8	+1.1		−3.1	+0.9
PTO	B3PW	+3.51	+6.89	+3.07	+3.12	−8.13	+5.32
	LDA [151]		−4.2	+2.6		−4.4	+3.1
CTO	B3PW	+7.89	−9.43	+1.12	+1.61	−4.46	+2.75
	GGA [67]	+0.37	−0.44	+0.22	+0.13	−0.41	+0.33
BZO	B3PW	+3.07	−4.77	+0.48	+0.09	−3.73	+1.97
SZO	B3LYP	+6.77	−8.49	+2.39	−0.72	−4.19	+2.85
	LDA [154]	+7.9	−9.1	+3.2	−0.7	−6.1	+4.2
	GGA [154]	+7.8	−9.3	+3.3	+0.3	−7.4	+4.9
CZO	B3LYP	+9.22	−11.12	+3.71	+1.01	−5.53	+4.28

According to our B3PW computations, the STO surface rumpling s for the SrO-terminated STO (001) surface (5.66% of a_0) is 2.67 times larger than the relevant surface rumpling s for the TiO_2-terminated STO (001) surface (Table 6). As we can see from Table 6, our B3LYP-computed surface rumpling (1.36% of a_0) for the BaO-terminated BSO (001) surface is almost two times larger than for the SnO_2-terminated BSO (001) surface (+0.70% of a_0). The systematic trend is (Table 6) that for all our computed ABO perovskites, the surface rumpling for AO-terminated (001) surfaces is larger than for BO_2-terminated (001) surfaces. The single exception from this systematic trend is our B3PW computed BTO perovskite, where the surface rumpling for the TiO_2-terminated BTO (001) surface (2.73% of a_0) is approximately two times larger than for the BaO-terminated BTO (001) surface (1.37% of a_0) (Table 6). As it is possible to see from Table 6, there is always a contraction between the first and second layer metal atoms (Δd_{12}) for both AO- and BO_2-terminated (001) surfaces of all our computed ABO perovskites. The sole exception from this systematic trend is the expansion of the PbO-terminated PTO perovskite (001) surface first and second layer metal atoms by (6.89% of a_0) (Table 6). Just the opposite, there is always an expansion between the second- and third-layer metal atoms (Δd_{23}) for both AO- and BO_2-terminated (001) surfaces of all our computed ABO perovskites [151–154] (Table 6). Our B3PW computations for AO- and TiO_2-terminated STO (001) surfaces are in fair agreement with previous ab initio LDA computations performed by Meyer et al. [151]. Also, our B3PW computed surface rumpling s amplitudes for SrO- and TiO_2-terminated (001) surfaces of STO perovskite are in fair agreement with available LEED [152] and RHEED [153] experiments. It is worth noting that for TiO_2-terminated STO (001) surfaces, our B3PW computer surface rumpling s value (2.12% of a_0) is in almost perfect agreement with available LEED [152] experimental data (2.1 ± 2% of a_0).

As we can see from Tables 2 and 7 and Figure 9, the B atom charge on BO_2-terminated (001) surfaces of BSO (+1.957e), BTO (+2.307e), STO (2.291e), PTO (2.279e), and CTO (2.278e) are smaller than in their bulk (+2.122e, +2.367e, +2.351e, +2.341e, and +2.330e, respectively). On the contrary, for BZO, SZO, and CZO perovskites, the B atom charges on their ZrO_2-terminated (001) surfaces (+2.173e, +2.196e, and +2.172e) (Table 7) are larger than in their bulk (+2.134e, +2.174e, and +2.144e, respectively) (Table 2). The largest charge difference is for the Sn atom in the BSO bulk (+2.122e) and on its SnO_2-terminated (001) surface (+1.957e) (Figure 9), equal to (+0.165e). As we can see from Figure 10, the absolutely largest chemical bond population is between the Sn and O atoms on the SnO_2-terminated BSO perovskite (001) surface (+0.298e) (Table 7). The chemical bond population between the Sn and O atoms is also very large in the BSO bulk (+0.284e) (Table 2). Without doubt, the systematic trend is that for all eight of our ab initio computed ABO perovskites, the B-O atom chemical bond population on their BO_2-terminated (001) surface always is larger than in the ABO perovskite bulk (Figure 10).

Table 7. Our computed magnitudes of atomic displacements are D (in Å), the effective atomic charges Q (in e), and the bond populations P (in e) between the nearest Me-O atoms for the BO_2-terminated ABO perovskite (001) surfaces.

			BO_2-Terminated ABO Perovskite (001) Surfaces							
			BSO	**BTO**	**STO**	**PTO**	**CTO**	**BZO**	**SZO**	**CZO**
	Functional		**B3LYP**	**B3PW**	**B3PW**	**B3PW**	**B3PW**	**B3PW**	**B3LYP**	**B3LYP**
Layer	**Prop.**	**Ion**	**SnO_2**	**TiO_2**	**TiO_2**	**TiO_2**	**TiO_2**	**ZrO_2**	**ZrO_2**	**ZrO_2**
1	*D*	B	−0.04	−0.123	−0.088	−0.111	−0.066	−0.076	−0.058	−0.054
	Q		+1.957	+2.307	+2.291	+2.279	+2.278	+2.173	+2.196	+2.172
	P		+0.298	+0.126	+0.118	+0.114	+0.114	+0.132	+0.114	+0.102
	D	O	0.011	−0.014	−0.005	+0.012	−0.004	−0.072	−0.088	−0.096
	Q		−1.089	−1.280	−1.296	−1.184	−1.267	−1.239	−1.277	−1.258
	P		−0.022	−0.038	−0.014	+0.044	+0.016	−0.018	−0.002	+0.018
2	*D*	A	+0.038	+0.101	+0.139	+0.209	+0.106	+0.082	+0.118	+0.176
	Q		+1.803	+1.767	+1.850	+1.275	+1.754	+1.797	+1.869	+1.772
	P		−0.026	−0.030	−0.008	+0.008	+0.006	−0.010	+0.002	+0.012
	D	O	−0.002	+0.015	+0.022	+0.050	+0.041	+0.036	+0.038	+0.052
	Q		−1.333	−1.343	−1.365	−1.167	−1.324	−1.273	−1.287	−1.235
	P		+0.268	+0.090	+0.080	+0.080	+0.086	+0.106	+0.094	+0.090
3	*D*	B	−0.005	-	-	-	-	−0.001	−0.002	−0.002
	Q		+2.105	+2.365	+2.348	+2.335	+2.326	+2.133	+2.172	+2.14
	P		+0.290	+0.104	+0.096	+0.108	+0.090	+0.116	+0.102	+0.098
	D	O	+0.001	-	-	-	-	0	−0.002	−0.004
	Q		−1.303	−1.371	−1.384	−1.207	−1.354	−1.30	−1.331	−1.286
	P		−0.030	−0.034	−0.010	+0.018	+0.008	−0.012	+0.002	+0.014

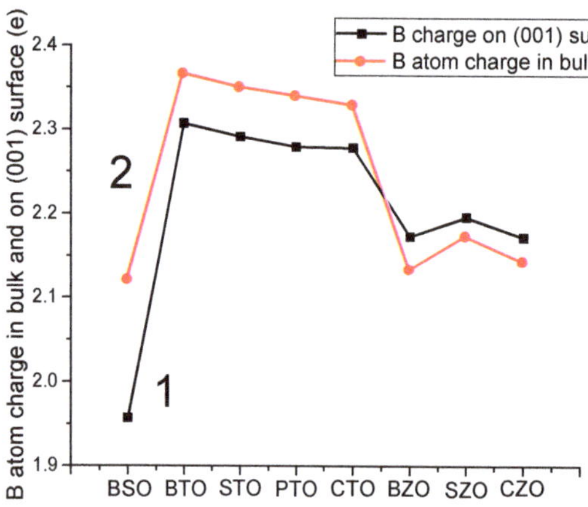

Figure 9. Our ab initio computed BO_2-terminated ABO perovskite (001) surface (line 1) upper layer B atom as well as ABO perovskite bulk (line 2) B atom charges (in e).

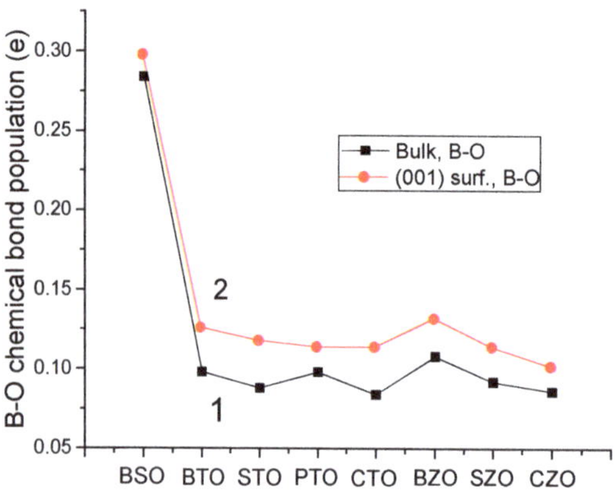

Figure 10. Our ab initio computed B-O chemical bond populations P (in e) for the ABO perovskite bulk (1) as well as BO_2-terminated (001) surfaces (2).

Our B3LYP computed BSO bulk Γ-Γ band gap (3.65 eV) is reduced near the SnO_2-terminated BSO (001) surface (1.90 eV) (Table 8). It is intriguing to observe that our B3LYP-computed BaO-terminated BSO (001) surface Γ-Γ band gap (3.81 eV) is approximately two times larger than the SnO_2-terminated (001) band gap (1.90 eV) (Table 8). For STO perovskite, our computed Γ-Γ bulk band gap (3.96 eV) [56] almost coincides with the TiO_2-terminated STO (001) surface band gap (3.95 eV) [56]. In general, the ABO perovskite bulk Γ-Γ band gaps, as a rule, are larger than the AO or BO_2-terminated (001) surface band gaps (Table 8). The single exception from this general rule is the BSO perovskite, where the AO-terminated (001) surface band gap (3.81 eV) is larger than the bulk Γ-Γ band gap (3.65 eV) (Table 8). Our B3PW-computed BTO (001) surface band structures are depicted in Figure 11. As we can wee from Figure 11a, our B3PW-computed BaO-terminated BTO (001) surface Γ-Γ band gap is equal to 3.49 eV [56]. Even smaller is our B3PW computed Γ-Γ band gap for the TiO_2-terminated BTO (001) surface (2.96 eV) (Figure 11b). Figure 12 exhibits our B3PW computed PTO (001) surface band structures [56]. Namely, the Γ-Γ band gap for the PbO-terminated PTO (001) surface (Figure 12a) is equal to 3.58 eV [56], whereas the Γ-Γ band gap for the PTO TiO_2-terminated (001) surface is equal to 3.18 eV [56] (Figure 12b). Finally, our B3PW-computed band structures for the BZO (001) surfaces [131] are displayed in Figure 13. Our B3PW-computed Γ-Γ band gap for the BaO-terminated BZO (001) surface is equal to 4.82 eV (Figure 13a), while the ZrO_2-terminated BZO (001) surface Γ-Γ band gap is even smaller and equal to 4.48 eV [131] (Figure 13b).

Table 8. Our B3LYP- or B3PW-computed ABO perovskite bulk as well as AO- and BO_2-terminated (001) surface Γ-Γ band gaps (in eV).

Material, Method	Bulk Γ-Γ Band Gap	AO-Termin., (001)	BO_2-Termin., (001)
BSO, B3LYP	3.65	3.81	1.90
BTO, B3PW	3.55 [56]	3.49 [56]	2.96 [56]
STO, B3PW	3.96 [56]	3.72 [56]	3.95 [56]
PTO, B3PW	4.32 [56]	3.58 [56]	3.18 [56]
CTO, B3PW	4.18 [131]	3.87 [131]	3.30 [131]
BZO, B3PW	4.93 [131]	4.82 [131]	4.48 [131]
SZO, B3LYP	5.31 [125]	5.04	4.91
CZO, B3LYP	5.40 [126]	5.00	5.22

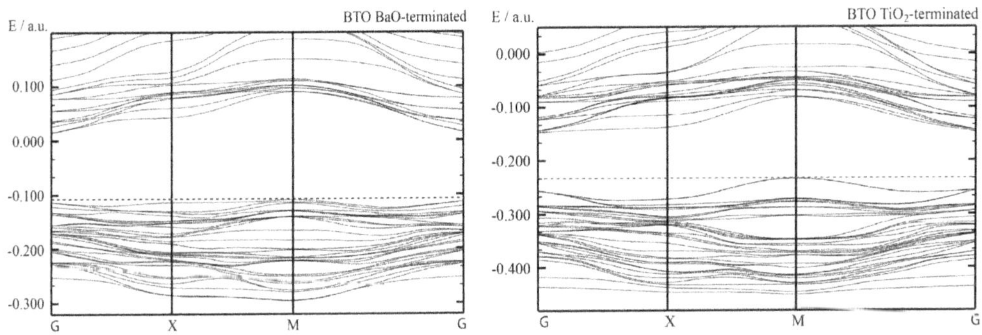

(**a**) (**b**)

Figure 11. Our B3PW-computed band structures for BaO- and TiO_2-terminated BTO (001) surfaces ((**a**) and (**b**), respectively). The dashed lines correspond to the bulk valence band maximum.

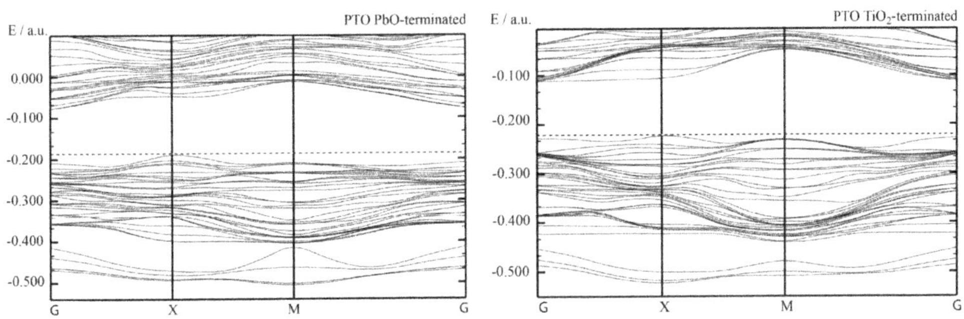

(**a**) (**b**)

Figure 12. Our B3PW-computed band structures for PbO- and TiO_2-terminated PTO (001) surfaces ((**a**) and (**b**), respectively). The dashed lines correspond to the valence band maximum.

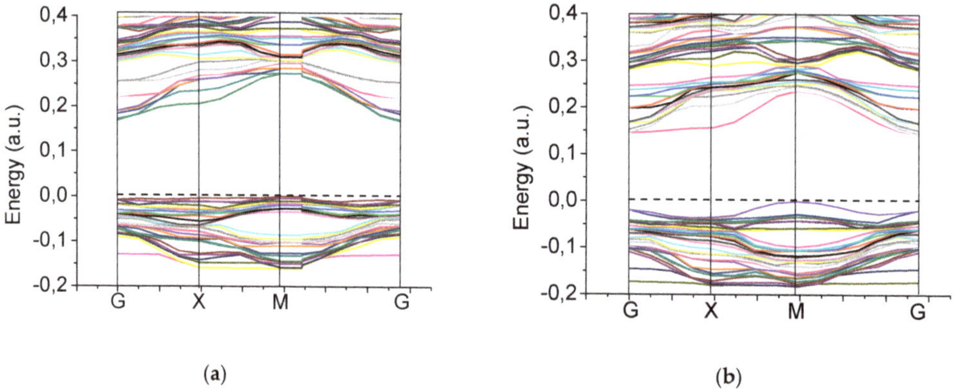

Figure 13. Our B3PW-omputed band structures for BaO- and ZrO_2-terminated BZO (001) surfaces ((**a**) and (**b**), respectively). The dashed lines correspond to the valence band maximum.

Our B3LYP- or B3PW-computed eight ABO perovskite bulk as well as AO- and BO_2-terminated (001) surface Γ-Γ band gaps are pictured in Figure 14. As we can see from Figure 14, the BSO perovskite SnO_2-terminated (001) surface has the absolutely smallest Γ-Γ band gap, equal to 1.90 eV. In contrast, the largest Γ-Γ band gap, between all eight of our computed ABO perovskite BO_2-terminated (001) surfaces, is for the ZrO_2-terminated CZO (001) surface (5.22 eV) (Table 8 and Figure 14). The smallest Γ-Γ band gap, between all eight of our computed ABO perovskite AO-terminated (001) surfaces (Figure 14), is for the BaO-terminated BTO (001) surface (3.49 eV), whereas the largest is for the SrO-terminated SZO (001) surface (5.04 eV) (Table 8 and Figure 14). As it is possible to see from Table 8 and Figure 14, in 15 cases, that our computed ABO perovskite Γ-Γ band gaps for the AO- and BO_2-terminated (001) surfaces are smaller than the respective ABO perovskite bulk Γ-Γ band gaps. The single exception from this systematic trend is our B3LYP computed BSO perovskite BaO-terminated (001) surface Γ-Γ band gap (3.81 eV), which is slightly larger than the BSO perovskite bulk Γ-Γ band gap (3.65 eV).

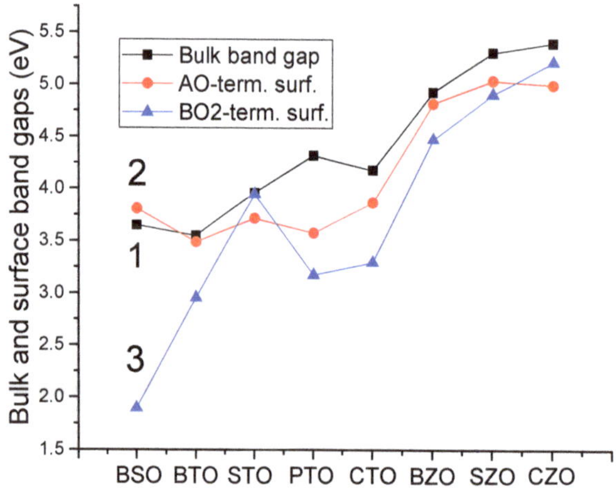

Figure 14. Our B3PW- or B3LYP-computed bulk (l) as well as AO- (2) and BO_2-terminated (3) eight ABO perovskite (001) surface Γ-Γ band gaps.

Table 9 lists our B3PW- or B3LYP-calculated eight ABO perovskite (001) surface energies. As we can see from Table 9, the absolute largest surface energies, according to our B3LYP computations, are for BSO perovskite BaO (1.38 eV) and SnO_2-terminated (1.39 eV) (001) surfaces (Table 9). The energy difference between these two BSO perovskite (001) surface terminations as well as BZO perovskite BaO (1.30 eV) and ZrO_2-terminated (1.31 eV) (001) surfaces are extremely small, at only 0.01 eV (Table 9). The absolute smallest ABO perovskite (001) surface energies, according to our B3PW computations, are for PTO AO- (0.83 eV) and TiO_2-terminated (0.74 eV) (001) surfaces (Table 9). The unconditionally largest surface energy difference between ABO perovskite AO- and BO_2-terminated (001) surfaces is for CZO perovskite CaO- (0.87 eV) and ZrO_2-terminated (1.33 eV) (001) surfaces, equal to 0.46 eV.

Table 9. Our B3LYP- or B3PW-computed (001) surface energies (in eV per surface cell) for BSO, BTO, STO, PTO, CTO, BZO, SZO, and CZO perovskites [2–4,124–126].

ABO Perovskite	**(001) Surface Energies**	
Termination	**AO**	**BO_2**
BSO, B3LYP	1.38 [This paper]	1.39 [This paper]
BTO, B3PW	1.19 [3]	1.07 [3]
STO, B3PW	1.15 [4]	1.23 [4]
PTO, B3PW	0.83 [3]	0.74 [3]
CTO, B3PW	0.94 [2]	1.13 [2]
BZO, B3PW	1.30 [124]	1.31 [124]
SZO, B3LYP	1.13 [125]	1.24 [125]
CZO, B3LYP	0.87 [126]	1.33 [126]

3.3. ABO Perovskite (111) Surface Atomic and Electronic Structure

We performed first in the world ab initio computations for the BSO perovskite (111) surfaces (Tables 10 and 11). As we can see from the performed B3LYP computations for the Sn-terminated BSO (111) surface (Table 10), the upper layer Sn atom very slightly relaxes inward by (0.50% of a_0). Also, both second-layer atoms, Ba and O, relax inward by a rather small relaxation magnitude of (0.51% and 0.30% of a_0, respectively). In contrast, the third layer Sn atom moves upward by a slightly larger relaxation magnitude (0.73% of a_0) (Table 10). The largest upper layer atom inward relaxation magnitudes are for Ti (11.19% of a_0) and Zr (11.07% of a_0) atoms on the Ti- and Zr-terminated BTO and CZO (111) surfaces, respectively (Table 10). All eight ABO perovskite (111) surface upper layer B atoms relax inward (Table 10). Also, all eight send layer metal atoms on the B-terminated ABO perovskite (111) surface relax inward. The largest second-layer metal atom inward relaxation magnitude is for the Ti atom (14.02% of a_0) on the Ti-terminated CTO (111) surface (Table 10). Contrarily, six of eight second-layer oxygen atoms on the B-terminated ABO perovskite (111) surface relax upward (Table 10). Finally, the metal atom relaxation directions on the third layer of B-terminated ABO perovskite (111) surfaces are rather random. Namely, the BSO, STO, PTO, and SZO perovskite third-layer metal atoms relax upward, whereas the BTO, CTO, BZO, and CZO third-layer atoms relax inward (Table 10).

Table 10. Our B3LYP-computed relaxation of B-terminated BSO, BTO, STO, PTO, CTO, BZO, SZO, and CZO (111) surface upper three layer atoms (% of a_0).

Layer	Ion	BSO	BTO	STO	PTO	CTO	BZO	SZO	CZO
Displacement		(Δz)	(Δz)	(Δz)	(Δz)	(Δz)	(Δz)	(Δz)	(Δz)
Termination		Sn-ter.	Ti-ter.	Ti-ter.	Ti-ter.	Ti-ter.	Zr-ter.	Zr-ter.	Zr-ter.
1	B	−0.50	−11.19	−3.58	−7.57	−6.23	−8.03	−5.72	−11.07
2	A	−0.51	−6.22	−11.24	−10.09	−14.02	−9.73	−11.92	−11.31
	O	−0.30	+2.74	+1.53	−0.13	+1.30	+0.78	+0.79	+0.14
3	B	+0.73	−0.25	+0.26	+0.53	−0.26	−0.02	+1.53	−0.96

Table 11. Our B3LYP-computed relaxation of the AO_3-terminated BSO, BTO, STO, PTO, CTO, BZO, SZO, and CZO (111) surface upper three layer atoms (% of a_0).

Layer	Ion	BSO	BTO	STO	PTO	CTO	BZO	SZO	CZO
Displacement		(Δz)	(Δz)	(Δz)	(Δz)	(Δz)	(Δz)	(Δz)	(Δz)
Termination		BaO_3	BaO_3	SrO_3	PbO_3	CaO_3	BaO_3	SrO_3	CaO_3
1	A	+3.90	−1.24	+1.33	+1.01	−0.52	+1.70	−0.74	−3.61
	O	−0.35	−3.98	−0.03	−2.52	−0.81	−0.57	−0.52	−0.07
2	B	+0.47	+2.49	+1.81	+0.02	+2.13	+0.21	+0.74	+1.20
3	A	+0.45	+1.49	−0.03	+1.26	+2.60	+0.71	−0.02	−0.02
	O	−0.01	−0.25	−0.26	+1.26	−0.07	−0.01	−0.18	−0.07

As we can see from Table 11, according to our performed B3LYP computations, all eight upper layer O atoms on the AO_3-terminated ABO perovskite (111) surfaces always relax inward. Their inward relaxation magnitudes are in the range from 0.03% of a_0 for the STO perovskite to 3.98% of a_0 for the BTO perovskite (Table 11). In contrast, four upper-layer metal atoms relax upward as well as four metal atoms inward (Table 11). The largest upward relaxation magnitude is for the Ba atom (3.90% of a_0) on the BaO_3-terminated BSO (111) surface upper layer (Table 11). All second-layer metal atoms on the AO_3-terminated ABO perovskite (111) surface relax outward (Table 11). The largest upward relaxation magnitude on the AO_3-terminated ABO perovskite (111) surface second layer is for the BTO perovskite Ti atom (2.49% of a_0), whereas the smallest upward relaxation magnitude is for the PTO perovskite Ti atom, only (0.02% of a_0) (Table 11). On the third layer of the AO_3-terminated ABO perovskite (111) surface, most of the metal atoms (five) relax outward, whereas most of the O atoms (seven) relax inward (Table 11). The largest relaxation magnitude, between all third-layer atoms, is for the CTO perovskite Ca atom (2.60% of a0) (Table 11).

As a next step, we computed the Γ-Γ band gaps for seven ABO perovskite B- and AO_3-terminated (111) surfaces (Table 12). In most cases, the ABO perovskite bulk Γ-Γ band gaps are reduced near their ABO_3- and B-terminated (111) surfaces (Table 12). Namely, for the AO_3-terminated ABO perovskite (111) surfaces, only the BaO_3-terminated BTO (111) surface Γ-Γ band gap (3.60 eV) is larger than the BTO perovskite Γ-Γ bulk band gap (3.49 eV) (Table 12). The largest reduction in the Γ-Γ band gap on the AO_3-terminated (111) surface is for the SrO_3-terminated SZO (111) surface, where the surface band gap (4.75 eV) is reduced regarding the SZO bulk Γ-Γ band gap (5.31 eV) by 0.56 eV (Table 12). On the B-terminated ABO_3 perovskite (111) surfaces, the bulk Γ-Γ band gaps for STO, PTO, BZO, and SZO perovskites (3.99 eV, 4.15 eV, 4.79, and 5.31 eV, respectively) are reduced near their Ti-, Ti-, Zr-, and Zr-terminated (111) surfaces (3.98 eV, 3.93 eV, 4.47 eV, and 4.57 eV, respectively) (Table 12).

Table 12. Our B3LYP-computed Γ-Γ band gaps for ABO perovskite bulk as well as their B- and AO_3-terminated (111) surfaces (in eV).

Termin., Γ-Γ Gap	BSO	BTO	STO	PTO	CTO	BZO	SZO	CZO
Bulk, Γ-Γ band gap	3.65	3.49	3.99	4.15	4.20	4.79	5.31	5.40
B-term., (111) surf.	Sn	Ti	Ti	Ti	Ti	Zr	Zr	Zr
Γ-Γ band gap	4.19	4.14	3.98	3.93	4.44	4.47	4.57	-
AO_3-t., (111) surf.	BaO_3	BaO_3	SrO_3	PbO_3	CaO_3	BaO_3	SrO_3	CaO_3
Γ-Γ band gap	3.60	3.60	3.72	3.78	3.78	4.51	4.75	-

Our B3LYP-computed surface energies of polar B (line 3) and AO_3-terminated (line 4) ABO perovskite (111) surfaces are presented in Table 13 and plotted in Figure 15, together with ABO perovskite neutral BO_2 (line 2) and AO-terminated (line 1) (001) surface energies (Table 9). We computed the ABO perovskite (111) surface energies using Equations (4)–(6). Just opposite to the ABO perovskite (001) surfaces (Table 9), it is possible to see from Table 13 that different polar ABO perovskite (111) surface terminations B and AO_3, lead to quite different (111) surface energies (Table 13 and Figure 15). As we can see from Table 13 and Figure 15, the lowest ABO perovskite (111) surface energies are for Ti-terminated CTO (4.18 eV) and STO (4.99 eV) (111) surfaces. At the same time, the largest ABO perovskite (111) surface energies are for the SrO_3- and CaO_3-terminated SZO and CZO (111) surfaces (9.45 eV and 9.62 eV, respectively) (Table 13). The largest (111) surface energy difference between two different surface terminations for ABO perovskite is for Ti- (6.14 eV) and PbO_3-terminated (8.11 eV) PTO perovskite (111) surfaces (1.97 eV) (Table 13 and Figure 15). In contrast, the smallest energy difference is between the Sn- (5.20 eV) and BaO_3-terminated (5.13 eV) BSO perovskite (111) surfaces, i.e., only 0.07 eV (Figure 15). It is worth noting that for seven ABO perovskites, their AO_3-terminated (111) surface energy is larger than their B-terminated (111) surface energy (Table 13 and Figure 15). The only exception from this systematic trend is the BSO perovskite, where the Sn-terminated (111) surface energy (5.20 eV) is larger than the BaO_3-terminated (111) surface energy (5.13 eV).

Table 13. Our B3LYP-computed surface energies for 8 ABO perovskite B and AO_3-terminated (111) surfaces (in eV).

Term., Surf. Energy	BSO	BTO	STO	PTO	CTO	BZO	SZO	CZO
B-term., (111) surf.	Sn	Ti	Ti	Ti	Ti	Zr	Zr	Zr
Surf. Energy (111)	5.20	7.28	4.99	6.14	4.18	7.94	7.98	8.19
AO_3-t., (111) surf.	BaO_3	BaO_3	SrO_3	PbO_3	CaO_3	BaO_3	SrO_3	CaO_3
Surf. Energy (111)	5.13	8.40	6.30	8.11	5.86	9.33	9.45	9.62

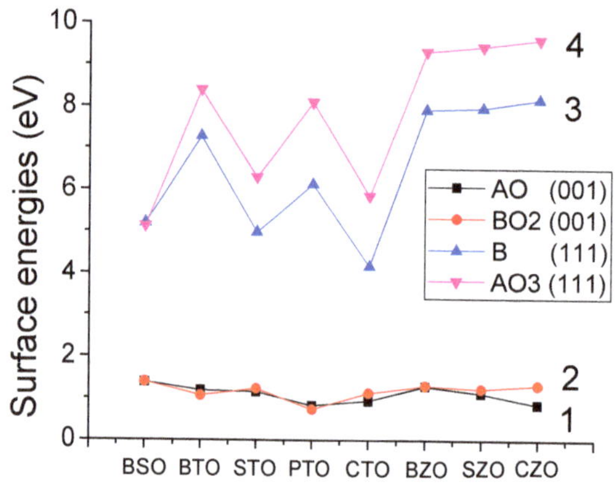

Figure 15. Our B3LYP-computed ABO perovskite surface energies for neutral AO- (line 1) and BO_2-terminated (line 2) (001) surfaces as well polar B- (line 3) and AO_3-terminated (line 4) (111) surfaces.

4. Conclusions

As we can see from Figure 8, the hybrid exchange-correlation functionals B3LYP and B3PW allow us to achieve a fair agreement with the experiments for ABO perovskite bulk Γ-Γ band gaps [2–4,9,56]. At the same time, the HF method considerably overestimates the bulk Γ-Γ band gaps, whereas the PWGGA method underestimates them [2–4,9,56]. This was the key reason why we choose the B3LYP and B3PW hybrid exchange-correlation functionals for our ABO perovskite bulk and surface computations [2–4,9,56].

We completed B3LYP and B3PW computations for AO and BO_2-terminated (001) as well as AO_3- and B-terminated (111) surfaces of BSO, BTO, STO, PTO, CTO, BZO, SZO, and CZO perovskites. We observed that most of the upper-layer atoms for AO- and BO_2-terminated ABO perovskite (001) surfaces relax inward. The two exceptions from this systematic trend are the upward relaxation of oxygen atoms on the TiO_2-terminated PTO (001) surface by (0.31% of a_0) (Table 4) as well as on the SrO-terminated STO (001) surface by (0.84% of a_0) (Table 5). In contrast, all second-layer metal atoms relax upward. Also, practically all second-layer oxygen atoms relax upward. Only the oxygen atoms on SnO_2- and BaO-terminated BSO as well as SrO-terminated SZO (001) surfaces relax inward by a very small relaxation magnitude of (0.04, 0.07, and 0.05% of a_0, respectively) (Tables 4 and 5). Lastly, almost all third-layer atoms, again, relax inward.

This tendency is less pronounced for atomic relaxation of first-, second-, and third-layer atoms for AO_3- and B-terminated ABO perovskite (111) surfaces (Tables 10 and 11). Namely, 20 from 24, or 83.33%, of ABO perovskite (111) surface upper layer atoms relax inward. Nevertheless, only 14 from 24, or 58.33%, of ABO perovskite (111) surface second layer atoms relax upward. Finally, again, only 58.33% of the third layer (111) surface atoms relax inward. For almost all ABO perovskites, their (001) surface rumplings s are considerably larger for AO-terminated compared to BO_2-terminated surfaces. For example, B3PW- or B3LYP-computed surface rumplings s for AO-terminated STO, BSO, and CTO (001) surfaces (5.66, 1.36, and 7.89) are considerably larger than the respective surface rumplings s for BO_2-terminated STO, BSO, and CTO (001) surfaces (2.12, 0.70, and 1.61, respectively) (Table 6).

On the contrary, the ABO perovskite (001) surface energies, for both AO- and BO_2-terminations, are essentially equivalent. The largest computed ABO perovskite (001) surface energy is for the BO_2-terminated BSO (001) surface (1.39 eV), whereas the smallest is for the TiO_2-terminated PTO (001) surface (0.74 eV) (Table 9). The ABO perovskite polar (111) surface energies always are substantially larger than their neutral (001) surface energies.

Namely, the largest ABO perovskite polar (111) surface energy is for the CaO_3-terminated CZO (111) surface (9.62 eV), whereas the smallest is for the Ti-terminated CTO (111) surface (4.18 eV) (Table 13 and Figure 15). In most cases, the surface energies of AO_3-terminated ABO perovskite polar (111) surfaces are considerably larger than their B-terminated surface energies. Specifically, for BTO, STO, PTO, CTO, BZO, SZO, and CZO perovskites, their AO_3-terminated (111) surface energy is considerably larger than their B-terminated (111) surface energy (Table 13 and Figure 15). The only omission from this systematic trend is BSO perovskite, where the Sn-terminated (111) surface energy (5.20 eV) is 0.07 eV larger than the BaO_3-terminated (111) surface energy (5.13 eV).

Our computations illustrate a noticeable boost in the B-O chemical bond covalency near the ABO perovskite BO_2-terminated (001) surface related to the bulk. Particularly, for BSO, BTO, STO, PTO, CTO, BZO, SZO, and CZO perovskites, their bulk B-O chemical bond covalency (+0.284*e*, +0.098*e*, +0.088*e*, +0.098*e*, +0.084*e*, +0.108*e*, +0.092*e*, +0.086*e*) increases near the BO_2-terminated ABO perovskite (001) surface (+0.298*e*, +0.126*e*, +0.118*e*, +0.114*e*, +0.114*e*, +0.132*e*, +0.114*e*, +0.102*e*). The absolutely largest B-O chemical bond covalency is near the SnO_2-terminated BSO perovskite (001) surface (+0.298*e*), whereas the smallest B-O chemical bond covalency is between Ti-O atoms (+0.084*e*) in the CTO perovskite bulk (Tables 2 and 7). Our computed ABO perovskite bulk Γ-Γ band gaps are almost always reduced near the AO- and BO_2-terminated neutral (001) surfaces as well as in most cases also near the AO_3- and B-terminated polar (111) surfaces.

We want to stress that our B3LYP and B3PW computations of ABO perovskite neutral (001) as well as polar (111) surface characteristics are very helpful in order to interpret processes, where surfaces exhibit a key role [155–157]. For instance, in the chemistry and physics of ABO perovskite surface reactions, (001) and (111) surface as well as interface aspect, and adsorption. Specifically, a wide variety of useful technological applications of ABO perovskites counting electrooptical and piezoelectrical devices as well as fuel cells and microelectrodes have inspired their theoretical research [155–157].

Author Contributions: Conceptualization, R.I.E. and J.P.; methodology, R.J.; software, S.P.K.; validation, S.W., R.I.E., and J.P.; formal analysis, R.J.; investigation, S.P.K.; resources, S.W.; data curation, R.I.E., J.P., R.J., S.P.K., and S.W.; writing—original draft preparation, R.I.E.; writing—review and editing, R.I.E., J.P., R.J., S.P.K., and S.W.; visualization, R.J.; supervision, S.W.; project administration, R.I.E.; funding acquisition, J.P. All authors have read and agreed to the published version of the manuscript.

Funding: This study was funded by the Latvian Council of Science Grant Number: LZP-2021/1-464. The Institute of Solid State Physics, University of Latvia (Latvia), as the Centre of Excellence, got funding from the European Union's Horizon 2020 Framework Program H2020-WIDESPREAD01-2016-2017-Teaming Phase2 under Grant Agreement No. 739508, project CAMART-2. S.P.K. acknowledges support by the National Academy of Science of Ukraine (Funding. No. 0116U002067). S.P.K. thanks the funding from Simons Foundation, under the Contract NUMBER-0121U109816. J.P. acknowledges SWEB project funded by the Horizon Europe programme (Number - GA101087367).

Institutional Review Board Statement: Not applicable.

Informed Consent Statement: Not applicable.

Data Availability Statement: Data are contained within the article.

Conflicts of Interest: The authors declare no conflicts of interest.

References

1. Dawber, M.; Rabe, K.M.; Scott, J.F. Physics of thin-film ferroelectric oxides. *Rev. Mod. Phys.* **2005**, *77*, 1083–1130.
2. Eglitis, R.I.; Vanderbilt, D. Ab initio calculations of the atomic and electronic structure of $CaTiO_3$ (001) and (011) surfaces. *Phys. Rev. B* **2008**, *78*, 155420. [CrossRef]

3. Eglitis, R.I.; Vanderbilt, D. Ab initio calculations of $BaTiO_3$ and $PbTiO_3$ (001) and (011) surface structures. *Phys. Rev. B* **2007**, *76*, 155439. [CrossRef]
4. Eglitis, R.I.; Vanderbilt, D. First-principles calculations of atomic and electronic structure of $SrTiO_3$ (001) and (011) surfaces. *Phys. Rev. B* **2008**, *77*, 195408. [CrossRef]
5. Chun, H.J.; Jeon, J.E.; Park, S.C. Facet-Dependent Ba Dissolution of Tetragonal $BaTiO_3$ Single Crystal Surfaces. *J. Phys. Chem. C* **2023**, *127*, 1848–1854. [CrossRef]
6. Guo, X.W.; Zou, M.J.; Wang, Y.J.; Tang, Y.L.; Zhu, Y.L.; Ma, X.L. Effects of anisotropic misfit strains on equilibrium phases and domain structures in (111)-oriented ferroelectric $PbTiO_3$ films. *Acta Mater.* **2021**, *206*, 116639. [CrossRef]
7. Erdman, N.; Poeppelmeier, K.R.; Asta, M.; Warschkov, O.; Ellis, D.E.; Marks, L.D. The structure and chemistry of the TiO_2-rich surface of $SrTiO_3$ (001). *Nature* **2002**, *419*, 55–58. [CrossRef]
8. Chun, H.J.; Lee, Y.; Kim, S.; Yoon, Y.; Kim, Y.; Park, S.C. Surface termination of $BaTiO_3$ (111) single crystal: A combined DFT and XPS study. *Appl. Surf. Sci.* **2022**, *578*, 152018. [CrossRef]
9. Eglitis, R.I.; Kleperis, J.; Purans, J.; Popov, A.I.; Jia, R. Ab initio calculations of $CaZrO_3$ (011) surfaces: Systematic trends in polar (011) surface calculations of ABO_3 perovskites. *J. Mater. Sci.* **2020**, *55*, 203–217. [CrossRef]
10. Costa-Amaral, R.; Gohda, Y. First-principles study of the adsorption of 3d transition metals on BaO- and TiO_2-terminated cubic-phase $BaTiO_3$ (001) surfaces. *J. Chem. Phys.* **2020**, *152*, 204701. [CrossRef]
11. Eglitis, R.I.; Jia, R. Review of Systematic Tendencies in (001), (011) and (111) Surfaces Using B3PW as Well as B3LYP Computations of $BaTiO_3$, $CaTiO_3$, $PbTiO_3$, $SrTiO_3$, $BaZrO_3$, $CaZrO_3$, $PbZrO_3$ and $SrZrO_3$ Perovskites. *Materials* **2023**, *16*, 7623. [CrossRef]
12. Muff, S.; Fanciulli, M.; Weber, A.P.; Pilet, N.; Ristić, Z.; Wang, Z.; Plumb, N.C.; Radović, M.; Dil, J.H. Observation of a two-dimensional electron gas at $CaTiO_3$ film surfaces. *Appl. Surf. Sci.* **2018**, *432*, 41–45. [CrossRef]
13. Tian, H.; Mao, A.J.; Zhao, H.J.; Cui, Y.; Li, H.; Kuang, X.Y. Large polarization and dielectric response in epitaxial $SrZrO_3$ films. *Phys. Chem. Chem. Phys.* **2016**, *18*, 7680–7687. [CrossRef]
14. Heifets, E.; Kotomin, E.A.; Maier, J. Semi-empirical simulations of surface relaxation for perovskite titanates. *Surf. Sci.* **2000**, *462*, 19–35. [CrossRef]
15. Enterkin, J.A.; Subramanian, A.K.; Russell, B.C.; Castell, M.R.; Poeppelmeier, K.R.; Marks, L.D. A homologous series of structures on the surface of $SrTiO_3$ (110). *Nat. Mater.* **2010**, *9*, 245–248. [CrossRef] [PubMed]
16. Blaess, C.; Matzen, S.; Lin, H.; Magnan, H.; Moussy, J.P.; Rountree, C.L.; Mocuta, C.; Silly, M.G.; Plantevin, O.; Charra, F.; et al. Nitrogen Doping in Epitaxial Self-Oxidized $BaTiO_3$ Ferroelectric Thin Films. *J. Phys. Chem. C* **2025**, *129*, 3849–3861. [CrossRef]
17. Celik, F.A. Electronic structure of two-dimensional-layered $PbTiO_3$ perovskite crystal: An extended tight-binding study based on DFT. *Bull. Mater. Sci.* **2022**, *45*, 108. [CrossRef]
18. Heifets, E.; Ho, J.; Merinov, B. Density functional simulation of the $BaZrO_3$ (011) surface structure. *Phys. Rev. B* **2007**, *75*, 155431. [CrossRef]
19. Wang, Z.; Hao, X.; Gerhold, S.; Schmid, M.; Franchini, C.; Diebold, U. Vacancy clusters at domain boundaries and band bending at the $SrTiO_3$ (110) surface. *Phys. Rev. B* **2014**, *90*, 035436. [CrossRef]
20. Eglitis, R.I.; Kotomin, E.A.; Borstel, G. Quantum chemical modelling of perovskite solid solutions. *J. Phys. Condens. Matter* **2000**, *12*, L431. [CrossRef]
21. Li, W.; Landis, C.M.; Demkov, A. Domain morphology and electro-optic effect in Si-integrated epitaxial $BaTiO_3$ films. *Phys. Rev. Mater.* **2022**, *6*, 095203. [CrossRef]
22. Setvin, M.; Reticcioli, M.; Poelzleitner, F.; Hulva, J.; Schmid, M.; Boatner, L.A.; Franchini, C.; Diebold, U. Polarity compensation mechanisms on the perovskite surface $KTaO_3$ (001). *Science* **2018**, *359*, 572–575.
23. Kruchinin, S.P.; Eglitis, R.I.; Novikov, V.E.; Oleś, A.M.; Wirth, S. Control of Strongly Nonequilibrium Coherently Correlated States and Superconducting Transition Temperature. *Symmetry* **2023**, *15*, 1732. [CrossRef]
24. Ellinger, F.; Shafiq, M.; Ahmad, I.; Reticcioli, M.; Franchini, C. Small polaron formation on the Nb-doped $SrTiO_3$ (001) surface. *Phys. Rev. Mater.* **2023**, *7*, 064602.
25. Eglitis, R.I.; Bocharov, D.; Piskunov, S.; Jia, R. Review of First Principles Simulations of STO/BTO, STO/PTO, and SZO/PZO (001) Heterostructures. *Crystals* **2023**, *13*, 799. [CrossRef]
26. Cohen, R.E. Origin of ferroelectricity in perovskite oxides. *Nature* **1992**, *358*, 136–138. [CrossRef]
27. Krainyukova, N.V.; Hamalii, V.O.; Peschanskii, A.V.; Popov, A.I.; Kotomin, E.A. Low temperature structure transformations on the (001) surface of $SrTiO_3$ single crystals. *Low. Temp. Phys.* **2020**, *46*, 740–750.
28. Bottin, F.; Finocchi, F.; Noguera, C. Stability and electronic structure of the (1×1) $SrTiO_3$ (110) polar surfaces by first-principles calculations. *Phys. Rev. B* **1999**, *68*, 179–198.
29. Millers, D.; Grigorjeva, L.; Pankratov, V.; Trepakov, V.A.; Kapphan, S.E. Pulsed electron beam excited transient absorption in $SrTiO_3$. *Nucl. Instr. Methods B* **2002**, *194*, 469–473. [CrossRef]
30. Hwang, H.Y.; Iwasa, Y.; Kawasaki, M.; Keimer, B.; Nagaosa, N.; Tokura, Y. Emergent phenomena at oxide interfaces. *Nat. Mater.* **2012**, *11*, 103–113.

31. Matsuda, T.; Yamanaka, S.; Kurosaki, K.; Kobayashi, S.I. High temperature phase transitions of $SrZrO_3$. *J. Alloys Compd.* **2003**, *351*, 43–46. [CrossRef]
32. Schober, T. Protonic conduction in $BaIn_{0.5}Sn_{0.5}O_{2.75}$. *Solid State Ion.* **1998**, *109*, 1–11. [CrossRef]
33. Elmahgary, M.G.; Mahran, A.M.; Ganoub, M.; Abdellatif, S.O. Optical investigation and computational modelling of $BaTiO_3$ for optoelectronic devices applications. *Sci. Rep.* **2023**, *13*, 4761. [CrossRef] [PubMed]
34. Li, R.; Zhang, C.; Liu, J.; Zhou, J.; Xu, L. A review on the electrical properties of doped $SrTiO_3$ as anode material for solid oxide fuel cells. *Mater. Res. Express* **2019**, *6*, 102006. [CrossRef]
35. Jiang, J.; Kato, K.; Fujimori, H.; Yamakata, A.; Sakata, Y. Investigation on the highly actice $SrTiO_3$ photocatalyst toward overall H_2O splitting by doping Na ion. *J. Catal.* **2020**, *390*, 81–89. [CrossRef]
36. Ikegami, S.; Ueda, I.; Nagata, T. Electromechanical properties of $PbTiO_3$ ceramics containing La and Mn. *J. Acoust. Soc. Am.* **1971**, *50*, 1060–1966. [CrossRef]
37. Passi, M.; Pal, B. A review on $CaTiO_3$ photocatalyst: Activity enhancement methods and photocatalytic applications. *Powder Technol.* **2021**, *388*, 274–304. [CrossRef]
38. Liu, Y.; Zhang, W.; Wang, B.; Sun, L.; Li, F.; Xue, Z.; Zhou, G.; Liu, B.; Nian, H. Theoretical and experimental investigations on high temperature mechanical and thermal properties of $BaZrO_3$. *Ceram. Int.* **2018**, *44*, 16475–16482. [CrossRef]
39. Nair, J.; Nair, P.; Doesburg, E.B.M.; Ommen, J.G.V.; Ross, J.R.H.; Burggraaf, A.J. Preparation and characterization of lanthanum zirconate. *J. Mater. Sci.* **1998**, *33*, 4517–4523. [CrossRef]
40. Longo, V.M.; Cavalcante, L.S.; Figueiredo, A.T.; Santos, L.P.S.; Longo, E.; Varela, J.A. High intense violet-blue emission at room temperature in structurally disordered $SrZrO_3$ powders. *Appl. Phys. Lett.* **2007**, *90*, 091906. [CrossRef]
41. Higuchi, T.; Tsukamoto, T.; Matsumoto, H.; Shimura, T.; Yashiro, K.; Kawada, T. Electronic structure of protonic conductor SrZrMO (M = Y, Sc) probed by soft-X-ray spectroscopy. *Solid State Ion.* **2005**, *176*, 2435–2438. [CrossRef]
42. Prasanth, C.S.; Kumar, H.P.; Pazhani, R.; Solomon, S.; Thomas, J.K. Synthesis, characterization and microwave dielectric properties of nanocrystalline $CaZrO_3$ ceramics. *J. Alloys Compd.* **2008**, *464*, 306–309.
43. Chang, Y.J.; Phark, S.H. Atomic-scale visualization of initial growth of perovskites on $SrTiO_3$ (001) using scanning tunneling microscope. *Curr. Appl. Phys.* **2017**, *17*, 640–656.
44. Wang, Y.; Zhang, Z.; Wang, Y.; Doan, E.; Yuan, L.; Tang, W.; Yang, K. First-principles investigation of structural, electronic, and energetic properties of $BaSnO_3$ (001) surfaces. *Vacuum* **2023**, *212*, 111977. [CrossRef]
45. Barret, N.; Dionot, J.; Martinotti, D.; Salje, E.K.H.; Mathieu, C. Evidence of surface anomaly during the cubic-tetragonal phase transition in $BaTiO_3$ (001). *Appl. Phys. Lett.* **2018**, *113*, 022901.
46. Chen, P.; Xu, Y.; Wang, N.; Oganov, A.R.; Duan, W. Effects of ferroelectric polarization on surface phase diagram: Evolutionary algorithm study of the $BaTiO_3$ (001) surface. *Phys. Rev. B* **2015**, *92*, 085432.
47. Saghayezhian, M.; Sani, S.M.R.; Zhang, J.; Plummer, E.W. Rumpling and enhanced covalency at the $SrTiO_3$ (001) surface. *J. Phys. Chem. C* **2019**, *123*, 8086–8091.
48. Sambrano, J.R.; Longo, V.M.; Longo, E.; Taft, C.A. Electronic and structural properties of the (001) $SrZrO_3$ surface. *J. Mol. Struct. THEOCHEM* **2007**, *813*, 49–56. [CrossRef]
49. Lee, T.T.; Ekerdt, J.G. Epitaxial growth of high-*k* $BaxSr_{1-x}TiO_3$ thin films on $SrTiO_3$ (001) substrates by atomic layer deposition. *J. Vac. Sci. Technol. A* **2020**, *38*, 032401.
50. Heifets, E.; Eglitis, R.I.; Kotomin, E.A.; Maier, J.; Borstel, G. Ab initio modeling of surface structure for $SrTiO_3$ perovskite crystals. *Phys. Rev. B* **2001**, *64*, 235417.
51. Zhang, R.; Hwang, G.S. First-principles mechanistic study of the initial growth of SrO by atomic layer deposition on TiO_2-terminated $SrTiO_3$ (001). *J. Phys. Chem. C* **2020**, *124*, 28116. [CrossRef]
52. Kasai, M.; Dohi, H. Surface structure and electrochemical properties of platinum films grown on $SrTiO_3$ (100) substrates. *Surf. Sci.* **2017**, *666*, 14–22. [CrossRef]
53. Borstel, G.; Eglitis, R.I.; Kotomin, E.A.; Heifets, E. Modelling of defects and surfaces in perovskite ferroelectrics. *Phys. Status Solidi B* **2003**, *236*, 253–264. [CrossRef]
54. Wang, Y.X. Surface properties of the (001) surface of cubic $PbZrO_3$ and $PbTiO_3$. *Phys. Status Solidi B* **2007**, *244*, 602–609.
55. Zhong, M.; Zeng, W.; Liu, F.S.; Tang, B.; Liu, Q.J. First-principles study of the atomic structures, electronic properties, and surface stability of $BaTiO_3$ (001) and (011) surfaces. *Surf. Interface Anal.* **2019**, *51*, 1021–1032.
56. Piskunov, S.; Kotomin, E.A.; Heifets, E.; Maier, J.; Eglitis, R.I.; Borstel, G. Hybrid DFT calculations of the atomic and electronic structure for ABO_3 perovskite (001) surfaces. *Surf. Sci.* **2005**, *575*, 75–88. [CrossRef]
57. Brik, M.G.; Ma, C.G.; Krasnenko, V. First-principles calculations of the structural and electronic properties of the cubic $CaZrO_3$ (001) surfaces. *Surf. Sci.* **2013**, *608*, 146–153. [CrossRef]
58. Eglitis, R.I.; Purans, J.; Popov, A.I.; Bocharov, D.; Chekhovska, A.; Jia, R. Ab initio computations of O and AO as well as ReO_2, WO_2 and BO_2-terminated ReO_3, WO_3, $BaTiO_3$, $SrTiO_3$ and $BaZrO_3$ (001) surfaces. *Symmetry* **2022**, *14*, 1050. [CrossRef]

59. Alam, N.N.; Malik, N.A.; Samat, M.H.; Hussin, N.H.; Jaafar, N.K.; Radzwan, A.; Mohyedin, M.Z.; Haq, B.U.; Ali, A.M.M.; Hassan, O.H.; et al. Underlaying mechanism of surface (001) cubic $ATiO_3$ (A = Pb,Sn) in enhancing thermoelectric performance of thin-film applications using density functional theory. *Surf. Interfaces* **2021**, *27*, 101524.
60. Zhao, X.; Selloni, A. Structure and stability of $NaTaO_3$ (001) and $KTaO_3$ (001) surfaces. *Phys. Rev. Mater.* **2019**, *3*, 015801.
61. Kolpak, A.M.; Li, D.; Shao, R.; Rappe, A.M.; Bonnell, D.A. Evolution of the surface structure and thermodynamic stability of the $BaTiO_3$ (001) surface. *Phys. Rev. Lett.* **2008**, *101*, 036102. [CrossRef]
62. Rondinelli, J.A.; Spaldin, N.A. Electron-lattice instabilities suppress cuprate-like electronic structures in $SrFeO_3/SrTiO_3$ superlattices. *Phys. Rev. B* **2010**, *81*, 085109. [CrossRef]
63. Wang, J.; Neaton, J.B.; Zheng, H.; Nagarjan, V.; Ogale, S.B.; Liu, B.; Viehland, D.; Vaithyanathad, V.; Schlom, D.G.; Vagmare, U.V.; et al. Epitaxial $BiFeO_3$ multiferroic thin film heterostructures. *Science* **2003**, *299*, 1719–1722. [CrossRef] [PubMed]
64. Piskunov, S.; Eglitis, R.I. First principles hybrid DFT calculations of $BaTiO_3/SrTiO_3$ (001) interface. *Solid State Ion.* **2015**, *274*, 29–33. [CrossRef]
65. Waldow, S.P.; Souza, R.A.D. Computational study of oxygen diffusion along a [100] dislocations in the perovskite oxide $SrTiO_3$. *ACS Appl. Mater. Interfaces* **2016**, *8*, 1224–1256. [CrossRef] [PubMed]
66. Vivek, M.; Goniakowski, J.; Santander-Syro, A.; Gabay, M. Octahedral rotations and defect-driven metallicity at the (001) surface of $CaTiO_3$. *Phys. Rev. B* **2023**, *107*, 045101. [CrossRef]
67. Wang, Y.X.; Arai, M.; Sasaki, T.; Wang, C.L. First-principles study of the (001) surface of cubic $CaTiO_3$. *Phys. Rev. B* **2006**, *73*, 035411. [CrossRef]
68. Luo, B.; Wang, X.; Tian, E.; Li, G.; Li, L. Structural and electronic properties of cubic $KNbO_3$ (001) surfaces: A first-principles study. *Appl. Surf. Sci.* **2015**, *351*, 558–564. [CrossRef]
69. Krasnenko, V.; Platonenko, A.; Liivand, A.; Rusevich, L.L.; Mastrikov, Y.A.; Zvejnieks, G.; Sokolov, M.; Kotomin, E.A. Modeling of the Lattice Dynamics in Strontium Titanate Films of Various Thicknesses: Raman Scattering Studies. *Materials* **2023**, *16*, 6207. [CrossRef]
70. Iles, N.; Finocchi, F.; Khodja, K.D. A systematic study of ideal and double layer reconstruction of ABO_3 (001) surfaces (A = Sr, Ba: B = Ti, Zr) from first principles. *J. Phys. Condens. Matter* **2010**, *22*, 305001. [CrossRef]
71. Heifets, E.; Dorfman, S.; Fuks, D.; Kotomin, E. Atomistic simulation of the [001] surface structure in $BaTiO_3$. *Thin Solid Film.* **1997**, *296*, 76–78. [CrossRef]
72. Wang, Y.; Zhao, H.; Zhang, L.; Chen, J.; Xing, X. $PbTiO_3$-based perovskite ferroelectric and multiferroic thin films. *Phys. Chem. Chem. Phys.* **2017**, *19*, 17493–17515. [CrossRef]
73. Kim, J.S.; Yang, J.H.; Kim, B.K.; Kim, Y.C. Proton conduction at BaO-terminated (001) $BaZrO_3$ surface using density functional theory. *Solid State Ion.* **2015**, *275*, 19–22. [CrossRef]
74. Shimada, T.; Wang, J.; Araki, Y.; Mrovec, M.; Elsässer, C.; Kitamura, T. Multiferroic vacancies at ferroelectric $PbTiO_3$ surfaces. *Phys. Rev. Lett.* **2015**, *115*, 107202. [CrossRef]
75. Zhang, Y.; Li, G.P.; Shimada, T.; Wang, J.; Kitamura, T. Disappearance of ferroelectric critical thickness in epitaxial ultrathin $BaZrO_3$ films. *Phys. Rev. B* **2014**, *90*, 184107. [CrossRef]
76. Eglitis, R.I.; Purans, J.; Popov, A.I.; Jia, R. Tendencies in ABO_3 Perovskite and SrF_2, BaF_2 and CaF_2 Bulk and Surface *F*-Center Ab initio Computations at High Symmetry Cubic Structure. *Symmetry* **2021**, *13*, 1920. [CrossRef]
77. Cord, B.; Courths, R. Photoemission study of $BaTiO_3$ (100) surfaces. *Surf. Sci.* **1985**, *152*, 1141–1146. [CrossRef]
78. Eglitis, R.I.; Piskunov, S.; Popov, A.I.; Purans, J.; Bocharov, D.; Jia, R. Systematic Trends in Hybrid-DFT Computations of $BaTiO_3/SrTiO_3$, $PbTiO_3/SrTiO_3$ and $PbZrO_3/SrZrO_3$ (001) Hetero Structures. *Condens. Matter* **2022**, *7*, 70. [CrossRef]
79. Morales, E.H.; Bonnell, D.A. On the relationship between surface reconstructions and step edge stability on $BaTiO_3$. *Surf. Sci.* **2013**, *609*, 62–66. [CrossRef]
80. Eglitis, R.I.; Piskunov, S. First principles calculations of $SrZrO_3$ bulk and ZrO_2-terminated (001) surface *F* centers. *Comput. Condens. Matter* **2016**, *7*, 1–6. [CrossRef]
81. Berlich, A.; Strauss, H.; Langheinrich, C.; Chassé, A.; Morgner, H. Surface termination of $BaTiO_3$ (001) single crystals: A combined electron spectroscopic and theoretical study. *Surf. Sci.* **2011**, *605*, 158–165.
82. Sokolov, M.; Eglitis, R.I.; Piskunov, S.; Zhukovskii, Y.F. Ab initio hybrid DFT calculations of $BaTiO_3$ bulk and BaO-terminated (001) surface *F*-centers. *Int. J. Mod. Phys. B* **2017**, *31*, 1750251.
83. Hagendorf, C.; Schindler, K.M.; Doege, T.; Neddermeyer, H. An STM, XPS and LEED investigation of the $BaTiO_3$ (111) surface. *Surf. Sci.* **1998**, *402*, 581–585.
84. Hagendorf, C.; Schindler, K.M.; Doege, T.; Neddermeyer, H. A scanning tunneling microscopy, X-ray photoelectron spectroscopy and low-energy electron diffraction investigation of the $BaTiO_3$ (111) surface. *Surf. Sci.* **1999**, *436*, 121–130. [CrossRef]
85. Eglitis, R.I. Ab initio hybrid DFT calculations of $BaTiO_3$, $PbTiO_3$, $SrZrO_3$ and $PbZrO_3$ (111) surfaces. *Appl. Surf. Sci.* **2015**, *358*, 556–562.
86. Pojani, A.; Finocchi, F.; Noguera, C. Polarity on the $SrTiO_3$ (111) and (110) surfaces. *Surf. Sci.* **1999**, *442*, 179–198.

87. Biswas, A.; Rossen, P.B.; Yang, C.H.; Siemons, W.; Jung, M.H.; Yang, I.K.; Ramesh, R.; Jeong, Y.H. Universal Ti-rich termination of atomically flat $SrTiO_3$ (001), (011), and (111) surfaces. *Appl. Phys. Lett.* **2011**, *98*, 051904. [CrossRef]
88. Sekiguchi, S.; Fujimoto, M.; Kang, M.G., Koizumi, S.; Chao, S.B.; Tanaka, J. Structure Analysis of $SrTiO_3$ (111) Polar Surfaces. *Jpn. J. Appl. Phys.* **1998**, *37*, 4140.
89. Tanaka, H.; Kawai, T. Surface structure of reduced $SrTiO_3$ (111) observed by scanning tunneling microscopy. *Surf. Sci.* **1996**, *365*, 437–442.
90. Eglitis, R.I. Comparative ab initio calculations of $SrTiO_3$ and $CaTiO_3$ polar (111) surfaces. *Phys. Status Solidi B* **2015**, *252*, 635–642.
91. Pang, Q.; Zhang, J.M.; Xu, K.W.; Ji, V. Structural, electronic properties and stability of the (1 × 1) $PbTiO_3$ (111) polar surfaces by first-principles calculations. *Appl. Surf. Sci.* **2009**, *255*, 8145–8152. [CrossRef]
92. Liu, W.; Wang, C.; Cui, J.; Man, Z.Y. Ab initio calculations of the $CaTiO_3$ (111) polar surfaces. *Solid State Commun.* **2009**, *149*, 1871–1876.
93. Eglitis, R.I. Comparative First-Principles Calculations of $SrTiO_3$, $BaTiO_3$, $PbTiO_3$ and $CaTiO_3$ (001), (011) and (111) Surfaces. *Ferroelectrics* **2015**, *483*, 53–67. [CrossRef]
94. Eglitis, R.I. Ab initio calculations of the atomic and electronic structure of $BaZrO_3$ (111) surfaces. *Solid State Ion.* **2013**, *230*, 43–47.
95. Eglitis, R.I. Ab initio calculations of $CaZrO_3$, $BaZrO_3$, $PbTiO_3$ and $SrTiO_3$ (001), (011) and (111) surfaces as well as their (001) interfaces. *Integr. Ferroelectr.* **2019**, *196*, 7–15.
96. Eglitis, R.I.; Kotomin, E.A.; Popov, A.I.; Kruchinin, S.P.; Jia, R. Comparative ab initio calculations of $SrTiO_3$, $BaTiO_3$, $PbTiO_3$ and $SrZrO_3$ (001) and (111) surfaces as well as oxygen vacancies. *Low. Temp. Phys.* **2022**, *48*, 80–88.
97. Janifer, M.A.; Anand, S.; Prabagar, C.J.; Pauline, S. Structural and optical properties of $BaSnO_3$ ceramics by solid state reaction method. *Mater. Today Proc.* **2021**, *47*, 2067–2070.
98. Kim, H.J.; Kim, V.; Kim, T.H.; Kim, J.; Kim, H.M.; Jeon, B.G.; Lee, W.J.; Mun, H.S.; Hong, K.T.; Yu, J.; et al. Physical properties of transparent perovskite oxides $(Ba,La)SnO_3$ with high electrical mobility at room temperature. *Phys. Rev. B* **2012**, *86*, 165205.
99. Evarestov, R.A.; Bandura, A.V. First-principles calculations on the four phases of $BaTiO_3$. *J. Comput. Chem.* **2012**, *33*, 1123–1130.
100. Goudochnikov, P.; Bell, A.J. Correlations between transition temperature, tolerance factor and cohesive energy in 2+:4+ perovskites. *J. Phys. Condens. Matter* **2007**, *19*, 176201.
101. Wemple, S.H. Polarization Fluctuations and the Optical-Absorption Edge in $BaTiO_3$. *Phys. Rev. B* **1970**, *2*, 2679–2689. [CrossRef]
102. Benrekia, A.R.; Benkhettou, N.; Nassour, A.; Driz, M.; Sahnoun, M.; Lebéque, S. Structural, electronic and optical properties of cubic $SrTiO_3$ and $KTaO_3$: Ab initio and GW calculations. *Phys. B Condens. Matter* **2012**, *407*, 2632–2636. [CrossRef]
103. Benthem, K.; Elsässer, C.; French, R.H. Bulk electronic structure of $SrTiO_3$: Experiment and theory. *J. Appl. Phys.* **2001**, *90*, 6156–6164.
104. Waghmare, U.V.; Rabe, K.M. Ab initio statistical mechanics of the ferroelectric phase transition in $PbTiO_3$. *Phys. Rev. B* **1997**, *55*, 6161–6173. [CrossRef]
105. Robertson, J. Band offsets of wide-band-gap oxides and implications for future electronic devices. *J. Vacuum. Sci. Technol.* **2000**, *18*, 1785–1791. [CrossRef]
106. Ali, R.; Yashima, M. Space group and crystal structure of the perovskite $CaTiO_3$ from 296 to 1720K. *J. Solid State Chem.* **2005**, *178*, 2867–2872.
107. Ueda, K.; Yanagi, H.; Noshiro, R.; Hosono, H.; Kawazoe, H. Vacuum ultraviolet reflectance and electron energy loss spectra of $CaTiO_3$. *J. Phys. Condens. Matter* **1998**, *10*, 3669–3677.
108. Knight, K.S. Low-temperature thermophysical and crystallographic properties of $BaZrO_3$ perovskite. *J. Mater. Sci.* **2020**, *55*, 6417–6428.
109. Hasegawa, S.; Sugimoto, T.; Hashimoto, T. Investigation of structural phase transition behavior of $SrZrO_3$ by thermal analysis and high-temperature X-ray diffraction. *Solid State Ion.* **2010**, *181*, 1091–1097. [CrossRef]
110. Lee, Y.S.; Lee, J.S.; Noh, T.W.; Byun, D.Y.; Yoo, K.S.; Yamaura, K.; Takayama-Muromachi, E. Systematic trends in the electronic structure parameters of the 4d transition-metal oxides $SrMO_3$ (M = Zr, Mo, Ru, and Rh). *Phys. Rev. B* **2003**, *67*, 113101. [CrossRef]
111. Koopmans, H.J.A.; Velde, G.M.H.; Gellings, P.J. Powder neutron diffraction study of the perovskites $CaTiO_3$ and $CaZrO_3$. *Actra Crystallogr.* **1983**, *C39*, 1323–1325. [CrossRef]
112. Stoch, P.; Szczerba, J.; Lis, J.; Madej, D.; Pedzich, Z. Crystal structure and ab initio calculations of $CaZrO_3$. *J. Eur. Ceram. Soc.* **2012**, *32*, 665–670. [CrossRef]
113. Rosa, I.L.V.; Oliveira, M.C.; Assis, M.; Ferrer, M.; André, R.S.; Longo, E.; Gurgel, M.F.C. A theoretical investigation of the structural and electronic properties of orthorhombic $CaZrO_3$. *Ceram. Int.* **2015**, *41*, 3069–3074. [CrossRef]
114. Eom, K.; Chung, B.; Oh, S.; Zhou, H.; Seo, J.; Oh, S.H.; Jang, J.; Choi, S.Y.; Choi, M.; Seo, I.; et al. Surface triggered stabilization of metastable charge ordered phase in $SrTiO_3$. *Nat. Commun.* **2024**, *15*, 1180. [CrossRef] [PubMed]
115. Kubo, T.; Nozoye, H. Surface structure of $SrTiO_3$. *Surf. Sci.* **2003**, *542*, 177–191. [CrossRef]
116. Maldonado, F.; Rivera, R.; Villamagua, L.; Maldonado, J. DFT modelling of ethanol on $BaTiO_3$ (001) surface. *Appl. Surf. Sci.* **2018**, *456*, 276–289. [CrossRef]

117. Behera, R.K.; Hinojosa, B.B.; Sinott, S.B.; Asthagiri, A.; Philpott, S.R. Coupling of surface relaxation and polarization in $PbTiO_3$ from atomistic simulation. *J. Phys. Condens. Matter* **2008**, *20*, 395004.
118. Si, Y.; Zhang, T.; Liu, C.; Das, S.; Xu, B.; Burkovsky, R.G.; Wei, X.K.; Chen, Z. Antiferroelectric oxide thin-films: Fundamentals, properties and applications. *Prog. Mater. Sci.* **2024**, *142*, 101231. [CrossRef]
119. Porotnikova, N.; Osinkin, D. Segregation and interdiffusion processes in perovskites: A review of recent advances. *J. Mater. Chem. A* **2024**, *12*, 2620–2646.
120. Magagnin, G.; Lubin, C.; Escher, M.; Weber, N.; Tortech, L.; Barrett, N. Ferroelastic Twin Angles at the Surface of $CaTiO_3$ Quantified by Photoemission Electron Microscopy. *Phys. Rev. Lett.* **2024**, *132*, 056201. [CrossRef]
121. Dovesi, R.; Saunders, V.R.; Roetti, C.; Orlando, R.; Zicovich-Wilson, C.M.; Pascale, F.; Civalleri, B.; Doll, K.; Harrison, N.M.; Bush, I.J.; et al. *CRYSTAL-2017 User Manual*; University of Torino: Torino, Italy, 2017.
122. Dovesi, R.; Orlando, R.; Roetti, C.; Pisani, C.; Saunders, V.R. The Periodic Hartree-Fock Method and Its Implementation in the Crystal Code. *Comput. Simul. Mater. At. Level* **2000**, *217*, 63–88.
123. Piskunov, S.; Heifets, E.; Eglitis, R.I.; Borstel, G. Bulk properties and electronic structure of $SrTiO_3$, $BaTiO_3$, $PbTiO_3$ perovskites: An ab initio HF/DFT study. *Comput. Mater. Sci.* **2004**, *29*, 165–178.
124. Eglitis, R.I. First-principles calculations of $BaZrO_3$ (001) and (011) surfaces. *J. Phys. Condens. Matter* **2007**, *19*, 356004. [CrossRef]
125. Eglitis, R.I.; Rohlfing, M. First-principles calculations of the atomic and electronic structure of $SrZrO_3$ and $PbZrO_3$ (001) and (011) surfaces. *J. Phys. Condens. Matter* **2010**, *22*, 415901. [CrossRef] [PubMed]
126. Eglitis, R.I. Theoretical Modelling of the Energy Surface (001) and Topology of $CaZrO_3$ Perovskite. *Ferroelectrics* **2015**, *483*, 75–85.
127. Perdew, J.P.; Wang, Y. Accurate and simple density functional for the electronic exchange energy: Generalized gradient approximation. *Phys. Rev. B* **1986**, *33*, 8800–8802, Erratum in *Phys. Rev. B* **1989**, *40*, 3399.
128. Perdew, J.P.; Wang, Y. Accurate and simple analytic representation of the electron-gas correlation energy. *Phys. Rev. B* **1992**, *45*, 13244–13249.
129. Lee, C.; Yang, W.; Parr, R.G. Development of the Colle-Salvetti correlation-energy formula into a functional of the electron density. *Phys. Rev. B* **1988**, *37*, 785–789.
130. Monkhorst, H.J. Special points for Brillouin-zone integrations. *Phys. Rev. B* **1976**, *13*, 5188.
131. Eglitis, R.I.; Popov, A.I. Systematic trends in (001) surface ab initio calculations of ABO_3 perovskites. *J. Saudi Chem. Soc.* **2018**, *22*, 459–468.
132. Pojani, A.; Finocchi, F.; Noguera, C. A theoretical study of the unreconstructed polar (111) face of $SrTiO_3$. *Appl. Surf. Sci.* **1999**, *142*, 177–181. [CrossRef]
133. Tasker, P.W. The stability of ionic crystal surfaces. *J. Phys. C Solid State Phys.* **1979**, *12*, 4977. [CrossRef]
134. Noguera, C. Polar oxide surfaces. *J. Phys. Condens. Matter* **2000**, *12*, R367.
135. Eglitis, R.; Purans, J.; Popov, A.I.; Jia, R. Systematic trends in $YAlO_3$, $SrTiO_3$, $BaTiO_3$, $BaZrO_3$ (001) and (111) surface ab initio calculations. *Int. J. Mod. Phys. B* **2019**, *33*, 1950390. [CrossRef]
136. Farfán, J.C.; Rodríguez, J.A.; Fajardo, F.; López, E.V.; Landínez Téllez, D.A.; Roa-Rojas, J. Structural properties, electric response and electronic feature of $BaSnO_3$ perovskite. *Phys. B Condens. Matter* **2009**, *404*, 2720–2722.
137. Hellwege, K.H.; Helwege, A.M. Ferroelectrics and Related Substances. In *Landolt-Bornstein, New Series, Group III*; Springer: Berlin, Germany, 1969; Volume 3.
138. Mabud, S.A.; Glazer, A.M. Lattice parameters and birefringence in $PbTiO_3$ single crystals. *J. Appl. Cryst.* **1979**, *12*, 49–53. [CrossRef]
139. Mathews, M.D.; Mirza, E.B.; Momin, A.C. High-temperature X-ray diffractometric studies of $CaZrO_3$, $SrZrO_3$ and $BaZrO_3$. *J. Mater. Sci. Lett.* **1991**, *10*, 305–306. [CrossRef]
140. Kennedy, B.J.; Howard, C.J.; Chakoumakos, B.C. High-temperature phase transitions in $SrZrO_3$. *Phys. Rev. B* **1999**, *59*, 4023–4027.
141. Slater, J.C. A Simplification of the Hartree-Fock Method. *Phys. Rev.* **1951**, *81*, 385.
142. Valatin, J.G. Generalized Hartree-Fock Method. *Phys. Rev.* **1961**, *122*, 1012. [CrossRef]
143. Pisani, C.; Dovesi, R. Exact-exchange Hartree-Fock calculations for periodic systems. I. Illustration of the method. *Int. J. Quantum Chem.* **1980**, *17*, 501–516. [CrossRef]
144. Hohenberg, P.; Kohn, W. Inhomogeneous Electron Gas. *Phys. Rev.* **1964**, *136*, B864. [CrossRef]
145. Kohn, W.; Sham, L.J. Self-Consistent Equations Including Exchange and Correlation Effects. *Phys. Rev.* **1965**, *140*, A1133. [CrossRef]
146. Ziesche, P.; Kurth, S.; Perdew, J.P. Density functionals from LDA to GGA. *Comput. Mater. Sci.* **1998**, *11*, 122–127. [CrossRef]
147. Davidson, E.R.; Clark, A.E. A viewpoint on population analyses. *Int. J. Quantum Chem.* **2022**, *122*, e26860. [CrossRef]
148. Zulueta, B.; Tulyani, S.V.; Westmoreland, P.R.; Frisch, M.J.; Petersson, E.J.; Petersson, G.A.; Keith, J.A. A Bond-Energy/Bond-Order and Populations Relationship. *J. Chem. Theory Comput.* **2022**, *18*, 4774–4794. [CrossRef]
149. Kaneko, T.; Sodeyama, K. First-principles molecular dynamics study for S-O bond dissociation of sulfolane on Li-metal negative electrode. *Chem. Phys. Lett.* **2021**, *762*, 138199. [CrossRef]

150. Hu, J.; Jian, X.; Yang, T.; Peng, X. Investigation on the interface characteristic between WC (001) and diamond (111) by first-principles calculation. *Diam. Relat. Mater.* **2022**, *123*, 108864. [CrossRef]
151. Meyer, B.; Padilla, J.; Vanderbilt, D. Theory of $PbTiO_3$, $BaTiO_3$ and $SrTiO_3$ surfaces. *Faraday Discuss.* **1999**, *114*, 395–405. [CrossRef]
152. Bickel, N.; Schmidt, G.; Heinz, K.; Müller, K. Ferroelectric relaxation of the $SrTiO_3$ (100) surface. *Phys. Rev. Lett.* **1989**, *62*, 2009–2011. [CrossRef]
153. Hikita, T.; Hanada, T.; Kudo, M.; Kawai, M. Structure and electronic state of the TiO_2 and SrO terminated $SrTiO_3$ (100) surfaces. *Surf. Sci.* **1993**, *287*, 377–381. [CrossRef]
154. Wang, Y.X.; Arai, M. First-principles study of the (001) surface of cubic $SrZrO_3$. *Surf. Sci.* **2007**, *601*, 4092–4096. [CrossRef]
155. Scott, J.F. *Ferroelectric Memories*; Springer: Berlin, Germany, 2000.
156. Noguera, C. *Physics and Chemistry at Oxide Surfaces*; Cambridge University Press: New York, NY, USA, 1996.
157. Fleig, J. Microelectrodes in solid state ionics. *Solid State Ion.* **2003**, *161*, 279–289. [CrossRef]

Disclaimer/Publisher's Note: The statements, opinions and data contained in all publications are solely those of the individual author(s) and contributor(s) and not of MDPI and/or the editor(s). MDPI and/or the editor(s) disclaim responsibility for any injury to people or property resulting from any ideas, methods, instructions or products referred to in the content.

inorganics

MDPI

Article

Improved Photothermal Heating of $NaNdF_4$ Microcrystals via Low-Level Doping of Sm^{3+} for Thermal-Responsive Upconversion Luminescence Anti-Counterfeiting

Ronghua Jian and Tao Pang *

Huzhou Key Laboratory of Materials for Energy Conversion and Storage, College of Science, Huzhou University, Huzhou 313000, China; jianronghua1982@126.com
* Correspondence: tpang@126.com

Abstract: This work reports the light-to-heat conversion (LHC) behavior of $NaNdF_4$ doped with Sm^{3+}. Due to the cross-relaxation between Nd^{3+} and Sm^{3+}, the improved LHC is obtainable by introducing 5% Sm^{3+}. When the laser power density is only 1.72 W/cm^2, the spot temperature of $NaNdF_4$:5%Sm^{3+} powder reaches as high as 138.7 ± 4.04 °C. More importantly, the photoheating response to the pump laser has favorable linear characteristics within a specific power range. A simple physical model is applied to analyze the relationship between photothermal heating and pump power. Finally, the temperature-responsive luminescence anti-counterfeiting is designed by combining the LHC material with the $NaYF_4$:Yb^{3+}/Ho^{3+}/Ce^{3+} microcrystals. This novel strategy only requires two laser beams, and thus is more convenient to apply.

Keywords: photothermal heating; $NaNdF_4$:Sm^{3+}; upconversion luminescence; anti-counterfeiting

1. Introduction

Photothermal heating (i.e., electron−phonon coupling under radiation excitation) has been considered harmful to luminescence [1]. However, it has recently attracted increasing attention because of its potential applicability in tumor photothermal therapy [2–5].

In principle, the ion centers for light-to-heat conversion (LHC) should have a high state density to allow for the fast multi-phonon relaxation processes. Moreover, it is significant to incorporate a high concentration of LHC ions into the host materials since the cross-relaxation between LHC ions is another effective channel to quench luminescence. Due to meager quantum yield, Yb^{3+}-sensitized upconversion luminescence (UCL) materials have been reported for optical heaters [6–8]. However, this strategy is not particularly ideal because, on the one hand, the Yb^{3+} ions produce strong ~1000 nm emission [9], while on the other hand, the energy gap between some levels of Er^{3+}, Tm^{3+}, and Ho^{3+} is relatively large, which makes it difficult for the multi-phonon relaxation to occur. To solve the issues, Chen et al. selected Dy^{3+} and Sm^{3+}, with a high state density in the energy range of 0–10,000 cm^{-1}, as LHC ions, thus obtaining favorable photothermal heating properties [10,11]. Compared to Yb^{3+}, Nd^{3+} possesses a larger absorbing cross-section, and its relatively abundant energy levels allow it to be a self-sensitized LHC ion [12,13]. Notably, the pure Nd^{3+}-based materials (e.g., $NaNdF_4$) emit efficient near-infrared light when excited at ~800 nm radiation [14,15], which means the LHC is improvable via quenching the luminescence. Recently, Yu et al. reported that designing Prussian blue-coated $NaNdF_4$ nanoparticles is an effective strategy [16]. In contrast, Nd^{3+}/Sm^{3+} codoping is much simpler in the preparation procedure. Nevertheless, the previous reports doped relatively low concentrations of Nd^{3+} or high concentrations of Sm^{3+}, resulting in low radiation absorption at ~800 nm [12,17]. A more reasonable doping should be to fully quench the luminescence of Nd^{3+} ions without significantly affecting the light absorption. That is, the content of Nd^{3+} should be high, while that of Sm^{3+} is relatively low.

On the other hand, some studies have shown that the thermal-responsive UCL color tuning is potentially applicable in advanced luminescence anti-counterfeiting [18–21]. Nevertheless, external heating prevents its real applications. Applying Nd^{3+}-sensitized LHC materials to tune the upconversion luminescence seems a promising strategy. But as far as we know, there have been no relevant reports.

In this work, 5% Sm^{3+} is introduced into the $NaNdF_4$ lattice to quench the near-infrared emission of Nd^{3+} under 808 nm excitation. As a result, the improved LHC property is obtainable. Mechanism studies indicate that the absorbed excitation energy is completely converted into thermal energy through the self-quench of Nd^{3+} and the cross-relaxation between Nd^{3+} and Sm^{3+}. Impressively, when the change of dissipation temperature with power is ignorable, the spot temperature of $NaNdF_4$:5%Sm^{3+} has a favorable linear relationship with the pump power density. Finally, this LHC material is applied to regulate the UCL color of $NaYF_4$:20%Yb^{3+}/2%Ho^{3+}/5%Ce^{3+} under 980 nm excitation. Compared to the external temperature control, this novel strategy only requires two laser beams and thus is more convenient to use.

2. Results and Discussion

Figure 1a,b shows the Rietveld refinement of XRD data for the samples obtained by adjusting the ratio of Ln^{3+} and F^-. R_{wp} and R_p below 10% suggest that the phase purity is high. More interestingly, the phase structure of fluorides is controllable. When the molar ratio of Ln^{3+} to F^- is 4:1 and 12:1, the as-prepared samples are indexed to hexagonal NdF_3 (JCPDS No. 09-0416) and $NaNdF_4$ (JCPDS No. 72-1532), respectively. After introducing 5% Sm^{3+} into the host, no impurities are observable, which suggests that the introduction of Sm^{3+} ions does not destroy the crystal structure. Based on Hume-Rothery rules for atomic substitution [22], it can be inferred that the doped Sm^{3+} ions occupy the Nd^{3+} sites. SEM images, shown in Figure 1c,d, reveal that NdF_3:5%Sm^{3+} and $NaNdF_4$:5%Sm^{3+} present different preferred orientation growth, which can be attributed to the difference in surface density of the lattice [18]. Furthermore, EDX mappings show that the Sm^{3+} ions are uniformly distributed in the lattice of the $NaNdF_4$:5%Sm^{3+} sample, as shown in Figure 1e–i. The concentration ratio of Nd to Sm is estimated to be about 19.8, which is very close to the designed molar ratio.

Figure 2 shows the heating characteristics of NdF_3:5%Sm^{3+} and $NaNdF_4$:5%Sm^{3+} samples under 808 nm laser irradiation. Both samples show good LHC behavior, but the latter is better due to its stronger light absorption. When the laser power density is 1.72 W/cm^2, the spot temperature of $NaNdF_4$:5%Sm^{3+} powder reaches 138.7 ± 4.04 °C, higher than $NaGdF_4$:Yb^{3+}/Er^{3+} [6] and $NaNdF_4$ (~128.4°C, in this work) under the same excitation conditions. More interestingly, the photoheating response of $NaNdF_4$:5%Sm^{3+} to the pump laser has favorable linear characteristics within a specific power range. Still, overall, the responsiveness (i.e., the slope of the linear fit) is on a downward trend. In addition, due to high energy dissipation, increased measurement errors are observed when the pump power density is beyond 6 W/cm^2.

Nd^{3+} and Sm^{3+} are both excellent luminescence centers [23–25], especially, the proportion between an electric dipole to magnetic dipole transition (i.e., symmetry ratio) of Sm^{3+} ions can be used to examine its local environment [26]. Herein, to understand the LHC behavior of NdF_3:5%Sm^{3+} and $NaNdF_4$:5%Sm^{3+}, we tested the photoluminescence spectra of the samples with and without Sm^{3+} ions under 808 nm excitation. As shown in Figure 3a, NdF_3 and $NaNdF_4$ produce typical near-infrared emission in the range of 850–910 nm, corresponding to the $^4F_{3/2}\rightarrow{}^4I_{11/2}$ and $^4F_{3/2}\rightarrow{}^4I_{13/2}$ transitions of Nd^{3+}, respectively. However, the luminescence quenching occurs after being doped with Sm^{3+}. When the Sm^{3+} concentration is 5%, the emission peaks are completely quenched. Continuing to increase the concentration of Sm^{3+} (e.g., 40%), would result in little observable change in the spectrum.

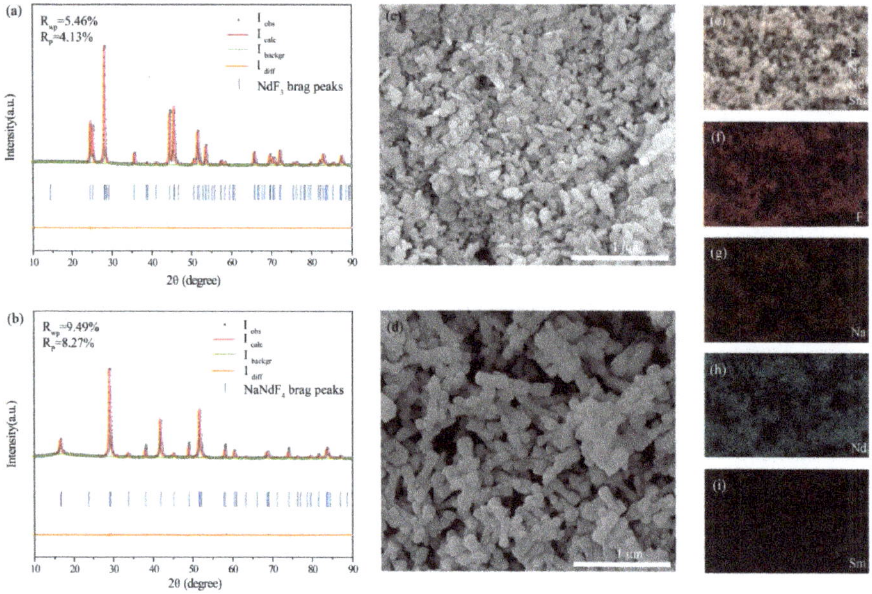

Figure 1. (**a**,**b**) Rietveld refinement of XRD data and (**c**,**d**) SEM image of $NaNdF_4$:5%Sm^{3+} and NdF_3:5%Sm^{3+} as well as (**e**–**i**) EDX mapping of F, Na, Nd and Sm elements in $NaNdF_4$:5%Sm^{3+} sample.

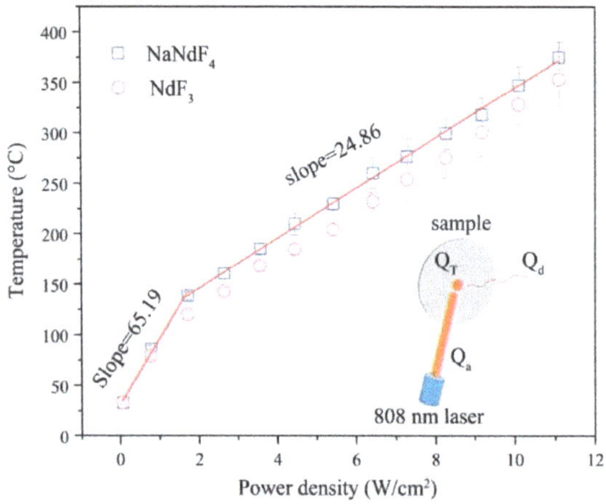

Figure 2. Photothermal heating of NdF_3:5%Sm^{3+} and $NaNdF_4$:5%Sm^{3+} flake samples, with a diameter of 10 mm and thickness of 1 mm, under 808 nm excitation. The inset gives a simple model for analyzing the relationship between photoheating and pump laser power, in which Q_a, Q_d, and Q_T represent the absorbed excitation energy, dissipated thermal energy, and residual thermal energy, respectively.

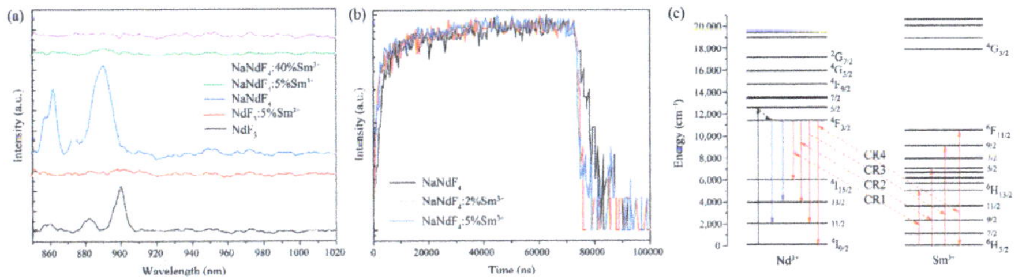

Figure 3. (**a**) Photoluminescence spectra of various Sm^{3+}-doped $NaNdF_4$ and NdF_3 under 808 nm excitation; (**b**) decay curves of 891 nm emission in various samples; (**c**) energy level of Nd^{3+} and Sm^{3+} and the proposed cross-relaxation channels between Nd^{3+} and Sm^{3+}.

No emissions belonging to Sm^{3+} ions are observed in the Sm^{3+}-doped sample, indicating the doped Sm^{3+} acts only as the quenching center of luminescence. Direct evidence can be found in Figure 3b, in which the emission lifetime of Nd^{3+} decreases with increasing the Sm^{3+} concentration. Taken together, Figure 3c illustrates the proposed mechanism of photon absorption, radiation transition, and non-radiative transition. When excited at 808 nm, Nd^{3+} first transitions from the $^4I_{9/2}$ level to the $^4F_{5/2}$ state and then rapidly decays to the $^4F_{3/2}$ level via the multi-phonon relaxation. Since the energy gap (~5380 cm^{-1}) from this level to the $^4I_{15/2}$ level far exceeds the phonon energy of NdF_3 and $NaNdF_4$ [15,27,28], the electrons at the $^4I_{15/2}$ level mainly return to $^4I_{11/2}$ and $^4I_{13/2}$ levels through radiation transition. However, after introducing Sm^{3+}, the following cross-relaxation processes occur and quench the luminescence of Nd^{3+}.

CR1: $^4F_{3/2}(Nd^{3+}) + {^6H_{5/2}}(Sm^{3+}) \rightarrow {^4I_{15/2}}(Nd^{3+}) + {^6H_{13/2}}(Sm^{3+})$;

CR2: $^4F_{3/2}(Nd^{3+}) + {^6H_{5/2}}(Sm^{3+}) \rightarrow {^4I_{13/2}}(Nd^{3+}) + {^6F_{5/2}}(Sm^{3+})$;

CR3: $^4F_{3/2}(Nd^{3+}) + {^6H_{5/2}}(Sm^{3+}) \rightarrow {^4I_{11/2}}(Nd^{3+}) + {^6F_{9/2}}(Sm^{3+})$;

CR4: $^4F_{3/2}(Nd^{3+}) + {^6H_{5/2}}(Sm^{3+}) \rightarrow {^4I_{9/2}}(Nd^{3+}) + {^6F_{11/2}}(Sm^{3+})$.

In summary, herein the absorbed excitation energy is completely converted into thermal energy through the self-quench of Nd^{3+} [12,13] and the cross-relaxation between Nd^{3+} and Sm^{3+}.

The inset in Figure 2 shows a simple model for analyzing the relationship between LHC and pump laser power. As shown, Q_a, Q_d, and Q_T represent the absorbed excitation energy, dissipated thermal energy, and residual thermal energy, respectively. According to the principle of conservation, $Q_T = Q_a - Q_d$. Therefore, the temperature function is written as

$$T_s = \frac{\rho \sigma N_0}{C} - \frac{Q_d}{C} + T_0 \quad (1)$$

where T_s is the radiation spot temperature, ρ represents the 808 nm laser power density, C represents the heat capacity of the material, σ represents the transition cross-section from the Nd^{3+} ground state to $^4F_{5/2}$, N_0 represents the population of Nd^{3+} ground state, and T_0 is the ambient temperature. Furthermore, according to the dimensions of physical quantities, the first and second terms on the right side of Equation (1) can be defined as the light absorption temperature (T_a) and the heat dissipation temperature (T_d), respectively. In this case, Equation (1) is rewritten as the following simplified form:

$$T_s = T_a - T_d + T_0 \quad (2)$$

It is evident that T_a and T_d are increasing functions of pump power. However, T_a is always greater than T_d, and thus T_s increases monotonically with the increase in pump power. Impressively, when T_d does not change much with power, T_s can be approximately regarded as a linear function of pump power. The analysis is consistent with the experimental results in Figure 2, indicating that the physical model we proposed is reasonable.

As a proof of concept, the LHC of $NaNdF_4$:5%Sm^{3+} will be applied to regulate the UCL color of $NaYF_4$:20%Yb^{3+}/2%Ho^{3+}/5%Ce^{3+} microcrystals. Figure 4a,b shows the XRD pattern and SEM image of $NaYF_4$:20%Yb^{3+}/2%Ho^{3+}/5%Ce^{3+}, respectively. It is clear that the as-prepared UCL material belongs to the hexagonal structure, and its particles appear as short rods. Furthermore, the temperature-dependent UCL properties, shown in Figure 4c, indicate that the $NaYF_4$:20%Yb^{3+}/2%Ho^{3+}/5%Ce^{3+} is applicable in thermal-responsive UCL color regulation. The possible mechanism is given in Figure 4d. Since heating improves the phonon-assisted CR1 and CR2 processes, the UCL color tuning from green to red is observed [29]. As shown in Figure 4e, a luminescence anti-counterfeiting is designed by using $NaNdF_4$:5%Sm^{3+} and $NaYF_4$:20%Yb^{3+}/2%Ho^{3+}/5%Ce^{3+}, in which the Chinese character "Light" is printed on a transparent glass substrate, and $NaNdF_4$:5%Sm^{3+} and silica gel are mixed and evenly coated on the Chinese character to form a dense heating film (referred to as NNS film). As expected, when irradiated from the glass substrate side with a 980 nm laser, clear green fonts can be observed (Figure 4f), followed by the spot temperature of 35 °C. However, when 808 nm is used to excite the NNS film, the color of the font quickly turns to red, corresponding to the spot temperature as high as 180 °C. Keeping the 808 nm radiation but turning off the 980 nm laser, the temperature of the irradiated area remains almost unchanged, but the luminous font is no longer visible. Notably, herein 808 nm and 980 nm lasers are used as a thermal- and optical-switch, respectively, and thus the external temperature-control device is abandoned.

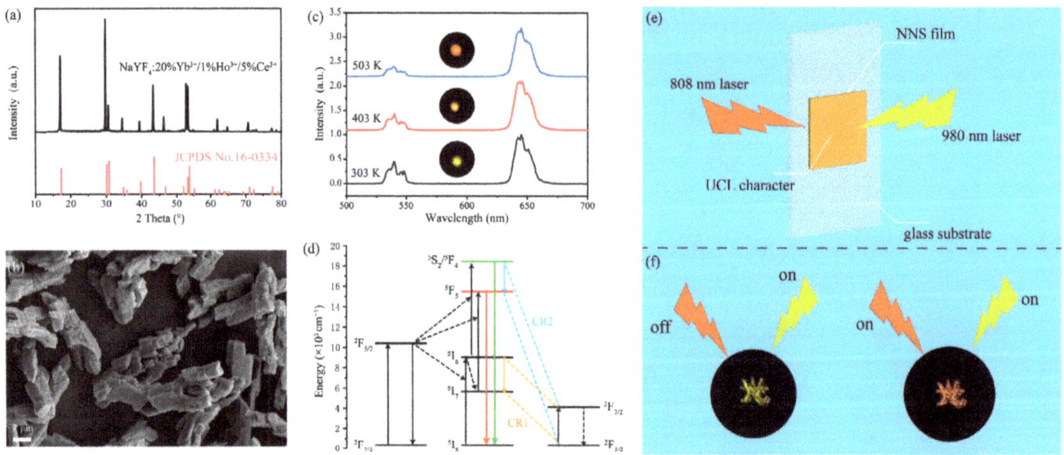

Figure 4. XRD (**a**), SEM (**b**), UCL spectra (**c**), and luminescence mechanism (**d**) of $NaYF_4$:20%Yb^{3+}/2%Ho^{3+}/5%Ce^{3+} as well as the photothermal-responsive anti-counterfeiting structure designed using $NaNdF_4$:5%Sm^{3+} and $NaYF_4$:20%Yb^{3+}/2%Ho^{3+}/5%Ce^{3+} (**e**) and its application display (**f**).

3. Experimental Section

3.1. Preparation of Sm^{3+}-Doped NdF_3 and $NaNdF_4$ Microcrystals

A hydrothermal method was used to prepare the NdF_3 and $NaNdF_4$ microcrystals doped with different concentrations of Sm^{3+}. Typically, $LnNO_3$ (Ln^{3+} = 95 mol% Nd^{3+} + 5 mol% Sm^{3+}) solution (20 mL) was obtained by dissolving $Nd(NO_3)_3 \cdot 6H_2O$ (99.9%, Aladdin Scientific Corp., Riverside, CA, USA) and $Sm(NO_3)_3 \cdot 6H_2O$ (99.9%, Aladdin Scientific Corp.) with deionized water, and then 1.7295 g $C_6H_8O_7 \cdot H_2O$ (A.C.S. grade, Aladdin Scientific Corp.) (mole ratio, citrate/Ln^{3+} = 2/1) was added under vigorous stirring. Subsequently, another 20 mL aqueous solution containing 2.0741 g NaF (A.C.S. grade, Sinopharm

Chemical Reagent Co., Ltd., Shanghai, China)was added slowly drop by drop (mole ratio, F^-/Ln^{3+} = 12/1). After 20 min, the resulting precursor solution was transferred to a 50 mL autoclave. The autoclave was heated at 200 °C for 12 h and then allowed to cool down to room temperature naturally. The product was separated from the reaction media by centrifugation and then washed three times with deionized water. After being dried at 50 °C for 12 h, the white phosphors were obtained.

3.2. Preparation of $NaYF_4$:20%Yb^{3+}/2%Ho^{3+}/5%Ce^{3+} Upconversion Materials

A similar procedure was used to prepare $NaYF_4$:20%Yb^{3+}/2%Ho^{3+}/5%Ce^{3+} upconversion microcrystals, in which Ln = 73 mol% Y + 20 mol% Yb + 2 mol% Ho + 5 mol% Ce, citrate/Ln^{3+} = 2/1, and F^-/Ln^{3+} = 12/1.

3.3. Characterization

X-ray diffraction (XRD) analysis was performed at 40 kV and 15 mA using a Miniflex600 X-ray generator (Rigaku Corp., Tokyo, Japan) with Cu K_α radiation (λ = 1.5406 Å). The 2θ scan range was 10–90° with a step size of 0.02°. Sample morphology was determined using a JSM-7800F (JEOL Ltd., Tokyo, Japan) scanning electric microscope (SEM). Energy dispersive X-ray (EDX) mapping was determined using a Phenom Pharos G2 (Phenom-World B.V., Eindhoven, The Netherlands) desktop scanning electron microscope equipped with elemental mapping. Photoluminescence spectra were collected by a portable spectrometer (Maya2000Pro, Ocean Optics Co., Hong Kong) using a continuous 980 nm diode laser as the excitation source. The photothermal heating was evaluated by a Fotric 288 infrared thermal imaging camera (Feichuke Intelligent Technology Co., Ltd., Shanghai, China).

4. Conclusions

The light-to-heat conversion (LHC) behavior of $NaNdF_4$ doped with Sm^{3+} is reported. Due to the cross-relaxation between Nd^{3+} and Sm^{3+}, the improved LHC is obtained after introducing the 5% Sm^{3+}. The principle of conservation is applied to understand the relationship between photothermal heating and pump laser power. Results show that when the change of dissipation temperature with power is ignorable, the spot temperature of $NaNdF_4$:5%Sm^{3+} has a favorable linear relationship with the pump power density. Combining the $NaNdF_4$:5%Sm^{3+} with the classic $NaYF_4$:Yb^{3+}/Ho^{3+}/Ce^{3+}, dynamic luminescence anti-counterfeiting is developed. This novel strategy only requires two laser beams and thus, is more convenient to apply compared to the external temperature control.

Author Contributions: Conceptualization, T.P.; Methodology, R.J.; Formal analysis, T.P.; Data curation, R.J.; Writing—original draft, R.J.; Writing—review & editing, T.P.; Funding acquisition, R.J. and T.P. All authors have read and agreed to the published version of the manuscript.

Funding: A project supported by the Scientific Research Fund of Zhejiang Provincial Education Department (Y202248378), and the Natural Science Foundation of Huzhou City (No. 2022YZ06).

Data Availability Statement: The original contributions presented in the study are included in the article, further inquiries can be directed to the corresponding author.

Conflicts of Interest: The authors declare no conflicts of interest.

References

1. Zhou, J.; Wen, S.; Liao, J.; Clarke, C.; Tawfik, S.A.; Ren, W.; Mi, C.; Wang, F.; Jin, D. Activation of the surface dark-layer to enhance upconversion in a thermal field. *Nat. Photonics* **2018**, *12*, 154–158. [CrossRef]
2. Pásciak, A.; Marin, R.; Abiven, L.; Pilch-Wróbel, A.; Misiak, M.; Xu, W.; Prorok, K.; Bezkrovnyi, O.; Marciniak, L.; Chaneac, C.; et al. Quantitative comparison of the light-to-heat conversion efficiency in nanomaterials suitable for photothermal therapy. *ACS Appl. Mater. Interfaces* **2022**, *14*, 33555–33566. [CrossRef] [PubMed]

3. Jaque, D.; Maestro, M.; Rosal, B.D.; Haro-Gonzalez, P.; Benayas, A.; Plaza, J.L.; Rodriguez, E.M.; Solé, J.G. Nanoparticles for photothermal therapies. *Nanoscale* **2014**, *6*, 9494–9530. [CrossRef]
4. Zhu, X.; Feng, W.; Chang, J.; Tan, Y.; Li, J.; Chen, M.; Sun, Y.; Li, F. Temperature-feedback upconversion nanocomposite for accurate photothermal therapy at facile temperature. *Nat. Commun.* **2015**, *7*, 10437. [CrossRef] [PubMed]
5. Pang, T.; Jian, R.; Xie, J.; Lu, W. Up-conversion luminescence and photo-thermal effect of KY_3F_{10}:Yb^{3+},Ho^{3+} nanocrystals. *J. Phys. D Appl. Phys.* **2018**, *51*, 355301. [CrossRef]
6. Pang, T.; Peng, W.; Yang, M.; Xie, J.; Lu, W. Synthesis, upconversion luminescence and optical heating of hexagonal $NaGdF_4$:Yb^{3+},Er^{3+}. *J. Rare Earth.* **2018**, *36*, 1136–1140. [CrossRef]
7. Shao, Q.; Ouyang, L.; Jin, L.; Jiang, J. Multifunctional nanoheater based on $NaGdF_4$:Yb^{3+},Er^{3+} upconversion nanoparticles. *Opt. Express* **2015**, *23*, 30057–30066. [CrossRef] [PubMed]
8. Zhang, Y.; Xu, S.; Li, X.; Zhang, J.; Sun, J.; Xia, H.; Hua, R.; Chen, B. Temperature sensing, excitation power dependent fluorescence branching ratios, and photothermal conversion in $NaYF_4$:Er^{3+}/Yb^{3+}@$NaYF_4$:Tm^{3+}/Yb^{3+} core-shell particles. *Opt. Mater. Express* **2018**, *8*, 368–384. [CrossRef]
9. Wang, H.; Xu, Y.; Pang, T.; Chen, B.; Xin, F.; Xing, M.; Tian, M.; Fu, Y.; Luo, X.; Tian, Y. Engineering Er^{3+}-sensitized nanocrystals to enhance NIR II-responsive upconversion luminescence. *Nanoscale* **2022**, *14*, 962–968. [CrossRef]
10. Li, Y.; Chen, B.; Tong, L.; Zhang, X.; Xu, S.; Li, X.; Zhang, J.; Sun, J.; Wang, X.; Zhang, Y.; et al. A temperature self-monitoring $NaYF_4$:Dy^{3+}/Yb^{3+}@$NaYF_4$:Er^{3+}/Yb^{3+} core-shell photothermal converter for photothermal therpy application. *Results Phys.* **2019**, *15*, 102704. [CrossRef]
11. Zheng, H.; Chen, B.; Yu, H.; Li, X.; Zhang, J.; Sun, J.; Tong, L.; Wu, Z.; Zhong, H.; Hua, R.; et al. Rod-shaped $NaY(MoO_4)_2$:Sm^{3+}/Yb^{3+} nanoheaters for photothermal conversion: Influence of doping concentration and excitation power density. *Sens. Actuators B Chem.* **2016**, *234*, 286–293. [CrossRef]
12. Pásciak, A.; Miśiak, M.; Trejgis, K.; Elzbieciak-Piecka, K.; Bezkrovnyi, O.; Marciniak, Ł.; Bednarkiewicz, A. Highly-doped lanthanide nanomaterials for efficient photothermal conversion-selection of the most promising ions and matrices. *J. Alloys Compd.* **2023**, *934*, 167900. [CrossRef]
13. Rocha, U.; Kumar, K.U.; Jacinto, C.; Ramiro, J.; Caamano, A.J.; Solé, J.G.; Jaque, D. Nd^{3+} doped LaF_3 nanoparticles as self-monitored photo-thermal agents. *Appl. Phys. Lett.* **2014**, *104*, 053703. [CrossRef]
14. Yu, S.; Cao, R.; Li, J.; Meng, L. Controlled synthesis of NdF_3 and $NaNdF_4$ micro- or nanocrystals by one-pot microwave-assisted hydrothermal reaction. *J. Fluorine Chem.* **2015**, *178*, 286–290. [CrossRef]
15. Li, M.; Hao, Z.H.; Peng, X.N.; Li, J.B.; Yu, X.F.; Wang, Q.Q. Controllable energy transfer in fluorescence upconversion of NdF_3 and $NaNdF_4$ nanocrystals. *Opt. Express* **2010**, *18*, 3364–3369. [CrossRef] [PubMed]
16. Yu, Z.; Hu, W.; Zhao, H.; Miao, X.; Guan, Y.; Cai, W.; Zeng, Z.; Fan, Q.; Tan, T.T.Y. Generating new cross-relaxation pathway by coating prussian blue on $NaNdF_4$ to fabricate enhanced photothermal agents. *Angew. Chem. Int. Ed.* **2019**, *58*, 8536–8540. [CrossRef]
17. Xu, S.; Xiang, S.; Zhang, Y.; Zhang, J.; Li, X.; Sun, J.; Cheng, L.; Chen, B. 808 nm laser induced photothermal effect of Sm^{3+}/Nd^{3+} doped $NaY(WO_4)_2$ microstructures. *Sens. Actuators B Chem.* **2017**, *240*, 386–391. [CrossRef]
18. Xu, L.; Zheng, H.; Pang, T.; Mao, J. Multicolor luminescence of hexagonal $NaYF_4$:Yb^{3+}/Ho^{3+}/Ce^{3+} microcrystals with tunable morphology under 940 nm excitation for temperature-responsvie anti-counterfeiting. *J. Rare Earth.* **2022**, *40*, 406–414. [CrossRef]
19. Wang, Y.; Lei, L.; Ye, R.; Jia, G.; Hua, Y.; Deng, D.; Xu, S. Integrating positive and negative thermal quenching effect for ultrasensitive ratiometric temperature sensing and anti-counterfeiting. *ACS Appl. Mater. Interfaces* **2021**, *13*, 23951–23959. [CrossRef] [PubMed]
20. Hu, Y.; Shao, Q.; Deng, X.; Song, D.; Han, S.; Dong, Y.; Jiang, J. Thermally induced multicolor emissions of upconversion hybrids with large color shifts for anticounterfeiting applications. *J. Mater. Chem. C* **2019**, *7*, 11770–11775. [CrossRef]
21. Shao, Q.; Zhang, G.; Ouyang, L.; Hu, Y.; Dong, Y.; Jiang, J. Emission color tuning of core/shell upconversion nanoparticles by modulating the laser power or temperature. *Nanoscale* **2017**, *9*, 12132–12141. [CrossRef] [PubMed]
22. Hu, T.; Ning, L.; Gao, Y.; Qiao, J.; Song, E.; Chen, Z.; Zhou, Y.; Wang, J.; Molokeev, M.S.; Ke, X.; et al. Glass crystallization making red phosphors for high-power warm white lighting. *Light Sci. Appl.* **2021**, *10*, 56. [CrossRef]
23. Benayas, A.; Rosal BDPérez-Delgado, A.; Santacruz-Gómez, K.; Jaque, D.; Hirata, G.A.; Vetrone, F. Nd:YAG near-infrared luminescent nanothermometers. *Adv. Opt. Mater.* **2015**, *3*, 687–694. [CrossRef]
24. Xia, Z.; Luo, Y.; Guan, M.; Liao, L. Near-infrared luminescence and energy transfer studies of LaOBr:Nd^{3+}/Yb^{3+}. *Opt. Expresses* **2012**, *20*, A722–A728. [CrossRef]
25. Kindrat, I.I.; Padlyak, B.V.; Drzewiecki, A. Luminescence properties of the Sm-doped borate glasses. *J. Lumin.* **2015**, *166*, 264–275. [CrossRef]
26. Madej, D.; Kruk, A. Classical and new insights into the methodology for characterizing the hydration of calcium aluminate cements. *Cem. Wapno Beton* **2023**, *28*, 318–328. [CrossRef]
27. Carnall, W.T.; Fields, P.R.; Rajnak, K. Electronic energy levels in the trivalent lanthanide aquo ions. I. Pr^{3+}, Nd^{3+}, Pm^{3+}, Sm^{3+}, Dy^{3+}, Ho^{3+}, Er^{3+}, and Tm^{3+}. *J. Chem. Phys.* **1968**, *49*, 4424. [CrossRef]

28. Kong, M.; Gu, Y.; Chai, Y.; Ke, J.; Liu, Y.; Xu, X.; Li, Z.; Feng, W.; Li, F. Luminescence interference-free lifetime nanothermometry pinpoints in vivo temperature. *Sci. China Chem.* **2021**, *64*, 974–984. [CrossRef]
29. Pang, T.; Wang, J. Controllable upconversion luminescence and temperature sensing behavior in $NaGdF_4$:Yb^{3+}/Ho^{3+}/Ce^{3+} nano-phosphors. *Mater. Res. Express* **2018**, *5*, 015049. [CrossRef]

Disclaimer/Publisher's Note: The statements, opinions and data contained in all publications are solely those of the individual author(s) and contributor(s) and not of MDPI and/or the editor(s). MDPI and/or the editor(s) disclaim responsibility for any injury to people or property resulting from any ideas, methods, instructions or products referred to in the content.

inorganics

MDPI

Article

Enhanced Transparency and Resistive Switching Characteristics in AZO/HfO_2/Ti RRAM Device via Post Annealing Process

Yuseong Jang , Chanmin Hwang, Sanggyu Bang and Hee-Dong Kim *

Department of Semiconductor Systems Engineering, Convergence Engineering for Intelligent Drone, and Institute of Semiconductor and System IC, Sejong University, 209, Neungdong-ro, Gwangjin-gu, Seoul 05006, Republic of Korea; nr7782942@sju.ac.kr (Y.J.); hcm4808@sju.ac.kr (C.H.); qkdtkdrb7125@sju.ac.kr (S.B.)
* Correspondence: khd0708@sejong.ac.kr

Abstract: As interest in transparent electronics increases, ensuring the reliability of transparent RRAM (T-RRAM) devices, which can be used to construct transparent electronics, has become increasingly important. However, defects and traps within these T-RRAM devices can degrade their reliability. In this study, we investigated the improvement of transparency and reliability of T-RRAM devices with an AZO/HfO_2/Ti structure through rapid thermal annealing (RTA) at 450 °C for 60 s in a nitrogen atmosphere. The device without RTA exhibited a low transmittance of 30%, whereas the device with RTA showed a significantly higher transmittance of over 75%. Furthermore, the device operated at lower current levels after RTA, which resulted in a reduction in its operating voltages, and the forming, setting, and reset voltages changed from 3.3, 2.4, and −5.1 V, respectively, to 2, 1, and −2.7 V. This led to an improvement in the endurance characteristics of the device, which thereby suggests that these improvements can be attributed to a reduction in the defects and trap density within the T-RRAM device caused by RTA.

Keywords: T-RRAM; TCO; AZO; RTA; trap

1. Introduction

With the growing interest in transparent electronics, such as transparent displays, transparent RRAM (T-RRAM) devices based on metal oxide semiconductors (MOS), or transparent conductive oxides (TCO) like indium tin oxide (ITO), indium gallium zinc oxide (IGZO), and aluminum zinc oxide (AZO) are emerging with significant industrial potential [1]. Additionally, recent studies on photosensitive and optoelectronic neuromorphic memristors, which utilize the photogenerated charge carriers and optoelectronic properties of these MOS and TCO materials [2–4], are further expanding the potential application fields of these materials. ITO and IGZO are representative TCO materials that are widely studied in T-RRAM due to their high conductivity and transmittance [5]. However, the limited availability of indium, its extraction being geographically restricted, its high cost, and its toxicity [6] necessitate research into alternative materials for T-RRAM devices. AZO is, in particular, a TCO material doped with Al in order to compensate for the low conductivity of ZnO, which makes it a strong candidate to replace indium-based TCO. AZO offers a promising alternative, which is due to its lower cost and abundance compared to indium-based materials, but the trade-offs with regard to conductivity and performance stability present challenges in its widespread adoption. Recent efforts have focused on optimizing the properties of AZO via various techniques, such as doping concentration adjustments and post-deposition treatments [7], which are shown in Table 1, but the most recent studies heavily rely on the simultaneous use of AZO and ITO. It is, therefore, difficult to regard these studies as focused on replacing ITO in T-RRAM.

Furthermore, even if the transparency of the device is achieved, a large number of defects and traps within the device can degrade its reliability [8], significantly reducing its commercial viability as a transparent device. This is especially problematic in applications requiring long-term stability, where defects and trap formation can lead to premature device failure or inconsistent performance. Addressing these reliability concerns is critical for ensuring the commercialization of transparent electronic products. Therefore, efforts to ensure the stability of devices through annealing under various conditions are ongoing to guarantee the reliability of RRAMs based on transparent materials. For example, Zhao et al. reported improved reliability of TaO_x-based RRAM after rapid thermal annealing (RTA) at 300 °C for 120 s in an oxygen atmosphere [9].

Table 1. Summary of the recently reported AZO-based transparent RRAM devices.

Device	Switching Type	Threshold Voltage	ON/OFF Ratio	Transparency	Ref.
AZO/SiOx/ITO	bipolar	7.2 V	16	85%	[4]
AZO/ZnO/ITO	bipolar	4.5 V	14	80%	[5]
ITO/SiCN/AZO	bipolar	6.4 V	>1000	85	[6]
ITO/SiCN/AZO	bipolar	6.4 V	>1000	85	[10]
AZO/CeO_2/ITO	bipolar	−7 V	>10	92.5	[11]

RTA can be a key technology in regard to enhancing the performance of T-RRAM among the various annealing methods. RTA exposes thin films to high temperatures for short durations, which promotes crystallization, improves electrical conductivity, and enhances optical transmittance [12]. The crystallization process increases the order within the thin films and also reduces the number of grain boundaries, which are often the sites for defect and trap formations [13]. RTA can greatly minimize electron scattering and improve the overall current flow within the device by reducing grain boundaries and defects [14]. Moreover, the ability of RTA to be precisely controlled in terms of temperature and duration allows for the fine-tuning of material properties, which ensures that the process can be adapted in order to meet specific performance requirements across various applications [15]. In other words, the short processing time of RTA minimizes thermal damage as well as quickly improves the device's performance, which makes it a crucial factor in T-RRAM processing [16].

In this paper, we propose a T-RRAM device using an AZO/HfO_2/Ti structure. The proposed device initially showed a low transmittance of 30% in the as-deposited state. However, after RTA at 450 °C for 60 s in a nitrogen atmosphere (hereafter referred to as 'after RTA' for short), the transmittance increased to over 75%. Furthermore, the T-RRAM after RTA exhibited improved switching operation and stability characteristics. The X-ray diffraction (XRD) results confirmed that crystallization occurred, with a significant increase in grain size for both the HfO_2 and Ti films after RTA. Additionally, X-ray photoelectron spectroscopy (XPS) analysis revealed a transformation of the Ti top electrode into TiN following RTA. Subsequently, we investigated how RTA impacts the conduction mechanism and impedance characteristics of the proposed T-RRAM device. Although all devices were found to be governed by the SCLC conduction mechanism, slight variations in behavior were observed, which we attributed to the reduction in defects and trap density due to RTA. Additionally, we established equivalent physical models for both the low resistance state (LRS) and high resistance state (HRS), and we compared the parameters of their components.

2. Results and Discussion

First, our proposed T-RRAM structure is illustrated in Figure 1a, and as shown in Figure 1b, the AZO/HfO_2/Ti structure was confirmed through field-emission scanning electron microscopy (FE-SEM). After RTA, the thickness of the Ti layer decreased from 10 nm to 7 nm, and the thickness of the HfO_2 layer decreased from 20 nm to 13 nm. This reduction in thickness is likely due to RTA-induced crystallization, which reduced the grain boundary density in the films [17]. After that, to evaluate whether the proposed

$AZO/HfO_2/Ti$ structured T-RRAM is suitable for transparent applications, we measured its transmittance, as shown in Figure 1c. The proposed T-RRAM device features a top Ti layer with a circular pattern applied using a shadow mask. As a result, in the case of the device without RTA, the regions without Ti deposition exhibit high transmittance, while the localized areas with Ti dots exhibit transmittance below 30% due to the presence of the metal layer. On the other hand, the device after RTA exhibited transmittance of over 75% across all areas of the device. This marked improvement in optical transparency highlights the effectiveness of RTA in enhancing the structural and optical qualities of the T-RRAM, making it more applicable to transparent electronics [18]. The improved transmittance of our device may be attributed to the transformation of the Ti top electrode to TiN through RTA in a nitrogen atmosphere and the crystallization of the film layers. TiN has higher transmittance than Ti, and previous studies have shown that thin TiN films can be effectively used as transparent electrodes in devices [19,20]. Additionally, RTA-induced crystallization reduces film thickness and grain boundary density, minimizing optical scattering and enhancing transmittance [17]. To investigate this further, we performed XPS analysis on the Ti top electrode before and after RTA, as well as XRD analysis on the Ti and HfO_2 films. Further details on these analyses are provided in the following sections.

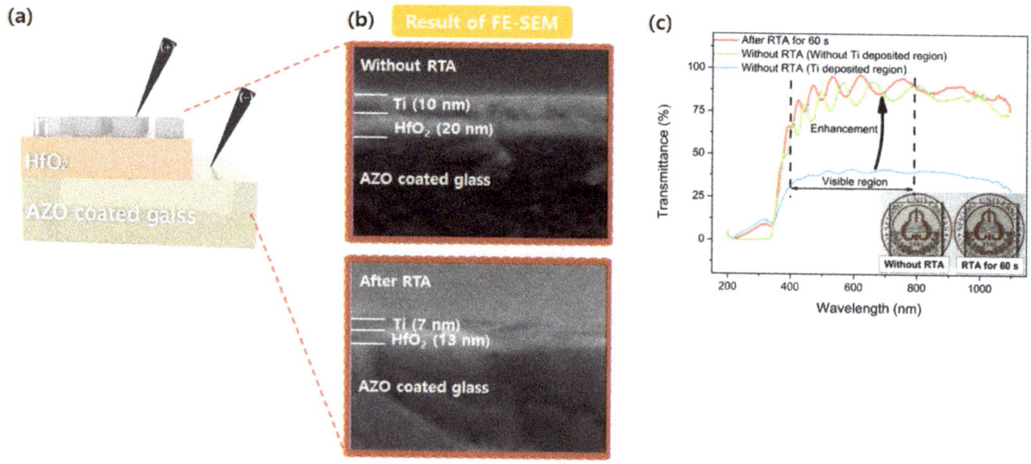

Figure 1. (**a**) Schematic structure, (**b**) cross-section FE-SEM images, and (**c**) transmittance at wavelengths of 200 nm to 1100 nm of the proposed T-RRAM.

In order to analyze the effect of RTA on the resistance switching characteristics, we analyzed the I–V curve of T-RRAM without RTA and after RTA. Figure 2a,b show the I–V curves of the T-RRAM without RTA and after RTA condition, respectively. In both cases, the log(I)–V curves were plotted to capture the distinct changes in resistance as the voltage bias was swept across both positive and negative directions. The DC voltage sweep was applied in the following sequence: $0 \rightarrow 4 \rightarrow 0 \rightarrow -8 \rightarrow 0$ V (for the device after RTA, the negative voltage was limited to -4 V). Each device shows bipolar RS characteristics, where the resistance of HfO_2 changes depending on the direction of applied bias [21]. This bipolar behavior is a hallmark of HfO_2-based RRAM, demonstrating its ability to switch between LRS and HRS depending on the applied voltage polarity [22]. The initial device shows HRS since the HfO_2 layer acts as an insulator with high resistance. When a voltage is applied to HfO_2-based RRAM, the conduction filament (CF) formation in the HfO_2 layer at a specific voltage condition causes an abrupt increase in current. This process is well known as forming, and it typically needs a higher voltage than the setting voltage. The forming voltage is critical for initiating the resistive switching mechanism, as it enables the initial formation of conductive paths through the insulating HfO_2 layer, which can later be modulated by subsequent voltage sweeps [23].

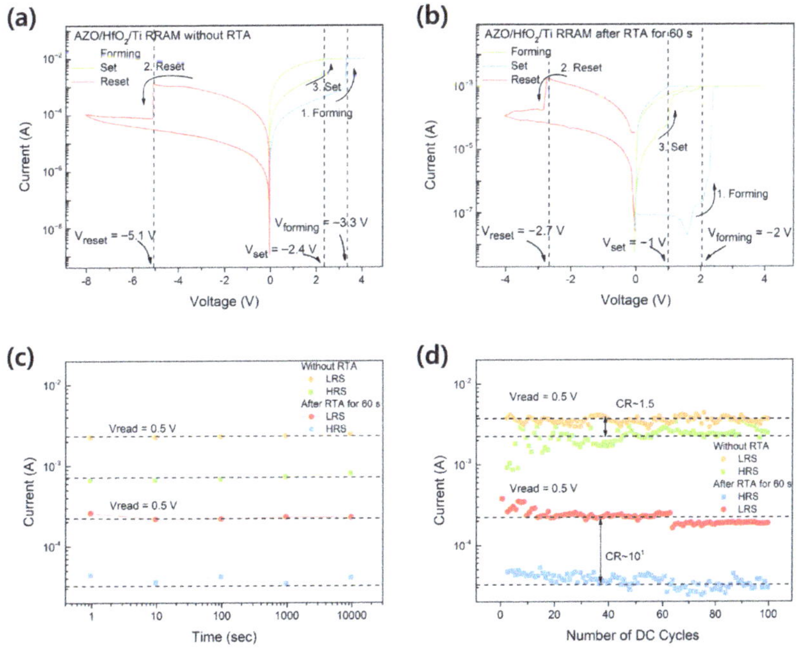

Figure 2. Resistive switching characteristics of proposed T-RRAM (**a**) without RTA and (**b**) after RTA at 450 °C for 60 s in a nitrogen atmosphere. (**c**) Retention and (**d**) endurance characteristics of T-RRAM without RTA and after RTA.

In the first positive voltage sweep, the forming voltage was 3.3 V and 2 V without and after RTA, respectively. After the initial forming process, in order to observe the RS characteristics, we applied a DC bias sweep from a negative to a positive direction. When a bias sweep was applied in the negative direction, the reset point was observed at −5.1 V for the device without RTA and at −2.7 V for the device after RTA, indicating a transition from the LRS to the HRS due to the partial or complete rupture of the CF [24]. Following that, to change HRS to LRS, the bias sweep was applied in a positive direction, and the setting point was observed at 2.4 V and 1 V, which shows an abrupt increase in current due to the reformation of the CF [25]. After that, we applied a positive voltage sweep again and confirmed the changed resistance state with an increased current level.

This varying conductive state mechanism is mainly attributed to the formation and rupture of CF consisting of oxygen vacancies in the HfO_2 layer [26]. In the initial forming process, a high voltage is applied to the top electrode Ti, resulting in oxygen vacancies (Vo^{2+}) generated in the insulator layer. With the increased amount of Vo^{2+} within the HfO_2 layer, they can act as a conductive path between the Ti and AZO, resulting in LRS facilitating electron transport through the HfO_2 layer [27]. In the reset process, it needs a high negative voltage to the bottom electrode AZO, inducing the HRS with the rupture of the CF [28]. During the setting operation to LRS transition, a positive voltage is applied to the top electrode, leading to the reformation of CF with Vo^{2+} aggregation [29].

After RTA, we observed that each forming, setting, and reset voltage, as well as the overall current level, decreased. The forming, set, and reset voltages for each device are shown in Table 2 below. This could be due to the decrease in defects and trap density and enhanced crystallization with the annealing process [30]. The correlation between RTA, smaterial phases, and trap density is explained in detail in a later section.

Table 2. The forming, setting, and reset voltages for each device.

Device	Forming	Setting	Reset
Without RTA	3.3 V	2.4 V	−5.1 V
After RTA	2 V	1 V	−2.7 V

To evaluate our device as a non-volatile memory (NVM) device, we measured retention and endurance data, as shown in Figure 2c,d. The on/off ratio was compared with resistance differences between HRS and LRS at a V_{read} of 0.5 V for devices without and after RTA. Retention measurements were conducted for 10,000 s at V_{read} to assess the stability and reliability of both devices over time. The retention test is conducted to ensure that the device can maintain its state (either LRS or HRS) without significant drift in resistance, which is crucial for long-term data storage applications [31]. The retention was measured in both the LRS and HRS states, and to further investigate the retention capabilities of the two devices, V_{read} pulses were applied to each device 1000 times with a 10 s delay between pulses. This process simulates the read operations that would occur in a typical memory application, ensuring that the device can maintain its state even after multiple read cycles [32]. As a result, both devices were confirmed to perform reliably in terms of retention, maintaining stable resistance values over the testing period. The ability to retain distinct resistance levels over time and under repeated read conditions indicates that both without RTA and after RTA devices have the potential to function effectively as NVM devices [33].

After that, the endurance characteristics were evaluated by measuring 100 endurance cycles, as shown in Figure 2d. During the endurance measurement, the device without RTA exhibited unstable endurance characteristics along with an on/off current ratio (CR) of approximately 1.5. The lower CR indicates a smaller difference between the high HRS and LRS, making it more difficult to reliably distinguish between the two states during read operations. This unstable endurance further suggests that the device without RTA may suffer from wear and degradation over time, potentially leading to reduced reliability in memory applications [34]. Conversely, the device after RTA shows a significantly increased CR of approximately 10, along with improved endurance characteristics. This higher CR provides a clearer distinction between HRS and LRS, which is essential for reliable read/write operations in non-volatile memory devices [35]. The enhanced endurance indicates that the device after RTA can withstand more switching cycles, suggesting that the RTA process strengthens the device's structural and electrical stability [36]. By reducing defects and trap density while stabilizing CF, RTA enhances both the CR and the endurance of the device, making it more robust for long-term use [37]. In contrast, the device without RTA is more prone to errors or data loss due to its lower CR and unstable endurance, making it less suitable for reliable memory applications [38]. When voltage is applied to the device, it induces the formation and dissolution of CF, altering the material's resistance [39]. However, if defects and traps are present in the material, unwanted leakage current may occur, leading to degraded device performance [40]. In other words, the RTA process helps reduce defect and trap density in the T-RRAM [41], which in turn lowers leakage current, operating voltage, and current, thereby promoting stable operation. This is consistent with findings from other studies [42–44].

Figure 3 shows the Ti 2p spectra of the Ti top electrode without and after RTA. In the Ti 2p spectrum prior to RTA, two peaks at 458.5 eV and 464.3 eV, corresponding to the oxidized state Ti^{4+} (Ti-O), were identified, indicating that Ti underwent oxidation upon exposure to air. After RTA, the full width at half maximum (FWHM) of the Ti^{4+} peaks increased, and their intensity decreased. Additionally, new peaks associated with nitridation emerged: Ti^{2+} (Ti-O-N) at 457.1 eV and Ti^{3+} (Ti-N) at 454 eV. These results suggest that, during RTA in a nitrogen atmosphere, inhomogeneous defects and structural distortions formed within the top electrode, indicating the conversion of the Ti top electrode to a nitride state [45]. As noted earlier, this nitridation of Ti is believed to have contributed to the enhanced transmittance observed in our device.

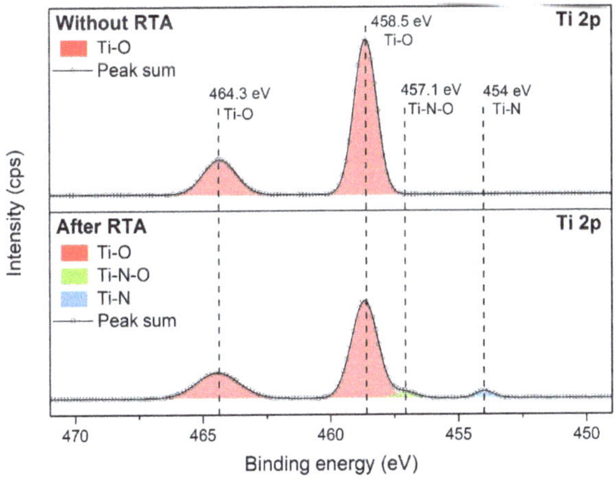

Figure 3. XPS spectra of the Ti 2p region of the Ti top electrode without RTA and after RTA.

As shown in Figure 4, we measured and analyzed the XRD patterns of the HfO_2 and Ti films before and after RTA to confirm the crystallization effect induced by RTA. For the HfO_2 film, no peaks were observed before RTA, but after RTA, peaks were found at 24.2°, 28.34°, 31.6°, 34.38°, 36.12°, 41.26°, 49.44°, 50.6°, 55.78° and 61.84°, corresponding to the (011), (−111), (111), (020), (200), (102), (211), (022), (−220), (221) and (113) planes of crystalline HfO_2 [46,47]. In the case of the Ti film, peaks were present both before and after RTA at 34.4°, 38.4°, 40.14°, 53.08°, 62.94°, 70.56°, and 76.3°, but sharper and narrower peaks were observed after RTA. These correspond to the (100), (002), (101), (102), (110), (103), and (112) planes of crystalline Ti [48,49]. This result is consistent with previous studies that show a significant improvement in diffraction intensity with increasing annealing temperature [50]. The sharpening and narrowing of the diffraction peaks with RTA indicate the formation of larger grains due to a high level of crystallinity and grain growth mechanisms. The average grain size (*D*) was calculated using Equation (1).

$$D = \frac{n\lambda}{\beta cos\theta} \quad (1)$$

Here, n represents the Scherrer constant (0.90), λ is the wavelength of the incident X-rays (0.15406 nm), β is the full width at half maximum (FWHM), and θ is the Bragg angle [51]. As a result, for HfO_2, no peaks were observed without RTA, making it impossible to calculate the grain size. However, after RTA, the grain size was calculated to be 22.88 nm. In the case of Ti, the grain size was found to be 19.03 nm without RTA and increased to 24.08 nm after RTA

Based on these results, it can be concluded that before RTA, the small grain size leads to the formation of numerous grain boundaries, causing a probabilistic process related to the distribution of oxygen vacancies, which results in uneven and high forming and operating voltages [52]. However, after RTA, grain growth reduces the grain boundaries, which can accelerate the diffusion of oxygen ions along these boundaries. Then, when an external bias is applied, the oxygen ions preferentially migrate along the grain boundaries, forming CF rich in oxygen vacancies [53]. In other words, the localization of grain boundaries through

RTA reduces variability in CF formation sites, thereby lowering the forming and operating voltages and improving device reliability [54,55].

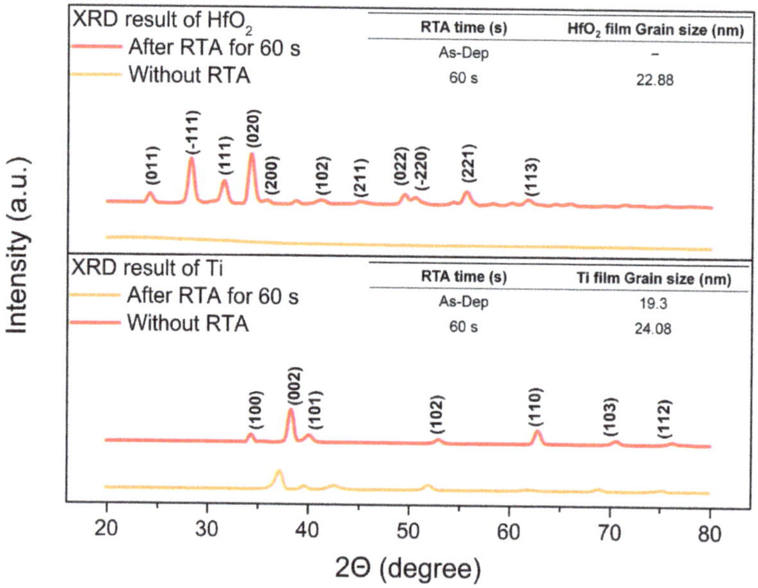

Figure 4. XRD patterns of as-deposited HfO_2 and Ti films and HfO_2 and Ti films after RTA, and (inset) average grain size of the HfO_2 and Ti films.

We plotted the I–V curves of the devices and analyzed them using various conduction mechanisms to gain a more detailed understanding of the crystallization effect induced by RTA on T-RRAM [56]. As a result, the I–V curves of both without and after RTA devices aligned well with the SCLC mechanism. Figure 5a,b show the I–V curves replotted on a double logarithmic scale for the positive voltage region of the device across cycles. The fitting results match the measured data, indicating that the conduction mechanism of the proposed T-RRAM is SCLC [57]. In the low-voltage region below 0.5 V, the current density follows Ohm's law, exhibiting a linear I–V relationship proportional to the applied voltage. However, in the high-voltage region above 0.5 V, the slope increases to 1.31 for the device without RTA and to 2.3 for the device after RTA, resulting in a nonlinear I–V relationship [58]. This behavior suggests that, as described above, the current in various voltage regions is well explained by the SCLC mechanism. The SCLC mechanism is modeled using the equation in Equation (2).

$$J = \frac{\theta}{\theta + 1} \times \frac{9}{8}\varepsilon\mu\frac{V^2}{L^3} \tag{2}$$

where $\theta = \left(\frac{N_c}{N_t}\right)\exp\left(\frac{E_c}{E_t}\right)/KT$ demonstrates the ration to free to trap electrons. Here, N_c and N_t are defined as the effective density of states in the conduction band and the number of effective electron traps, respectively, while ε and μ correspond to the permittivity and electron mobility of HfO_2. V represents the applied voltage, and L is the thickness of the HfO_2 film. Furthermore, in SCLC, the transition voltage (V_{TR}) and the trap-filled limit voltage (V_{TFL}) are two important parameters that can be derived from the I–V curve, providing information about the characteristics of the proposed T-RRAM. In the low-voltage region below V_{TR}, the electric field is uniformly distributed across the HfO_2, and since there is no space-charge region in HfO_2, the current density follows Ohm's law [59]. The energy band diagram illustrating the carrier transport behavior in this region is shown in Figure 6a. When a higher electric field is applied, the previously unoccupied trap sites in HfO_2 become filled, and a space-charge region is created. Once the applied voltage exceeds V_{TR}, the time required for externally injected charge carriers to pass through the HfO_2 becomes very short. At this point, the thermally generated charge carriers in the HfO_2 are no longer sufficient to mitigate or reduce the transit time of the externally injected carriers, as shown in Figure 6b [60]. As a result, the current density increases significantly because the externally injected carriers move through the material faster, encountering less interference from thermally generated carriers. Consequently, the Fermi level of HfO_2 rises above the trap level, and all traps in the dielectric film become filled, as shown in Figure 6c [61]. The voltage at this point is referred to as V_{TFL}, and it can be calculated using the equation in Equation (3) [62].

$$V_{TFL} = \frac{qN_tL^2}{2\varepsilon} \tag{3}$$

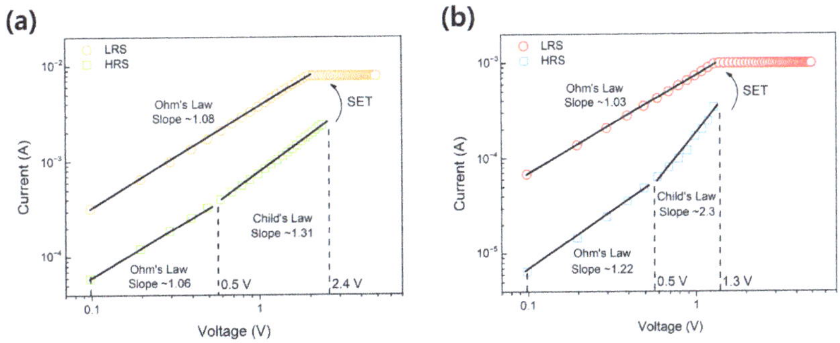

Figure 5. SCLC mechanism at the positive bias of proposed T-RRAM (**a**) without RTA and (**b**) after RTA.

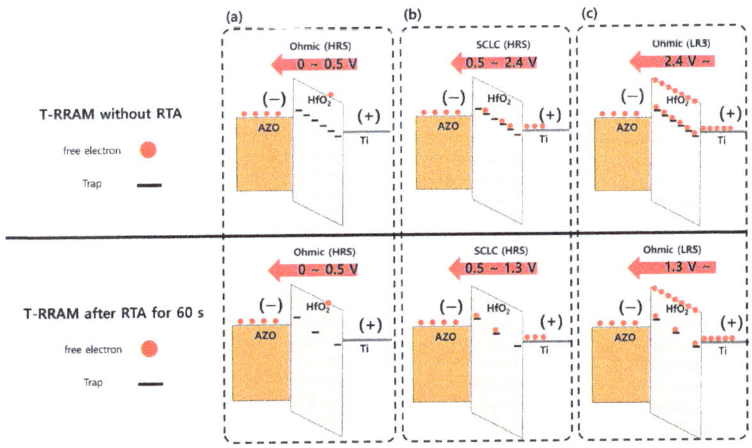

Figure 6. Band diagram of proposed T-RRAM (**a**) in the low-voltage region, (**b**) medium-voltage region and (**c**) high-voltage region. (The orange arrows indicate the direction of the electric field).

To calculate N_t, the static dielectric constant of amorphous HfO_2 was assumed to be 25 [63]. As a result, N_t values of $1.66 \times 10^{13}/\text{cm}^3$ for the device without RTA and $8.99 \times 10^{12}/\text{cm}^3$ for the device after RTA were obtained, confirming a reduction in both bulk and interface trap density due to RTA [64]. This phenomenon is also reflected in the energy band diagram shown in Figure 5.

Impedance spectroscopy (IS) is a highly effective characterization technique for investigating the electrical properties of dielectric thin films, where the films can be modeled using electrical components such as resistance, inductance, and capacitance [65]. By integrating IS analysis with DC I–V characteristics, we can gain a deeper understanding of the RS behavior and current conduction mechanisms in both the HRS and LRS [66]. Z can be defined as Equation (4).

$$Z(w) = Z' - jZ'' \tag{4}$$

Here, Z′ represents the real part and jZ″ represents the imaginary part of the complex impedance, while ω is the angular frequency. Figure 7a presents the Nyquist plots of the proposed T-RRAM in the HRS state, both without and after RTA. The semicircular shape observed in the HRS state indicates that the memory state resistance of the RRAM device can be modeled as a parallel combination of a resistor and a capacitor.

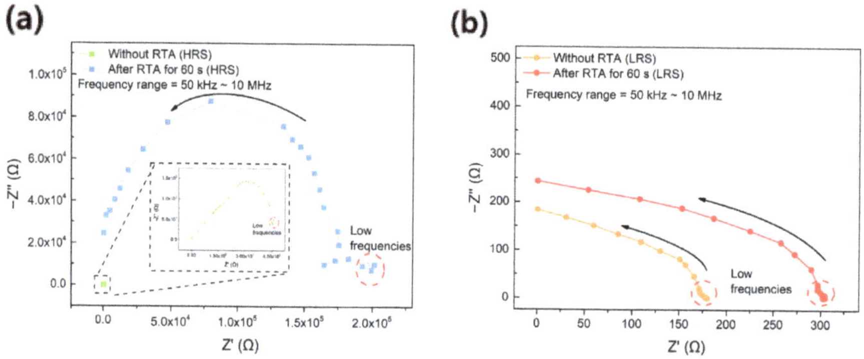

Figure 7. Nyquist plot of proposed T-RRAM at (**a**) HRS and (**b**) LRS.

Based on the Nyquist plots, the equivalent electrical circuits for the two devices in the HRS state can be represented as shown in Figure 8a [67]. The equivalent circuit can be equivalent to a parallel component (RC). Therefore, the Z in the HRS can be expressed by Equation (5).

$$Z_{HRS} = R_s + \frac{R_{HRS}}{1 + (wR_{HRS}C_{HRS})^2} - j\frac{wR_{HRS}^2C_{HRS}}{1 + (wR_{HRS}C_{HRS})^2} \tag{5}$$

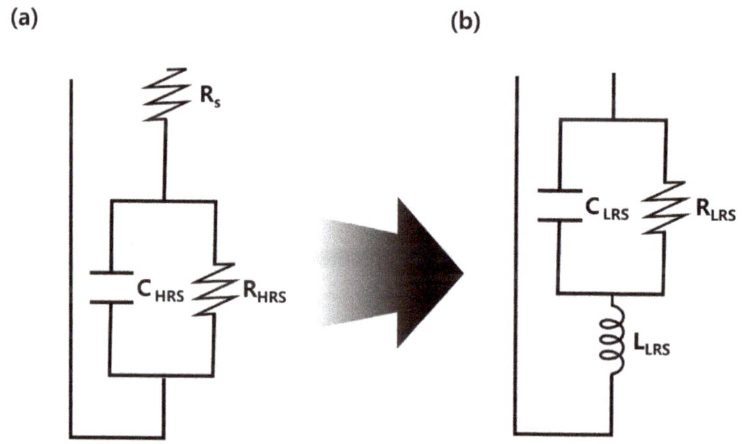

Figure 8. Equivalent circuit of proposed T-RRAM at (**a**) HRS and (**b**) LRS.

We assumed that Rs has a value of 0 to simplify the calculations. This is exclusively associated with all the resistance provided by the interfacial layer between the dielectric and the top electrode [67]. Furthermore, in the HRS, the relationship between R_{HRS} and C_{HRS} can be expressed by Equation (6).

$$C_{HRS} = \frac{1}{R_{HRS}w_{max}} \tag{6}$$

In the case of HRS, the R_{HRS} values for the device without RTA and the device after RTA were calculated as 453.89 Ω and 199.2 × 10^3 Ω, respectively, while the C_{HRS} values were calculated as 8.79 × 10^{-10} F and 3.99 × 10^{-13} F, respectively.

Figure 7b shows the Nyquist plots of the proposed T-RRAM in the LRS state, both with and without RTA. The fact that the semicircular curve shape is not entirely lost in the LRS state indicates that the RC parallel circuit continues to influence the device's behavior. However, the progressive increase in the positive Z" direction with increasing frequency suggests that inductance, caused by the formation of oxygen-vacancy-based CF, begins to play a significant role [68]. The equivalent circuit can be equivalent to a parallel component (RC) and L in series. Therefore, the Z in the LRS can be expressed by Equation (7) [67].

$$Z_{LRS} = \frac{R_{LRS}}{1+(wR_{LRS}C_{LRS})^2} + j(wL_{LRS} - \frac{wR_{LRS}^2C_{LRS}}{1+(wR_{LRS}C_{LRS})^2}) \tag{7}$$

In the case of LRS, the R_{LRS} values for the device without RTA and the device after RTA were calculated as 179.65 Ω and 301.99 Ω, respectively, while the C_{LRS} values were calculated as 1.84 × 10^{-5} F and 2.43 × 10^{-5} F, respectively. Additionally, in the case of LRS, the inductance component, L_{LRS}, was added and calculated as 8.49 × 10^{-8} H and 6.43 × 10^{-8} H, respectively. Compared to HRS, the formation of CF in LRS resulted in a reduction in R_{LRS}, an increase in C_{LRS}, and the appearance of the L_{LRS} component [69]. The presence of L_{LRS} could be attributed to the weak polarization caused by oxygen ions at the contact resistance between the HfO_2 film and the top electrode Ti due to CF formation [67]. Additionally, this contact effect is likely due to the small R_{LRS} and the broad frequency range [70]. Therefore, it can be considered that the conduction mechanism in LRS follows Ohmic behavior, which aligns well with our previous results [67]. Finally, the calculated values of the circuit parameters for HRS and LRS for each device are presented in Table 3 below. R, C, and L represent the parameters of the device without RTA, while R′, C′, and L′ denote the parameters of the device after RTA.

Table 3. Calculated parameter values of equivalent circuit elements.

Component	HRS	LRS
R	453.89 [Ω]	179.65 [Ω]
C	8.76 × 10^{-10} [Ω]	1.84 × 10^{-5} [F]
L	-	8.49 × 10^{-8} [H]
R′	199.2 × 10^3 [Ω]	301.99 [Ω]
C′	3.99 × 10^{-13} [F]	2.43 × 10^{-5} [F]
L′	-	6.43 × 10^{-8} [H]

3. Materials and Methods

3.1. Fabrication of AZO/HfO_2/Ti T-RRAM

First, the AZO-coated quartz substrate was sequentially cleaned with acetone for 10 min, methanol for 10 min, and deionized (DI) water for 10 min. After cleaning, a 20 nm-thick HfO_2 layer was deposited on AZO-coated quartz as a resistive switching layer using radio frequency (RF) sputtering (KVS-2000L) at 100 W in an Ar ambient of 20 sccm. After that, a 10 nm thick Ti top electrode (TE) was deposited using RF sputtering in an Ar ambient of 20 sccm with a metal shadow mask with a circular pattern. After deposition, the RTA process was carried out using a MILA-3000 (RTA, ULVAC, MILA-5000) for 60 s in a nitrogen atmosphere at 450 °C.

3.2. Analysis of Proposed T-RRAM Before and After RTA

To evaluate the optical properties of the proposed T-RRAM, transmittance was measured using a UV–Vis spectrophotometer (Cary 5000 UV–vis–NIR spectrophotometer, Agilent Technologies Inc., Santa Clara, CA, USA) over a spectral range of 200 nm to 1100 nm. The measurements were taken across this wide spectral range to ensure a comprehensive understanding of the device's transparency across both the ultraviolet and visible light regions. We also investigated changes in the grain size of the Ti and HfO_2 films without and after RTA using XRD. Following this, we utilized XPS to assess the chemical and material composition of the Ti top electrode and evaluate RTA-induced transformations. In addition, the electrical properties of the device were characterized. To measure the current–voltage (I–V) and impedance characteristics, a Keithley 4200A-SCS (current: ±0.025%, voltage: ±0.012%) parameter analyzer was utilized. The I–V measurements provided insights into the switching behavior and conduction mechanism of the device, and the impedance analysis provided insights into the frequency-dependent properties.

4. Conclusions

In this study, we demonstrated that RTA at 450 °C for 60 s in a nitrogen atmosphere significantly enhances both the transparency and reliability of AZO/HfO_2/Ti-based T-RRAM devices. The device without RTA exhibited a low transmittance of 30%, while the RTA-treated device showed a much higher transmittance of over 75%. Additionally, RTA resulted in the device operating at lower current levels, with reduced forming, set, and reset voltages from 3.3, 2.4, and −5.1 V to 2, 1, and −2.7 V, respectively. These changes improved the endurance characteristics of the device, indicating that the reduction in traps and defects within the T-RRAM structure, induced by RTA, plays a key role in the observed performance improvements.

Author Contributions: Conceptualization, H.-D.K. and Y.J.; methodology, Y.J. and H.-D.K.; validation, Y.J.; investigation, Y.J.; resources, H.-D.K.; data curation, H.-D.K., Y.J., S.B., and C.H.; writing—original draft preparation, Y.J.; writing—review and editing, H.-D.K.; visualization, Y.J., S.B., and C.H.; supervision, H.-D.K.; project administration, H.-D.K.; funding acquisition, H.-D.K. All authors have read and agreed to the published version of the manuscript.

Funding: This work was supported in part by the Basic Science Research Program via the National Research Foundation of Korea (NRF) funded by the Ministry of Education under grant NRF-2022R1F1A1060655 and in part by the Korea Institute for Advancement of Technology (KIAT) grant funded by the Korea Government (MOTIE) via the Competency Development Program for Industry Specialists under grant P0020966.

Data Availability Statement: The data are contained within the article.

Conflicts of Interest: The authors declare that there are no conflicts of interest.

References

1. Abbey, T.; Giotis, C.; Serb, A.; Stathopoulos, S.; Prodromakis, T. Thermal effects on initial volatile response and relaxation dynamics of resistive ram devices. *IEEE Electron Device Lett.* **2022**, *43*, 386–389. [CrossRef]
2. Chen, Q.; Zhang, Y.; Liu, S.; Han, T.; Chen, X.; Xu, Y.; Meng, Z.; Zhang, G.; Zheng, X.; Zhao, J. Switchable perovskite photovoltaic sensors for bioinspired adaptive machine vision. *Adv. Intell. Syst.* **2020**, *2*, 2000122. [CrossRef]
3. Park, H.; Ju, D.; Mahata, C.; Emelyanov, A.; Koo, M.; Kim, S. Long-and Short-Term Memory Characteristics Controlled by Electrical and Optical Stimulations in InZnO-Based Synaptic Device for Reservoir Computing. *Adv. Electron. Mater.* **2024**, *10*, 2300911. [CrossRef]
4. Fan, Z.-Y.; Tang, Z.; Fang, J.-L.; Jiang, Y.-P.; Liu, Q.-X.; Tang, X.-G.; Zhou, Y.-C.; Gao, J. Neuromorphic Computing of Optoelectronic Artificial BFCO/AZO Heterostructure Memristors Synapses. *Nanomaterials* **2024**, *14*, 583. [CrossRef]
5. Ismail, M.; Rana, A.M.; Talib, I.; Tsai, T.-L.; Chand, U.; Ahmed, E.; Nadeem, M.Y.; Aziz, A.; Shah, N.A.; Hussain, M. Room-temperature fabricated, fully transparent resistive memory based on ITO/CeO_2/ITO structure for RRAM applications. *Solid State Commun.* **2015**, *202*, 28–34. [CrossRef]

6. Ghosh, S.; Mallick, A.; Kole, A.; Chaudhury, P.; Garner, S.; Basak, D. Study on AZO coated flexible glass as TCO substrate. In Proceedings of the 2016 IEEE 43rd Photovoltaic Specialists Conference (PVSC), Portland, OR, USA, 5–10 June 2016; pp. 0634–0638.
7. Kim, H.; Osofsky, M.; Prokes, S.; Glembocki, O.; Piqué, A. Optimization of Al-doped ZnO films for low loss plasmonic materials at telecommunication wavelengths. *Appl. Phys. Lett.* **2013**, *102*, 171103. [CrossRef]
8. Zhang, L.; Hsu, Y.-Y.; Chen, F.T.; Lee, H.-Y.; Chen, Y.-S.; Chen, W.-S.; Gu, P.-Y.; Liu, W.-H.; Wang, S.-M.; Tsai, C.-H. Experimental investigation of the reliability issue of RRAM based on high resistance state conduction. *Nanotechnology* **2011**, *22*, 254016. [CrossRef]
9. Zhao, J.; Li, Y.; Li, J.; Zhou, L. Role and optimization of thermal rapid annealing in Ta/TaOx/Ru based resistive switching memory. *Vacuum* **2021**, *191*, 110392. [CrossRef]
10. Kumar, D.; Aluguri, R.; Chand, U.; Tseng, T.-Y. Conductive bridge random access memory characteristics of SiCN based transparent device due to indium diffusion. *Nanotechnology* **2018**, *29*, 125202. [CrossRef]
11. Ismail, M.; Kim, S. Negative differential resistance effect and dual bipolar resistive switching properties in a transparent Ce-based devices with opposite forming polarity. *Appl. Surf. Sci.* **2020**, *530*, 147284. [CrossRef]
12. Roy, S.; Ghosh, S.; Pradhan, D.; Sahu, P.; Kar, J. Investigation of morphological and electrical properties of RTA-processed TiO_2 for memristor application. *J. Sol-Gel Sci. Technol.* **2020**, *96*, 702–717. [CrossRef]
13. Jeong, D.G.; Park, E.; Jo, Y.; Yang, E.; Noh, G.; Lee, D.K.; Kim, M.J.; Jeong, Y.; Jang, H.J.; Joe, D.J. Grain boundary control for high-reliability HfO_2-based RRAM. *Chaos Solitons Fractals* **2024**, *183*, 114956. [CrossRef]
14. Ralls, K.; Skocpol, W.; Jackel, L.; Howard, R.; Fetter, L.; Epworth, R.; Tennant, D. Discrete resistance switching in submicrometer silicon inversion layers: Individual interface traps and low-frequency ($\frac{1}{f}$?) noise. *Phys. Rev. Lett.* **1984**, *52*, 228. [CrossRef]
15. Han, C.; Kwon, S.J.; Yim, J.; Kim, J.; Kim, S.; Jeong, S.; Park, E.C.; You, J.W.; Choi, R.; Kwon, D. Effects of RTA Rising Time on Ferroelectric Characteristics of $HfZrO_2$. *IEEE Trans. Electron Devices* **2022**, *69*, 3499–3502. [CrossRef]
16. Song, S.; Yang, T.; Liu, J.; Xin, Y.; Li, Y.; Han, S. Rapid thermal annealing of ITO films. *Appl. Surf. Sci.* **2011**, *257*, 7061–7064. [CrossRef]
17. Hsu, H.-H.; Yen, S.-S.; Chiu, Y.-C.; Chiou, P.; Chang, C.-Y.; Cheng, C.-H.; Lai, Y.-C.; Chang, C.-P.; Lu, H.-H.; Chuang, C.-S. Correlation of thermal annealing effect, crystallinity and electrical characteristics in c-axis crystallized InGaZnO thin-film transistors. *J. Alloys Compd.* **2015**, *643*, S187–S192. [CrossRef]
18. Hsu, C.-C.; Long, P.-X.; Lin, Y.-S. Enhancement of Resistive Switching Characteristics of Sol–Gel TiO x RRAM Using Ag Conductive Bridges. *IEEE Trans. Electron Devices* **2020**, *68*, 95–102. [CrossRef]
19. Kiuchi, M.; Chayahara, A. Titanium nitride for transparent conductors. *Appl. Phys. Lett.* **1994**, *64*, 1048–1049. [CrossRef]
20. Kim, G.; Cho, Y.; Kim, S. Short-term memory characteristics of TiN/WOX/FTO-based transparent memory device. *Chin. J. Phys.* **2024**, *88*, 1044–1052. [CrossRef]
21. Zhou, Q.; Zhai, J. HfOx bipolar resistive memory with robust endurance using ZrNx as buttom electrode. *Appl. Surf. Sci.* **2013**, *284*, 644–650. [CrossRef]
22. Lata, L.K.; Jain, P.K.; Chand, U.; Bhatia, D.; Shariq, M. Resistive switching characteristics of HfO_2 based bipolar nonvolatile RRAM cell. *Mater. Today Proc.* **2020**, *30*, 217–220. [CrossRef]
23. Shubhakar, K.; Mei, S.; Bosman, M.; Raghavan, N.; Ranjan, A.; O'Shea, S.J.; Pey, K.L. Conductive filament formation at grain boundary locations in polycrystalline HfO_2-based MIM stacks: Computational and physical insight. *Microelectron. Reliab.* **2016**, *64*, 204–209. [CrossRef]
24. Yuan, F.; Shen, S.; Zhang, Z.; Pan, L.; Xu, J. Interface-induced two-step RESET for filament-based multi-level resistive memory. *Superlattices Microstruct.* **2016**, *91*, 90–97. [CrossRef]
25. Balatti, S.; Larentis, S.; Gilmer, D.; Ielmini, D. Multiple memory states in resistive switching devices through controlled size and orientation of the conductive filament. *Adv. Mater.* **2013**, *25*, 1474–1478. [CrossRef]
26. Zhang, D.-l.; Wang, J.; Wu, Q.; Du, Y.; Holec, D. Ab initio study of oxygen vacancy filament formation at Ta/HfO_2 interface. *Surf. Interfaces* **2024**, *49*, 104418. [CrossRef]
27. Privitera, S.; Bersuker, G.; Lombardo, S.; Bongiorno, C.; Gilmer, D. Conductive filament structure in HfO_2 resistive switching memory devices. *Solid-State Electron.* **2015**, *111*, 161–165. [CrossRef]
28. Kumari, K.; Kar, S.; Thakur, A.D.; Ray, S. Role of an oxide interface in a resistive switch. *Curr. Appl. Phys.* **2022**, *35*, 16–23. [CrossRef]
29. Chen, Y.; Pourtois, G.; Wang, X.P.; Adelmann, C.; Goux, L.; Govoreanu, B.; Pantisano, L.; Kubicek, S.; Altimime, L.; Jurczak, M. Switching by Ni filaments in a HfO_2 matrix: A new pathway to improved unipolar switching RRAM. In Proceedings of the 2011 3rd IEEE International Memory Workshop (IMW), Monterey, CA, USA, 22–25 May 2011; pp. 1–4.
30. Zhang, X.; Zhang, T.-Y.; Wong, M.; Zohar, Y. Rapid thermal annealing of polysilicon thin films. *J. Microelectromech. Syst.* **1998**, *7*, 356–364. [CrossRef]
31. Wang, C.; Wu, H.; Gao, B.; Dai, L.; Deng, N.; Sekar, D.; Lu, Z.; Kellam, M.; Bronner, G.; Qian, H. Relaxation effect in RRAM arrays: Demonstration and characteristics. *IEEE Electron Device Lett.* **2015**, *37*, 182–185. [CrossRef]
32. Strenz, R. Review and outlook on embedded NVM technologies–from evolution to revolution. In Proceedings of the 2020 IEEE International Memory Workshop (IMW), Dresden, Germany, 17–20 May 2020; pp. 1–4.

33. Chin, A.; Lai, C.; Yang, H.; Chen, W.; Wu, Y.; Hwang, H. Extremely low voltage and high speed deep trapping MONOS memory with good retention. In Proceedings of the 2006 8th International Conference on Solid-State and Integrated Circuit Technology Proceedings, Shanghai, China, 23–26 October 2006; pp. 744–747.
34. Roy, T.; Kant, K. Enhancing endurance of ssd based high-performance storage systems using emerging nvm technologies. In Proceedings of the 2020 IEEE International Parallel and Distributed Processing Symposium Workshops (IPDPSW), New Orleans, LA, USA, 18–22 May 2020; pp. 1070–1079.
35. Chen, A. Emerging nonvolatile memory (NVM) technologies. In Proceedings of the 2015 45th European Solid State Device Research Conference (ESSDERC), Graz, Austria, 14–18 September 2015; pp. 109–113.
36. Kempen, T.; Waser, R.; Rana, V. 50x endurance improvement in taox rram by extrinsic doping. In Proceedings of the 2021 IEEE International Memory Workshop (IMW), Dresden, Germany, 16–19 May 2021; pp. 1–4.
37. Sedghi, N.; Li, H.; Brunell, I.; Dawson, K.; Guo, Y.; Potter, R.; Gibbon, J.; Dhanak, V.; Zhang, W.; Zhang, J. Enhanced switching stability in Ta_2O_5 resistive RAM by fluorine doping. *Appl. Phys. Lett.* **2017**, *111*, 092904. [CrossRef]
38. Lin, B.; Gao, B.; Pang, Y.; Yao, P.; Wu, D.; He, H.; Tang, J.; Qian, H.; Wu, H. A high-speed and high-reliability TRNG based on analog RRAM for IoT security application. In Proceedings of the 2019 IEEE International Electron Devices Meeting (IEDM), San Francisco, CA, USA, 7–11 December 2019; pp. 14.18.11–14.18.14.
39. Zahoor, F.; Azni Zulkifli, T.Z.; Khanday, F.A. Resistive random access memory (RRAM): An overview of materials, switching mechanism, performance, multilevel cell (MLC) storage, modeling, and applications. *Nanoscale Res. Lett.* **2020**, *15*, 90. [CrossRef] [PubMed]
40. Mao, L.-F. Investigating the effects of the interface defects on the gate leakage current in MOSFETs. *Appl. Surf. Sci.* **2008**, *254*, 6628–6632. [CrossRef]
41. Zhao, L.; Liu, H.-x.; Wang, X.; Fei, C.-x.; Feng, X.-y.; Wang, Y.-t. Effects of Annealing Ambient on the Characteristics of $LaAlO_3$ Films Grown by Atomic Layer Deposition. *Nanoscale Res. Lett.* **2017**, *12*, 108. [CrossRef] [PubMed]
42. Cai, W.; Zhu, Z.; Wei, J.; Fang, Z.; Ning, H.; Zheng, Z.; Zhou, S.; Yao, R.; Peng, J.; Lu, X. A simple method for high-performance, solution-processed, amorphous ZrO_2 gate insulator TFT with a high concentration precursor. *Materials* **2017**, *10*, 972. [CrossRef]
43. Ha, S.; Lee, H.; Lee, W.-Y.; Jang, B.; Kwon, H.-J.; Kim, K.; Jang, J. Effect of annealing environment on the performance of sol–gel-processed ZrO_2 RRAM. *Electronics* **2019**, *8*, 947. [CrossRef]
44. Modreanu, M.; Sancho-Parramon, J.; Durand, O.; Servet, B.; Stchakovsky, M.; Eypert, C.; Naudin, C.; Knowles, A.; Bridou, F.; Ravet, M.-F. Investigation of thermal annealing effects on microstructural and optical properties of HfO_2 thin films. *Appl. Surf. Sci.* **2006**, *253*, 328–334. [CrossRef]
45. Eid, K.; Sliem, M.H.; Abdullah, A.M. Tailoring the defects of sub-100 nm multipodal titanium nitride/oxynitride nanotubes for efficient water splitting performance. *Nanoscale Adv.* **2021**, *3*, 5016–5026. [CrossRef]
46. Park, M.H.; Chung, C.C.; Schenk, T.; Richter, C.; Opsomer, K.; Detavernier, C.; Adelmann, C.; Jones, J.L.; Mikolajick, T.; Schroeder, U. Effect of Annealing Ferroelectric HfO_2 Thin Films: In Situ, High Temperature X-Ray Diffraction. *Adv. Electron. Mater.* **2018**, *4*, 1800091. [CrossRef]
47. Fu, W.-E.; Chang, Y.-Q. Layer structure variations of ultra-thin HfO_2 films induced by post-deposition annealing. *Appl. Surf. Sci.* **2011**, *257*, 7436–7442. [CrossRef]
48. Li, Z.; Fu, L.; Fu, B.; Shan, A. Effects of annealing on microstructure and mechanical properties of nano-grained titanium produced by combination of asymmetric and symmetric rolling. *Mater. Sci. Eng. A* **2012**, *558*, 309–318. [CrossRef]
49. Potlog, T.; Dumitriu, P.; Dobromir, M.; Luca, D. XRD and XPS analysis of TiO_2 thin films annealed in different environments. *J. Mater. Sci. Eng. B* **2014**, *4*, 163–170.
50. Cui, X.; Tuokedaerhan, K.; Cai, H.; Lu, Z. Effect of annealing temperature on the microstructure and optical properties of lanthanum-doped hafnium oxide. *Coatings* **2022**, *12*, 439. [CrossRef]
51. Khan, M.I.; Neha, T.R.; Billah, M.M. UV-irradiated sol-gel spin coated AZO thin films: Enhanced optoelectronic properties. *Heliyon* **2022**, *8*, e08743. [CrossRef] [PubMed]
52. Ahmed, N.M.; Sabah, F.A.; Abdulgafour, H.; Alsadig, A.; Sulieman, A.; Alkhoaryef, M. The effect of post annealing temperature on grain size of indium-tin-oxide for optical and electrical properties improvement. *Results Phys.* **2019**, *13*, 102159. [CrossRef]
53. Tong, H.; Deng, Z.; Liu, Z.; Huang, C.; Huang, J.; Lan, H.; Wang, C.; Cao, Y. Effects of post-annealing on structural, optical and electrical properties of Al-doped ZnO thin films. *Appl. Surf. Sci.* **2011**, *257*, 4906–4911. [CrossRef]
54. Gan, K.-J.; Liu, P.-T.; Ruan, D.-B.; Chiu, Y.-C.; Sze, S.M. Annealing effects on resistive switching of IGZO-based CBRAM devices. *Vacuum* **2020**, *180*, 109630. [CrossRef]
55. Tsai, T.-L.; Chang, H.-Y.; Jiang, F.-S.; Tseng, T.-Y. Impact of post-oxide deposition annealing on resistive switching in HfO_2-based oxide RRAM and conductive-bridge RAM devices. *IEEE Electron Device Lett.* **2015**, *36*, 1146–1148. [CrossRef]
56. Hu, H.; Peng, C.; Krupanidhi, S. Effect of heating rate on the crystallization behavior of amorphous PZT thin films. *Thin Solid Film.* **1993**, *223*, 327–333. [CrossRef]
57. Liu, K.-C.; Tzeng, W.-H.; Chang, K.-M.; Chan, Y.-C.; Kuo, C.-C.; Cheng, C.-W. Transparent resistive random access memory (T-RRAM) based on Gd_2O_3 film and its resistive switching characteristics. In Proceedings of the 2010 3rd International Nanoelectronics Conference (INEC), Hong Kong, China, 3–8 January 2010; pp. 898–899.
58. Qi, Y.; Shen, Z.; Zhao, C.; Mitrovic, I.; Xu, W.; Lim, E.; Yang, L.; He, J.; Luo, T.; Huang, Y. Resistive switching behavior of solution-processed AlOx and GO based RRAM at low temperature. *Solid-State Electron.* **2020**, *168*, 107735. [CrossRef]

59. Wu, L.; Liu, H.; Li, J.; Wang, S.; Wang, X. A multi-level memristor based on Al-doped HfO_2 thin film. *Nanoscale Res. Lett.* **2019**, *14*, 177. [CrossRef]
60. Son, D.-I.; Park, D.-H.; Choi, W.K.; Cho, S.-H.; Kim, W.-T.; Kim, T.W. Carrier transport in flexible organic bistable devices of ZnO nanoparticles embedded in an insulating poly (methyl methacrylate) polymer layer. *Nanotechnology* **2009**, *20*, 195203. [CrossRef] [PubMed]
61. Liu, Y.; Luan, K.; He, Y. Effect of Annealing Temperature on the Resistive Characteristics of Ag/Hfo_2/P+-Si Storage Units. Available online: https://papers.ssrn.com/sol3/papers.cfm?abstract_id=4623563 (accessed on 20 November 2024).
62. Lim, Z.X.; Cheong, K.Y. Effects of drying temperature and ethanol concentration on bipolar switching characteristics of natural Aloe vera-based memory devices. *Phys. Chem. Chem. Phys.* **2015**, *17*, 26833–26853. [CrossRef] [PubMed]
63. Zhu, W.; Tamagawa, T.; Gibson, M.; Furukawa, T.; Ma, T. Effect of Al inclusion in HfO_2 on the physical and electrical properties of the dielectrics. *IEEE Electron Device Lett.* **2002**, *23*, 649–651. [CrossRef]
64. Lin, C.-L.; Chou, M.-Y.; Kang, T.-K.; Wu, S.-C. Electrical characteristics and TDDB breakdown mechanism of N2-RTA-treated Hf-based high-κ gate dielectrics. *Microelectron. Eng.* **2011**, *88*, 950–958. [CrossRef]
65. Gogoi, P.; Srinivas, P.; Sharma, P.; Pamu, D. Optical, dielectric characterization and impedance spectroscopy of Ni-substituted $MgTiO_3$ thin films. *J. Electron. Mater.* **2016**, *45*, 899–909. [CrossRef]
66. Sahu, V.K.; Das, A.K.; Ajimsha, R.; Misra, P. On origin of resistive and capacitive contributions to impedance of memory states in Cu/TiO_2/Pt RRAM devices by impedance spectroscopy. *Ceram. Int.* **2023**, *49*, 2215–2223. [CrossRef]
67. Bai, J.; Xie, W.; Zhang, W.; Yin, Z.; Wei, S.; Qu, D.; Li, Y.; Qin, F.; Zhou, D.; Wang, D. Conduction mechanism and impedance analysis of HfOx-based RRAM at different resistive states. *Appl. Surf. Sci.* **2022**, *600*, 154084. [CrossRef]
68. Wiśniewski, P.; Mazurak, A.; Jasiński, J.; Beck, R.B. Study of silicon-oxide RRAM devices based on complex impedance spectroscopy. *Solid-State Electron.* **2023**, *208*, 108732. [CrossRef]
69. Wiśniewski, P.; Jasiński, J.; Mazurak, A.; Stonio, B.; Majkusiak, B. Investigation of electrical properties of the Al/SiO_2/n++-Si resistive switching structures by means of static, admittance, and impedance spectroscopy measurements. *Materials* **2021**, *14*, 6042. [CrossRef]
70. Wang, Z.; Yu, H.; Tran, X.A.; Fang, Z.; Wang, J.; Su, H. Transport properties of HfO_{2-x} based resistive-switching memories. *Phys. Rev. B—Condens. Matter Mater. Phys.* **2012**, *85*, 195322. [CrossRef]

Disclaimer/Publisher's Note: The statements, opinions and data contained in all publications are solely those of the individual author(s) and contributor(s) and not of MDPI and/or the editor(s). MDPI and/or the editor(s) disclaim responsibility for any injury to people or property resulting from any ideas, methods, instructions or products referred to in the content.

inorganics

MDPI

Article

Some Aspects of Hot Carrier Photocurrent across GaAs *p-n* Junction

Steponas Ašmontas [1,*], Oleksandr Masalskyi [2,*], Ihor Zharchenko [1,2], Algirdas Sužiedėlis [1] and Jonas Gradauskas [1,2]

1 Laboratory of Electronic Processes, Center for Physical Sciences and Technology, LT-10257 Vilnius, Lithuania; ihor.zharchenko@ftmc.lt (I.Z.); algirdas.suziedelis@ftmc.lt (A.S.); jonas.gradauskas@ftmc.lt (J.G.)

2 Department of Physics, Vilnius Gediminas Technical University, LT-10223 Vilnius, Lithuania

* Correspondence: steponas.asmontas@ftmc.lt (S.A.); oleksandr.masalskyi@vilniustech.lt (O.M.)

Abstract: The photocurrent across crystalline GaAs *p-n* junction induced by Nd:YAG laser radiation was investigated experimentally. It is established that the displacement current is dominant at reverse and low forward bias voltages in the case of pulsed excitation. This indicates that hot carriers do not have enough energy to overcome the *p-n* junction until the forward bias significantly reduces the potential barrier. At a sufficiently high forward bias, the photocurrent is determined by the diffusion of hot carriers across the *p-n* junction. The current–voltage (*I-V*) characteristics measured at different crystal lattice temperatures show that the heating of carriers by laser radiation increases with a drop in crystal lattice temperature. This study proposes a novel model for evaluating carrier temperature based on the temperature coefficient of the *I-V* characteristic. It is demonstrated that the heating of carriers by light diminishes the conversion efficiency of a solar cell, not only through thermalisation but also because of the conflicting interactions between the hot carrier and conventional photocurrents, which exhibit opposite polarities. These findings contribute to an understanding of hot carrier phenomena in photovoltaic devices and may prompt a revision of the intrinsic losses in solar cells.

Keywords: GaAs; hot carriers; temperature coefficient; hot carrier temperature; *p-n* junction; solar cell

1. Introduction

The ambition to reduce the difference between the theoretically predicted and practically achieved efficiencies of a solar cell requires the introduction of new physical concepts. Hot carriers (HCs) are one of the modern candidates for solving this problem. Hot carriers are free carriers with excess energy higher than the crystal lattice temperature. In a semiconductor, light can heat the carriers in two cases. First, if the light photon energy $h\upsilon$ is lower than the bandgap E_g, heating results from intraband absorption. Such free carrier absorption is not spectrally selective and follows the classical Drude–Zener $\propto\lambda^2$ law (λ is the light wavelength) [1]. The HC photocurrent induced by below-bandgap photons was studied in Si, Ge, and GaAs *p-n* [2–4], and n-n^+ and p-p^+ [5] diodes exposed to CO_2 laser radiation ($h\upsilon$ = 0.12 eV). The polarity of the HC photocurrent indicates carrier flow up the potential barrier. Another possibility for creating hot carriers is the interband absorption of photons with energy $h\upsilon > E_g$. The excess energy, $h\upsilon - E_g$, is given to the generated electron or hole. Independent of excitation, HCs dissipate their energy through carrier–phonon and carrier–carrier scattering events. This process is defined by the energy relaxation time, which is typically on a picosecond scale [6].

The role of hot carriers in photovoltaics remains unclear. On the one hand, classical calculations of the possible efficiency of a solar cell ignore intraband absorption, which is treated as an intrinsic "*below E_g loss*" [7,8]. The indirect harmful impact of HCs on the operation of a cell is considered only as a "*thermalisation loss*"; that is, their excess energy heats the crystal lattice and hence leads to efficiency loss.

On the other hand, HCs are evidenced in solar cells, and despite their extremely short lifetime, they provide benefits. For example, excess energy in hot carriers can be optically eliminated by the emission of infrared photons [9], or used for photon upconversion [10]. Direct energy harvesting before thermalisation can be realised by impact ionisation [11,12], nanowire-modified phononic properties leading to increased relaxation times [13], or extraction of excess energy through contacts [14–16]. Extraction of hot electrons near the collector results in an increased open-circuit voltage of an ultrathin *p-i-n* cell [14]. If an energy-selective contact is made up of many contacts with particular energy differences, collecting HCs through it enhances the efficiency of a single-bandgap *p-i-n* solar cell both by using the extra energy of HCs and by avoiding lattice heating [15]. An efficient hot carrier solar cell, a cell based entirely on the HC phenomenon, was proposed [17]. According to theoretical calculations by Ross and Nozik, the conversion efficiency of HC solar cells can reach 60%. More recently, a large number of theoretical and experimental studies devoted to the development of HC solar cells were carried out [18–29]. But until now, no HC solar cell valuable for practical application has been created.

To understand and rule HC processes in semiconductors, knowledge of carrier temperature becomes a key parameter. Various techniques were developed to determine the temperature of the hot carriers. Ultrafast pump–probe spectroscopy [27,28], spontaneous photoemission spectroscopy [29], electrical measurements [30], steady-state photoluminescence spectroscopy [31], theoretical considerations [32] and other studies provide insights into the thermodynamic characteristics of the carriers.

This study presents the hot carrier photocurrent induced by close-to-bandgap laser light across a GaAs *p-n* diode. A model of the temperature coefficient of the voltage changes of the current–voltage characteristic is used to obtain the HC temperature. The distribution of 'cold' and hot carriers is analysed with regard to the potential barrier height of the *p-n* junction. The impact of HC on photovoltage formation in solar cells is discussed

2. Materials and Methods

The GaAs *p-n* junction was a 5 μm thick liquid-phase epitaxy-grown *p*-type layer on an *n*-type substrate with respective 5×10^{17} cm^{-3} and 3×10^{17} cm^{-3} hole and electron densities. For better ohmic contact, a thin heavily doped p^+-layer with a 2×10^{18} cm^{-3} hole density was additionally grown (see Figure 1a). The fabrication process involved traditional photolithography techniques finalised by thermal evaporation and standard annealing of the Au–Ge–Ni ohmic contacts. To avoid direct illumination and the rise of unwanted photosignals across the top p-p^+ and semiconductor–metal contacts, the p^+-layer was etched off, leaving it only below the metal contacts, and the contacts were positioned at the periphery of the 2.5 × 2.5 mm^2 cut sample (Figure 1a). Laser light of 1.064 μm wavelength, 17 ns long pulses, 50 Hz repetition rate, and 0.7 MW/cm^2 intensity was used for excitation (Nd:YAG laser fabricated by Ekspla Ltd., Vilnius, Lithuania). The material was chosen because of its bandgap E_g = 1.42 eV, which is close to the bandgap of a solar cell with maximum possible efficiency [8], and because single-crystal single-junction solar cells demonstrate higher efficiency as compared to multicrystalline or thin-film solar cells [33]. In addition, to evidence the hot carrier effect, the GaAs bandgap was wider than the laser light photon energy $h\nu$ = 1.17 eV.

The experiment was conducted under the photocurrent regime, where the photovoltage across the 50-Ohm load resistor was measured using a digital oscilloscope from Agilent Technologies DSO6102A (Santa Rosa, CA, USA) as shown in Figure 1a. The photoresponse consisted of two components of opposite polarity (see the inset in Figure 2a). The peak value of each photocurrent subpulse induced by pulsed laser light was measured and used for analysis. The research was carried out at 300 K and 80 K temperatures.

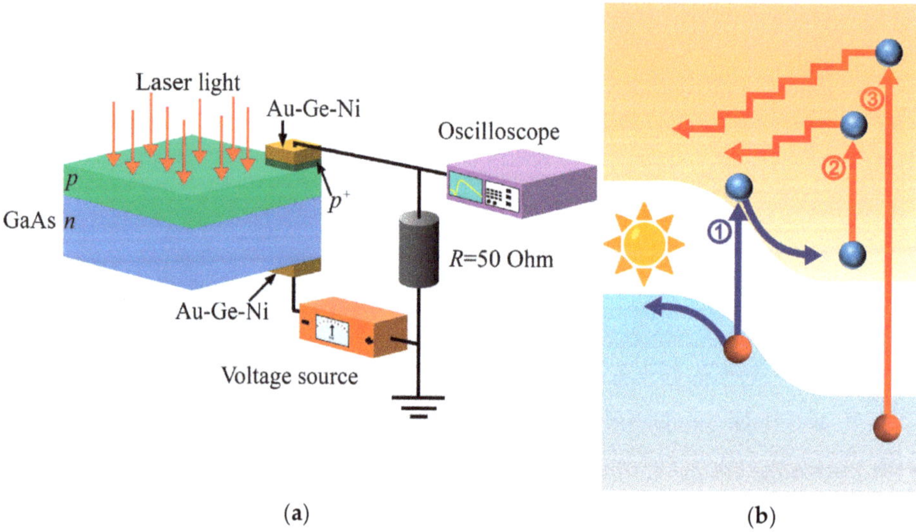

Figure 1. (**a**) Schematic of the sample and measurement circuit. (**b**) Schematic view of the formation of generation (blue) and HC (red) photocurrent across the *p-n* junction: 1—electron–hole pair generation by an equal-to-bandgap photon; 2—free electron heating; 3—generation of hot electron and hole pair. The stepped arrows indicate the cooling and diffusion of hot electrons. Analogous hot hole-related processes are omitted to avoid visual overloading.

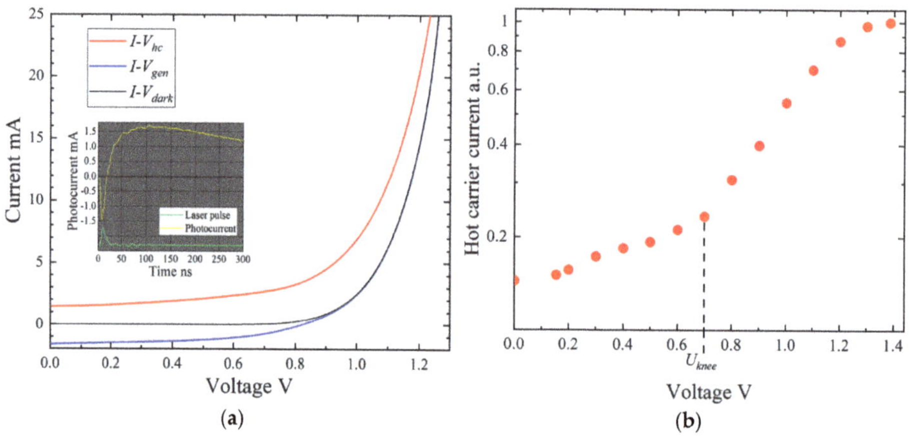

Figure 2. (**a**) *I-V* characteristics of the GaAs *p-n* junction in the dark (black line) and under illumination: the red line represents the HC photocurrent, and the blue line signifies the generation photocurrent. The inset shows a typical oscilloscope trace of a photocurrent pulse composed of two components, negative and positive, and a laser pulse below (not to scale). (**b**) Dependence of the HC photocurrent (normalised) on the bias voltage; U_{knee} = 0.7 V is the 'knee' voltage of the HC current.

3. Results

When the GaAs *p-n* junction is exposed to pulsed 1.064 μm laser light, the induced photocurrent, on the whole, consists of three components defined by carrier generation, hot carriers, and lattice heating [34]. Since the photon energy is lower than the semiconductor bandgap, the generated current results from interband two-photon absorption [35]. The polarity of the two heating-associated photocurrent components is opposite to that of the generation. Specifically, the diffusion of high-energy carriers across the junction against the built-in electric field is caused by the carrier density gradient, and it initiates the hot carrier photocurrent, as shown by the red stepped arrows in Figure 1b. At the same time, the heated lattice-induced change in the potential barrier results in the redistribution of charges and, thus, carrier transport in the same direction. The source of carrier heating is dual. First, it is determined by the free carrier absorption of the below-bandgap photons (Figure 1b, process 2). Second, they are heated by the energy $2h\nu - E_g$ = 0.91 eV left over from electron–hole generation (process 3). Until they dissipate their excess energy, the HCs can diffuse toward the junction barrier, thus forming the HC photocurrent. Calculation of the hot carrier diffusion length,

$$L_D = \sqrt{D \cdot \tau_\varepsilon}, \tag{1}$$

reveals that it is of the order of 100–150 nm; using approximate values of the diffusion coefficient $D \leq 200$ cm^2/s, and the carrier energy relaxation time τ_ε = 1 ps [6], we obtain L_D = 141 nm. This is comparable to the width of the depleted region of the junction W = 100 nm [36]. Thus, this version of hot carrier transport over the *p-n* junction, as previously believed [2], seems reasonable.

In this work, the negative (see inset in Figure 2a) photocurrent component is attributed to the hot carrier effect because of (a) its fast run determined by the carrier energy relaxation time, and (b) polarity corresponding to the carrier flow up the barrier of the *p-n* junction, as indicated by red stepped arrows in Figure 1b, and its rise with increasing forward bias. The role of the lattice heating-caused component was ignored because of its relatively small input into the net photocurrent, as shown earlier [34]. Due to their opposite polarities, competition between the generation (blue) and hot carrier (red) photocurrents is seen in the *I-V* characteristics, and both of them are influenced by the bias voltage (Figure 2a).

The HC photocurrent increases with the forward voltage (Figure 2b). At a particular voltage of 0.7 V (the 'knee' voltage), the dependence shows a typical turning point when the hot carrier photocurrent starts to grow sharply. This sudden exponential rise of the HC photocurrent was explained by the change in the photocapacitive nature of the pulsed displacement current at reverse and low-forward-bias voltage to the flow of hot carriers across the bias-lowered *p-n* junction barrier [37,38]. The displacement HC photocurrent is analogous to the AC current across a capacitor; it is present only in the case of pulsed carrier heating and resulting recharging of the junction. A similar HC-caused displacement current was also observed across an MOS structure [39]. A further increase in the forward bias leads to saturation of the HC photocurrent because of the reduced potential barrier and Joule's heating effect [2].

The potential barrier height is also influenced by the temperature of the diode. Figure 3 shows the *I-V* curves at room and liquid nitrogen temperatures. The shift of the *I-V*s with temperature change can be described by the temperature coefficient, which is a material-specific parameter and is associated with electronic components whose characteristics are temperature-dependent [15,28,40].

The temperature coefficient indicates the voltage change per temperature degree at a fixed current value [41]:

$$\alpha_T = \frac{\Delta U}{\Delta T}|i = const.|, \tag{2}$$

where ΔU is the voltage change, and ΔT is the corresponding change in temperature. Using the data in Figure 3, we obtain

$$\alpha_T = \frac{U_{N2}^{dark} - U_R^{dark}}{80\text{K} - 300\text{K}} = -2.14\ \frac{\text{mV}}{\text{K}}\bigg|i = 42.7\ \text{mA} \cdot \text{cm}^{-2}\bigg|. \tag{3}$$

The reason for choosing the 42.7 mA·cm^{-2} current density as a fixed one is described in the next paragraph. The obtained value of the temperature coefficient falls within the −(1.88–2.30) mV/K range announced for GaAs solar cells [42], and is close to −2.37 mV/K of GaAs diodes, as calculated using data from ref. [43]. In Equation (3) and elsewhere, the index "*dark*" indicates the voltage of the unilluminated diode, and the indices "*N2*" and "*R*" represent liquid nitrogen and room temperature, respectively.

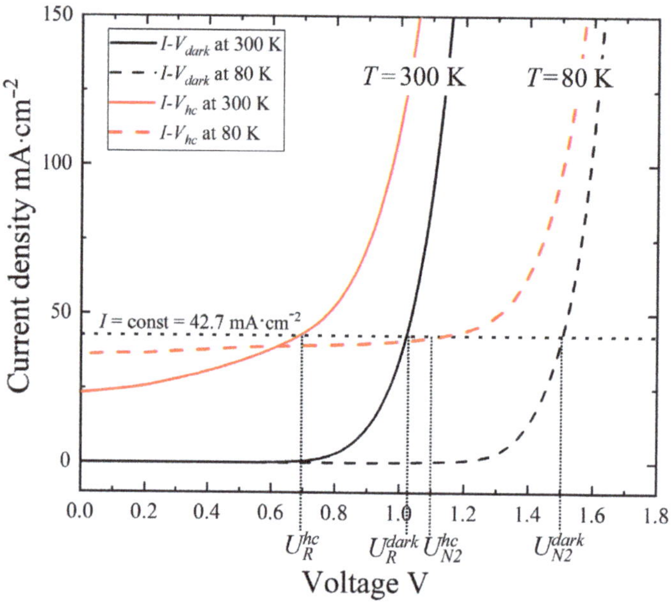

Figure 3. *I-V* characteristics in the dark (black) and with HC photocurrent (red) at temperatures of 300 K (solid lines) and 80 K (dashed lines). The bias voltages used for the calculation are U_R^{hc} = 0.70 V, U_R^{dark} = 1.03 V, U_{N2}^{hc} = 1.10 V, U_{N2}^{dark} = 1.50 V. Here, the index "*hc*" refers to the hot carriers, "*dark*" stands for the unilluminated diode, and "*N2*" and "*R*" represent liquid nitrogen and room temperature, respectively.

As Figure 3 shows, the dark current can be assumed to still be equal to zero at the mentioned 'knee' voltage of 0.7 V, and the corresponding 42.7 mA·cm^{-2} current density in the $I\text{-}V_{hc}$ is determined entirely by the hot carriers (Figure 3). This point of the $I\text{-}V_{hc}$ contains information about the carrier temperature. As the forward bias voltage increases further, the total current starts to consist of both the dark current and the HC photocurrent. Fixing this 42.7 mA·cm^{-2} value as a constant both for the illuminated and dark cases, and assuming that both these currents, the dark and HC, are of the same (recombination) nature, as explained below, we calculate the difference between the carrier and lattice temperatures, ΔT_C, under room temperature conditions using the obtained temperature coefficient α_T = −2.14 mV·K^{-1}:

$$\Delta T_C = \frac{U_R^{dark} - U_R^{hc}}{\alpha_T} = 154\ \mathrm{K}\left|i = 42.7\ \mathrm{mA}\cdot\mathrm{cm}^{-2}\right|, \tag{4}$$

where U_R^{hc} = 0.70 V, U_R^{dark} = 1.03 V, and the index "*hc*" refers to the $I\text{-}V_{hc}$ of the hot carriers. The same approach was used to calculate the HC temperature under the liquid nitrogen conditions assuming the same constant current density value of 42.7 mA·cm^{-2} at respective voltage values of U_{N2}^{hc} = 1.10 V (illuminated case, pure hot carrier current) and U_{N2}^{dark} = 1.50 V (dark current). As a result, ΔT_C = 187 K at T = 80 K. These findings indicate that the carrier heating process is more pronounced at lower temperatures, which is probably related to the increase in the carrier energy relaxation time constant at lower temperatures [6]. The proposed model for evaluating carrier temperature is supported by the following sequential reasoning: the stronger the heating, the stronger the hot carrier current, and the higher the upward shift of the current–voltage characteristic (see red lines in Figures 2a and 3). Higher shift is followed by a bigger difference in voltage values and, according to Equation (4), leads to a higher value of ΔT_C. The results obtained using the temperature coefficient method agree with the findings reported in other studies, as summarised in Table 1.

Table 1. Hot carrier temperature values at 300 K ambient temperature.

Absorber	T_C, K	Method	Excitation Details	Reference
GaAs	454	Temperature coefficient	700 kW/cm^2	This work
(In,Ga)As	430	Theoretical simulation	1 kW/cm^2	[44]
Bulk GaAs	300–680	Time-resolved luminescence	12.5 mW (at corresponding 100 ps to 1 ps time delay)	[45]
QW InGaAsP	400–1200	Photoluminescence	1.7–25 kW/cm^2	[46]

The estimated hot carrier temperature provides an additional tool to investigate the dynamics of hot carriers in a *p-n* junction. In general, the density of states occupied by free electrons in the conduction band is expressed as [36]

$$n(E) = N(E) \times f(E), \tag{5}$$

where $f(E)$ is the Fermi–Dirac probability function, and $N(E)$ is the density of states in the conduction band. After making all the necessary substitutions [36], Equation (5) becomes

$$n(E) = \frac{N_C\sqrt{E - E_C}}{4\pi^{3.5}k^{3.5}T^{1.5}} \times \frac{1}{1 + exp\left(\frac{E-E_F}{kT_C}\right)}, \tag{6}$$

where E_C= 0 eV is the energy of the conduction band bottom, N_C is the effective density of states in the conduction band, T is the lattice temperature, T_C is the carrier temperature, E is the energy (in eV), E_F is the Fermi level, and k is the Boltzmann constant.

The blue area in Figure 4 shows the electron distribution in the conduction band calculated in the case of no excitation, i.e., at T = T_C = 300 K. The data needed for the calculation were taken from ref. [6]. When the carriers are heated, in our case, we obtain a hot electron distribution using T_C = 454 K in the Fermi–Dirac probability function of Equation (6) at a T = 300 K lattice temperature (red area). Naturally, the 'tail' of the HC distribution is now extended to higher energies. However, its value is fading away at a level that is much lower than the unbiased potential barrier of the junction, 1.31 eV, and is still lower than the 0.7 V forwardly biased barrier of 1.31 eV − 0.7 eV = 0.61 eV. Statistically, only 10^{-4} % of the hot electrons, or of the total red area in Figure 4, are above the 0.61 eV energy level. Thus, the probability that a hot electron will form a current by overcoming the barrier is negligibly small. Similar results are obtained for the hot hole distribution in the valence band, and similar barrier-related considerations can be applied to the hot holes. This means that the hot carrier photocurrent across the *p-n* junction at reverse and low-forward-bias voltages cannot be a diffusion of the HCs over the barrier, as was supposed earlier [2] and is schematised in Figure 1b. Therefore, the hot carrier photocurrent changes its nature from a displacement character to a recombination character when the 'knee' voltage is exceeded. Still being well below the top of the barrier, the hot carriers do not manage to overcome it, as the calculated carrier distribution shows, but now, the heated electrons and holes can meet each other in the depleted region and recombine, thus forming the recombination current. However, at higher voltages, the HC photocurrent most probably turns into diffusion of the hot carriers over the substantially reduced potential barrier. To reach the same 42.7 mA·cm^{-2} recombinative photocurrent across the unbiased potential barrier, or, in other words, to obtain 10^{-4}% of hot electrons above the energy level of 1.31 eV, the electron temperature should be approximately 990 K.

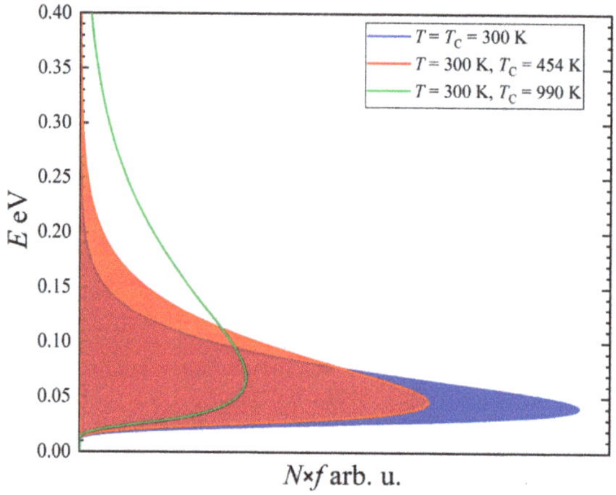

Figure 4. Electron distribution in the conduction band of the *n*-region of the *p-n* junction at a T = 300 K lattice temperature and electron temperatures of T_C = 300 K (blue area), T_C = 454 K (red area), and T_C = 990 K (green line).

In the case of sunlight-illuminated solar cells, the potential barrier of the *p-n* junction decreases due to generated electron–hole pairs, thereby creating favourable conditions for the HC current flow through the junction. Since the generation-related photocurrent and HC photocurrent flow in opposing directions, it is evident that carrier heating diminishes the overall current passing through the *p-n* junction, thereby decreasing the conversion efficiency of a solar cell.

4. Conclusions

In conclusion, we analysed the hot carrier photocurrent induced across the GaAs *p-n* junction by the 1.064 μm laser radiation. Based on earlier studies and this study, the hot carrier photocurrent, in contrast to traditional assumptions, has three stages: (1) displacement current at low bias voltage in the case of pulsed excitation, (2) recombination current at higher forward bias exceeding the 'knee' voltage, and (3) diffusion current, which is possible only in the case of a strong excitation level resulting in high carrier temperature and at a high enough forward bias allowing the hot carriers to overcome the potential barrier. On the basis of the temperature coefficient of the current–voltage characteristics, we propose a novel technique to estimate the hot carrier temperature. The result of the calculation falls within the reasonable range of carrier temperatures obtained by other methods. A more substantial change in the absolute hot carrier temperature at liquid nitrogen temperature compared with room temperature implies that carrier heating is stronger at lower lattice temperatures because of the longer energy relaxation time of the hot carriers.

Concerning solar photovoltaics, several points based on the results are worth noting. The hot carrier photocurrent flows in a direction opposite to the functional generation current, and this way, it can have a detrimental impact on the operation of a solar cell; it should be considered an additional intrinsic "*pre-thermalisational loss*". Absorption of the below-bandgap radiation also causes carrier heating and cannot be neglected. Although this study focuses on the hot carrier effect caused by single-wavelength radiation at intensities much higher than sunlight, hot carriers will still have a certain impact when considering the entire solar spectrum. These findings contribute to a deeper understanding of the hot carrier phenomena in single-junction solar cells and will probably facilitate a quicker alignment between theoretically predicted and practically achieved solar cell efficiencies.

Author Contributions: Conceptualisation, O.M., J.G. and S.A.; methodology, O.M., J.G., A.S. and S.A.; software, O.M. and I.Z.; validation, O.M., J.G. and S.A.; formal analysis, A.S., J.G. and S.A.; investigation, O.M. and I.Z.; resources, A.S.; data curation, O.M., J.G. and I.Z.; writing—original draft preparation, O.M.; writing—review and editing, J.G. and S.A.; visualisation, O.M. and J.G.; supervision, J.G. All authors have read and agreed to the published version of the manuscript.

Funding: This research received no external funding.

Data Availability Statement: The original contributions presented in the study are included in the article; further inquiries can be directed to the corresponding author.

Conflicts of Interest: The authors declare no conflicts of interest.

References

1. Murray, R. Infrared Absorption and Energy Levels Due to Impurities. In *Properties of Gallium Arsenide*, 3rd ed.; Brozel, M.R., Stillman, G.E., Eds.; The Institution of Electrical Engineers INSPEC: London, UK, 1996; pp. 225–256.
2. Umeno, M.; Sugito, Y.; Jimbo, T.; Hattori, H.; Amenixa, Y. Hot photo-carrier and hot electron effects in p-n junctions. *Solid-State Electron.* **1978**, *21*, 191–195. [CrossRef]
3. Encinas-Sanz, F.; Guerra, J.M. Laser-induced hot carrier photovoltaic effects in semiconductor junctions. *Prog. Quantum Electron.* **2003**, *27*, 267–294. [CrossRef]
4. Ašmontas, S.; Gradauskas, J.; Seliuta, D.; Širmulis, E. Photoelectrical properties of nonuniform semiconductor under infrared laser radiation. In Proceedings of the International Conference on Nonresonant Laser-Matter Interaction, St. Petersburg, Russia, 21–23 August 2000.
5. Ašmontas, S.; Gradauskas, J.; Seliuta, D.; Šilėnas, A.; Širmulis, E.; Marmur, I.Y. Photoelectrical properties of nonuniform GaAs structures under infrared laser illumination. In Proceedings of the International Conference on SPIE: Nonresonant Laser-Matter Interaction, St. Petersburg, Russia, 1–3 July 1996.
6. Dargys, A.; Kundrotas, J. Physical data for gallium arsenide. 143–188. In *Handbook on the Physical Properties of Ge, Si, GaAs and InP*, 1st ed.; Science and Encyclopedia: Vilnius, Lithuania, 1994.
7. Shockley, W.; Queisser, H. Detailed Balance Limit of Efficiency of p-n Junction Solar Cells. *J. Appl. Phys.* **1961**, *32*, 510–519. [CrossRef]
8. Hirst, L.C.; Ekins-Daukes, N.J. Fundamental losses in solar cells. *Prog. Photovolt. Res. Appl.* **2011**, *19*, 286–293. [CrossRef]
9. Saeed, S.; de Jong, E.M.L.D.; Dohnalova, K.; Gregorkiewicz, T. Efficient optical extraction of hot-carrier energy. *Nat. Commun.* **2014**, *5*, 4665. [CrossRef] [PubMed]
10. Torchynska, T.V.; Vivas Hernandez, A.; Goldstein, Y.; Jedrzejewskii, J.; Jimenez Sandoval, S. Photoluminescence of Si or Ge nanocrystallites embedded in silicon oxide. *J. Non-Cryst. Solids* **2006**, *352*, 1152–1155. [CrossRef]
11. Kolodinski, S.; Werner, J.H.; Wittchen, T.; Queisser, H.J. Quantum efficiencies exceeding unity due to impact ionization in silicon solar cells. *Appl. Phys. Lett.* **1993**, *63*, 2405–2407. [CrossRef]
12. Wurfel, P. Solar energy conversion with hot electrons from impact ionisation. *Sol. Energy Mater. Sol. Cells* **1997**, *46*, 43–52. [CrossRef]
13. Fast, J.; Aeberhard, U.; Bremner, S.P.; Linke, H. Hot-carrier optoelectronic devices based on semiconductor nanowires. *Appl. Phys. Rev.* **2021**, *8*, 021309. [CrossRef]
14. Kempa, K.; Naughton, M.J.; Ren, Z.F.; Herczynski, A.; Kirkpatrick, T.; Rybczynski, J.; Gao, Y. Hot electron effect in nanoscopically thin photovoltaic junctions. *Appl. Phys. Lett.* **2009**, *95*, 233121–2331214. [CrossRef]
15. Shayan, S.; Matloub, S.; Rostami, A. Efficiency enhancement in a single bandgap silicon solar cell considering hot-carrier extraction using selective energy contacts. *Opt. Express* **2021**, *29*, 5068–5080. [CrossRef] [PubMed]
16. Esgandari, M.; Barzinjy, A.A.; Rostami, A.; Rostami, G.; Dolatyari, M. Solar cells efficiency enhancement using multilevel selective energy contacts (SECs). *Opt. Quantum Electron.* **2022**, *54*, 1–9. [CrossRef]
17. Ross, R.T.; Nozik, A.J. Efficiency of hot-carrier solar energy converters. *J. Appl. Phys.* **1982**, *53*, 3813–3818. [CrossRef]
18. Neges, M.; Schwarzburg, K.; Willig, F. Monte Carlo simulation of energy loss and collection of hot charge carriers, first step towards a more realistic hot-carrier solar energy converter. *Sol. Energy Mater. Sol. Cells* **2006**, *90*, 2107–2128. [CrossRef]
19. Conibeer, G.; Ekins-Daukes, N.; Guillemoles, J.-F.; Konig, D.; Cho, E.-C.; Jiang, C.-W.; Shrestha, S.; Green, M.A. Progress on hot carrier cells. *Sol. Energy Mater. Sol. Cells* **2009**, *93*, 713–719. [CrossRef]
20. Le Bris, A.; Guillemoles, J.-F. Hot carrier solar cells: Achievable efficiency accounting for heat losses in the absorber and through contacts. *Appl. Phys. Lett.* **2010**, *97*, 113506. [CrossRef]
21. Kirk, A.P.; Fischetti, M.V. Fundamental limitations of hot-carrier solar cells. *Phys. Rev. B Condens. Matter.* **2012**, *86*, 165206. [CrossRef]

22. Rodiere, J.; Lombez, L.; Le Corre, A.; Durand, O.; Guillemoles, J.-F. Experimental evidence of hot carrier solar cell operation in multi-quantum wells heterostructures. *Appl. Phys. Lett.* **2015**, *106*, 183901. [CrossRef]
23. Conibeer, G.; Shrestha, S.; Huang, S.; Patterson, R.; Xia, H.; Feng, Y.; Zhang, P.; Gupta, N.; Tayebjee, M.; Smyth, S.; et al. Hot carrier solar cell absorber prerequisites and candidate material systems. *Sol. Energy Mater. Sol. Cells* **2015**, *135*, 124–129. [CrossRef]
24. Chung, S.; Wen, X.; Huang, S.; Gupta, N.; Conibeer, G.; Shrestha, S.; Harada, T.; Kee, T.W. Nanosecond long excited state lifetimes observed in hafnium nitride. *Sol. Energy Mater. Sol. Cells* **2017**, *169*, 13–18. [CrossRef]
25. Takeda, Y. Hot-carrier solar cells and improved types using wide-bandgap energy-selective contacts. *Prog. Photovolt. Res. Appl.* **2022**, *30*, 65–84. [CrossRef]
26. Sharma, A.S.; Hanif, M.; Bremner, S.P.; Nielsen, M.P.; Tayebjee, M.J.Y.; Rougieux, F.E.; Ekins-Daukes, N.J.; Pusch, A. Heat Flow through Nonideal Contacts in Hot-Carrier Solar Cells. *Phys. Rev. Appl.* **2023**, *20*, 034001. [CrossRef]
27. Li, M.; Bhaumik, S.; Goh, T.W.; Kumar, M.S.; Yantara, N.; Grätzel, M.; Mhaisalkar, S.; Mathews, N.; Sum, T.C. Slow cooling and highly efficient extraction of hot carriers in colloidal perovskite nanocrystals. *Nat. Commun.* **2017**, *8*, 14350. [CrossRef] [PubMed]
28. Yang, Y.; Ostrowski, D.P.; France, R.M.; Zhu, K.; Van De Lagemaat, J.; Luther, J.M.; Beard, M.C. Observation of a hot-phonon bottleneck in lead-iodide perovskites. *Nat. Photonics* **2016**, *10*, 53–59. [CrossRef]
29. Johannsen, J.C.; Ulstrup, S.; Cilento, F.; Crepaldi, A.; Zacchigna, M.; Cacho, C.; Turcu, I.E.; Springate, E.; Fromm, F.; Raidel, C.; et al. Direct View of Hot Carrier Dynamics in Graphene. *Phys. Rev. Lett.* **2013**, *111*, 027403. [CrossRef] [PubMed]
30. El Fatimy, A.; Myers-Ward, R.L.; Boyd, A.K.; Daniels, K.M.; Gaskill, D.K.; Barbara, P. Epitaxial graphene quantum dots for high-performance terahertz bolometers. *Nat. Nanotechnol.* **2016**, *11*, 335–338. [CrossRef] [PubMed]
31. Esmaielpour, H.; Lombez, L.; Giteau, M.; Delamarre, A.; Ory, D.; Cattoni, A.; Collin, S.; Guillemoles, J.-F.; Suchet, D. Investigation of hot carrier thermalization mechanisms in quantum well structures. In Proceedings of the SPIE, Online, 6–11 March 2021; Volume 11681. Physics, Simulation, and Photonic Engineering of Photovoltaic Devices X.
32. Gibelli, F.; Lombez, L.; Guillemoles, J.-F. Two carrier temperatures non-equilibrium generalized Planck law for semiconductors. *Phys. B Condens. Matter.* **2016**, *498*, 7–16. [CrossRef]
33. The National Renewable Energy Laboratory. Best Research-Cell Efficiency Chart. *Nrel.Gov.* Available online: https://www.nrel.gov/pv/cell-efficiency.html (accessed on 10 October 2023).
34. Gradauskas, J.; Ašmontas, S.; Sužiedėlis, A.; Šilėnas, A.; Vaičikauskas, V.; Čerškus, A.; Širmulis, E.; Žalys, O.; Masalskyi, O. Influence of Hot Carrier and Thermal Components on Photovoltage Formation across the p–n Junction. *Appl. Sci.* **2020**, *10*, 7483. [CrossRef]
35. Ašmontas, S.; Gradauskas, J.; Sužiedėlis, A.; Šilėnas, A.; Širmulis, E.; Švedas, V.; Vaičikauskas, V.; Žalys, O. Hot carrier impact on photovoltage formation in solar cells. *Appl. Phys. Lett.* **2018**, *113*, 071103. [CrossRef]
36. Sze, S.M. *Physics of Semiconductor Devices*, 1st ed.; John Wiley & Sons, Inc.: New York, NY, USA, 1969; pp. 25–38.
37. Ašmontas, S.; Širmulis, E.; Stonys, S. Investigation of the photovoltage across germanium p-n-junction under pulsed CO_2 laser irradiation. *Lith. J. Phys.* **1984**, *24*, 75–82.
38. Ašmontas, S.; Gradauskas, J.; Seliuta, D.; Suziedelis, A.; Silenas, A.; Valusis, G. GaAs/AlGaAs heterojunction: A promising detector for infrared radiation. In Proceedings of the 5th European Gallium Arsenide and related III–V compounds Applications Symposium, Bolonga, Italy, 3–5 September 1997.
39. Gradauskas, J.; Ašmontas, S. Hot Carrier Photocurrent through MOS Structure. *Appl. Sci.* **2021**, *11*, 7211. [CrossRef]
40. Green, M.A.; Emery, K.; Blakers, A.W. Silicon solar cells with reduced temperature sensitivity. *Electron. Lett.* **1982**, *18*, 97–98. [CrossRef]
41. Schaffner, J.S.; Shea, R.F. The Variation of the Forward Characteristics of Junction Diodes with Temperature. *Proc. IRE* **1955**, *43*, 101.
42. Singh, P.; Ravindra, N.M. Temperature dependence of solar cell performance—An analysis. *Sol. Energy Mater. Sol. Cells* **2012**, *101*, 36–45. [CrossRef]
43. Nootan, C.P. Voltage-Current Characteristics of GaAs Diodes as a Function of Temperature. Master's Thesis, Lehigh University, Bethlehem, PA, USA, 1964.
44. Cavassilas, N.; Makhfudz, I.; Daré, A.-M.; Lannoo, M.; Dangoisse, G.; Bescond, M.; Michelini, F. Theoretical Demonstration of Hot-Carrier Operation in an Ultrathin Solar Cell. *Phys. Rev. Appl.* **2022**, *17*, 064001. [CrossRef]
45. Pelouch, W.S.; Ellingson, R.J.; Powers, P.E.; Tang, C.L.; Szmyd, D.M.; Nozik, A.J. Comparison of hot-carrier relaxation in quantum wells and bulk GaAs at high carrier densities. *Phys. Rev. B* **1992**, *45*, 1450–1453. [CrossRef]
46. Nguyen, D.-T.; Lombez, L.; Gibelli, F.; Boyer-Richard, S.; Le Corre, A.; Durand, O.; Guillemoles, J.-F. Quantitative experimental assessment of hot carrier-enhanced solar cells at room temperature. *Nat. Energy* **2018**, *3*, 236–242. [CrossRef]

Disclaimer/Publisher's Note: The statements, opinions and data contained in all publications are solely those of the individual author(s) and contributor(s) and not of MDPI and/or the editor(s). MDPI and/or the editor(s) disclaim responsibility for any injury to people or property resulting from any ideas, methods, instructions or products referred to in the content.

inorganics

MDPI

Article

Composition-Dependent Phonon and Thermodynamic Characteristics of C-Based $X_xY_{1-x}C$ (X, Y ≡ Si, Ge, Sn) Alloys

Devki N. Talwar [1,2]

1 Department of Physics, University of North Florida, 1 UNF Drive, Jacksonville, FL 32224-7699, USA; devki.talwar@unf.edu; Tel.: +1-(724)-762-7719
2 Department of Physics, Indiana University of Pennsylvania, 975 Oakland Avenue, 56 Weyandt Hall, Indiana, PA 15705-1087, USA

Abstract: Novel zinc-blende (zb) group-IV binary XC and ternary $X_xY_{1-x}C$ alloys (X, Y ≡ Si, Ge, and Sn) have recently gained scientific and technological interest as promising alternatives to silicon for high-temperature, high-power optoelectronics, gas sensing and photovoltaic applications. Despite numerous efforts made to simulate the structural, electronic, and dynamical properties of binary materials, no vibrational and/or thermodynamic studies exist for the ternary alloys. By adopting a realistic rigid-ion-model (RIM), we have reported methodical calculations to comprehend the lattice dynamics and thermodynamic traits of both binary and ternary compounds. With appropriate interatomic force constants (IFCs) of XC at ambient pressure, the study of phonon dispersions $\omega_j(\vec{q})$ offered positive values of acoustic modes in the entire Brillouin zone (BZ)—implying their structural stability. For $X_xY_{1-x}C$, we have used Green's function (GF) theory in the virtual crystal approximation to calculate composition x, dependent $\omega_j(\vec{q})$ and one phonon density of states $g(\omega)$. With no additional IFCs, the RIM GF approach has provided complete $\omega_j(\vec{q})$ in the crystallographic directions for both optical and acoustical phonon branches. In quasi-harmonic approximation, the theory predicted thermodynamic characteristics (e.g., Debye temperature $\Theta_D(T)$ and specific heat $C_v(T)$) for $X_xY_{1-x}C$ alloys. Unlike SiC, the GeC, SnC and $Ge_xSn_{1-x}C$ materials have exhibited weak IFCs with low [high] values of $\Theta_D(T)$ [$C_v(T)$]. We feel that the latter materials may not be suitable as fuel-cladding layers in nuclear reactors and high-temperature applications. However, the XC and $X_xY_{1-x}C$ can still be used to design multi-quantum well or superlattice-based micro-/nano devices for different strategic and civilian application needs.

Keywords: C-based novel binary/ternary alloys; rigid-ion-model; lattice dynamics; Born's transverse effective charge e_T^*; Debye temperature and specific heat; Green's function method

1. Introduction

Since the invention of Si-based transistors [1], the evolution to include them in radio frequency (RF) devices for microelectronics has been remarkable [2–11]. With the low processing costs, Si-technology has led to many achievements by incorporating electrodes, dielectrics, and other elements in different integrated circuits (ICs). With truly monolithic optoelectronic functionality, Si has offered substantial cost benefits as well as long-term performance gains in the optoelectronic microsensor systems [12–16] and photovoltaic cells. These achievements have given tremendous opportunities for both the scientists and engineers to successfully incorporate different devices in energy harvesting and gas-sensing needs. Earlier accomplishments on RF devices and ICs have quickly grown to become valuable for wireless sensor networks (WSNs) [17,18]. These devices are now employed in a wide range of applications including photonics, optoelectronics, environment monitoring, medical diagnostics, forensic, spintronics, cellular base phone transceivers, amplifiers, Gigabit wireless local as well as personal area networks [12–32].

The progress in Si-based ICs and WSNs is the outcome of many scientific research reports published on different topics where materials' perspectives played increasingly important roles. New wavelength regimes are now extended from 1.55 μm to 5.0 μm for the operations of Si-based photonic and opto-electronic ICs [10]. While many multiple-quantum-wells (MQWs) and superlattices (SLs) based III-V semiconductors [19–22] have revealed extremely inspiring results, the high cost and incompatibility of III-Vs with the Si platform has been, and still is, the main impediment that prevented the large-scale commercial production of different opto-electronic devices [23–38].

Despite the technological achievements of group-IV materials, the IV-IV binary XC and ternary $X_xY_{1-x}C$ alloys with X, Y (≡Si, Ge, and Sn) have recently become quite attractive in preparing different heterostructures for bandgap and strain engineering [29–32]. At room temperature (RT), the C, Si, Ge, and α-Sn of diamond crystal structures are known to exhibit indirect band gap energies E_g [≡5.47 eV, 1.12 eV, 0.66 eV and 0 eV]. Higher mismatch between the lattice constants, a_o [≡C: 3.56 Å, Si: 5.43 Å, Ge: 5.66 Å, α-Sn: 6.49 Å] and low solubility of C has caused complications in the earlier designs of XC/Si MQWs and SLs [33–45]. However, the differences in their electronegativity now play a vital role in carefully optimizing parameters to achieve epitaxial growth of heterostructures apposite for diverse device applications. Recently, the use of Si/SiGe has gained considerable attention for designing heterojunction bipolar transistors with cut-off frequencies > 10 GHz [11].

Tremendous efforts have also been formulated in preparing novel C-based binary XC and/or ternary $X_xY_{1-x}C$ alloys by taking advantage of the unique and exciting properties of group-IV materials [12–37]. Good structural stability of zinc-blende (zb) binary and/or ternary compounds with different E_g, hardness, high stiffness, melting point, and high thermal conductivity is considered particularly favorable for applications [38–41] in blue/ultraviolet (UV) light-emitting diodes (LEDs), laser diodes (LDs), photodetectors and solar cells [12–16], etc. Due to large lattice mismatch and the differences in thermal expansion coefficients between $X_xY_{1-x}C$ epilayers and Si substrate, one would expect the possibilities of observing structural and/or intrinsic defects near the interfaces [29]. However, appropriate use of buffer layers acquiring load through the relaxation of mechanical stresses has helped improve the structural qualities of MQWs and SLs. While there remain a few intrinsic issues, which could constrain the design of opto-electronic device structures, solutions to these problems are not impossible and can be resolved by exploiting suitable experimental (e.g., growth, [42–93] characterization [94–107]) and theoretical [108–137] methods.

One must note that by employing the pulsed supersonic free jets techniques [42–45], an inverse heteroepitaxial growth of Si on SiC has been demonstrated to achieve multilayer structures [46–65]. A novel arc plasma C gun source is incorporated in the molecular beam epitaxial (MBE) methods to grow ultrathin MQWs and SLs [66–70]. Ultrahigh chemical vapor deposition (UH-CVD), reduced pressure RP-CVD and metalorganic (MOCVD) techniques are also successfully used [71–93] to prepare different $Si_{1-x}Ge_xC$/Si, $Ge_{1-x}Sn_xC$/Si, GeC/SiC epilayers. For commercial applications of these materials, the RP-CVD method has been preferred due to the balance between good epitaxial quality and relatively high growth rates [76]. Certainly, the progress made in the growth of complex and exotic C-based materials has set challenges for both the physicists and engineers in investigating their fundamental properties.

Although binary compounds are used in many technological applications, considerably less attention is paid to ternary alloys despite the successful growth of ultrathin epifilms. A variety of characterization techniques are also applied for analyzing/ monitoring their fundamental properties [94–107]. The classification of such methods includes reflection high-energy electron diffraction (RHEED) [94,95], Auger electron spectroscopy (AES) [96], He^+ Rutherford backscattering spectrometry (RBS) [97], atomic force microscopy (AFM) [98,99], high-resolution X-ray diffraction (HR-XRD) [100–102], transmission electron microscopy (XTEM) [103], photoluminescence (PL) [104], absorption, Fourier transform infrared (FTIR) spectroscopy [100–102], Raman scattering spectroscopy (RSS) [105–107] and spectroscopic ellipsometry (SE) techniques [107], etc. It is to be noted that not only have these techniques validated their crystal structures but also helped in assessing the film thickness, strain, intrinsic electrical, and optical traits. Despite the existence of numerous experimental studies on structural and electrical properties, there are limited IR absorption and RSS measurements [100–107] for evaluating their phonon characteristics.

From a theoretical standpoint, several calculations are performed to understand the structural, electronic, and optical properties of the binary compounds using full potential linear augmented plane wave (FP-LAPW), first-principles (ab initio) and molecular dynamical (MD) methods [108–126]. For the ternary $X_xY_{1-x}C$ alloys, however, no systematic studies are known for comprehending their lattice dynamical and/or thermodynamic characteristics. To accomplish major technological applications of the C-based heterostructures [e.g., Si/Si(Ge)C, GeSnC/GeC MQWs and $(SiC)_m/(Ge(Sn)C)_n$ SLs], such calculations are necessary to obtain the complete phonon dispersions of XC and $X_xY_{1-x}C$ materials from realistic lattice dynamical models. One reason for this requisite is that the dynamical response of polar lattices affects the key electronic properties, including the exciton-binding energies and charge-carrier mobilities. Examining the dynamical response of crystals and its impact on the dielectric environment provides a major step in realizing their structural characteristics. The other reason for its need is that, in MQWs and SLs, the phonon density of states (DOS) of binary/ternary alloys has played crucial roles in evaluating thermodynamic traits, including the thermal expansion coefficients α(T), Debye temperature $\Theta_D(T)$, heat capacity $C_v(T)$, Grüneisen constants γ(T), entropy, lattice thermal conductivity, etc.

The purpose of this work is to use a realistic rigid-ion model (RIM) [127] and report the methodical results of our comprehensive study to assess the structural, lattice dynamical (cf. Section 2.1), and thermodynamic (cf. Section 2.2) characteristics of binary XC and ternary (cf. Section 2.3) $X_{1-x}Y_xC$ alloys. For the tetrahedral (T_d^2: $F\bar{4}3m$) binary materials, we have carefully optimized the RIM interatomic force constants (IFCs) by exercising [128] (cf. Section 2.1.1) successive least-square fitting procedures. In these processes, the phonon frequencies at high critical points (Γ, X, and L) are used as input while exploiting their lattice—a_o, and elastic constants C_{ij} as constraints (see Table 1). For zb-SiC, the values of phonon frequencies are incorporated from inelastic X-ray scattering (IXS) measurements [129], while in GeC, SnC, we employed critical-point phonon frequencies from first-principles ABINIT results using a plane–wave pseudopotential method in density functional theory (DFT) and local-density perturbation approximation [125]. With the appropriate IFCs of binary XC materials, the dynamical matrix (cf. Section 3.1) is diagonalized to obtain phonon dispersions $\omega_j(\vec{q})$, at each wave vector $\vec{q}$ point in the Brillouin zone (BZ), as well as one phonon density of states (DOS), $g(\omega)$. For mixed $X_xY_{1-x}C$ alloys $0 \leq x \leq 1$, the composition-dependent phonon frequencies (cf. Section 3.2) are simulated by using a Green's function (GF) theory [130] in the virtual crystal approximation (VCA). Without considering additional IFCs, the RIM GF approach for the mixed alloys has offered a complete $\vec{q}$ dependent $\omega_j(\vec{q})$, in arbitrary crystallographic directions for both the optical and acoustical phonon branches. In $X_xY_{1-x}C$ materials, and using a quasi-harmonic approximation (QHA), we have systematically simulated the x-dependent $\omega_j(\vec{q})$, and $g(\omega)$ (cf. Section 3.2) to predict their thermodynamic characteristics (e.g., Debye temperature $\Theta_D(T)$ and specific heat $C_v(T)$). Born's transverse effective charges e_T^*, and Fröhlich coupling parameters α_F are also (cf. Sections 3.2.3 and 3.2.4) calculated. Theoretical results of $\omega_j(\vec{q})$, $g(\omega)$, and the thermo-dynamical traits for both binary XC and ternary $X_xY_{1-x}C$ alloys are carefully analyzed/contrasted (cf. Section 3.1) against the existing experimental [129] and theoretical [120–126] data (cf. Section 3) with concluding remarks presented in Section 4.

Table 1. Different parameters for XC (X ≡ Si, Ge and Sn) materials, viz., lattice constants a_0 (in); elastic constants c_{ij} $\left(\text{in } 10^{11} \frac{\text{dyn}}{\text{cm}^2}\right)$ and phonon frequencies (cm^{-1}) at Γ, X, and L critical points. These are required for evaluating the necessary interatomic force constants of the rigid-ion model (see: text).

Parameters	3C-SiC	zb-GeC	zb-SnC
a_0 (Å)	4.360 [(a)], 4.40 [(b)], 4.374 [(c)]	4.610 [(d)], 4.590 [(e)]	5.130 [(f)], 5.170 [(h)]
c_{11}	38.0 [(a)], 38.3 [(b)], 38.5 [(c)]	29.7 [(f)], 35.8 [(g)]	24.6 [(g)]
c_{12}	14.2 [(a)], 12.5 [(b)], 12.2 [(c)]	12.4 [(f)], 12.2 [(g)]	11.3 [(g)]
c_{44}	25.6 [(a)], 24.0 [(b)], 24.3 [(c)]	14.1 [(f)], 21.4 [(g)]	14.3 [(g)]
$\omega_{LO(\Gamma)}$	974 [(b)]	748 [(h)]	558 [(h)]
$\omega_{TO(\Gamma)}$	793 [(b)]	626 [(h)]	456 [(h)]
$\omega_{LO(X)}$	830 [(b)]	697 [(h)]	503 [(h)]
$\omega_{TO(X)}$	759 [(b)]	617 [(h)]	450 [(h)]
$\omega_{LA(X)}$	644 [(b)]	348 [(h)]	216 [(h)]
$\omega_{TA(X)}$	373 [(b)]	214 [(h)]	134 [(h)]
$\omega_{LO(L)}$	850 [(b)]	705 [(h)]	516 [(h)]
$\omega_{TO(L)}$	770 [(b)]	612 [(h)]	440 [(h)]
$\omega_{LA(L)}$	605 [(b)]	331 [(h)]	199 [(h)]
$\omega_{TA(L)}$	260 [(b)]	162 [(h)]	109 [(h)]

[(a)] Ref. [130]; [(b)] Ref. [129]; [(c)] Refs. [112,113]; [(d)] Ref. [115]; [(e)] Ref. [116]; [(f)] Ref. [119]; [(g)] Ref. [124]; [(h)] Ref. [125].

2. Computational Details

The vibrational properties of XC materials have played a valuable role in assessing their phase transitions, transport coefficients and other physical phenomena [130–137]. Understanding the electron–phonon interactions and polaron characteristics in polar compounds is important for evaluating their optical and transport properties [138]. In $X_{1-x}Y_xC$ alloys, the exploration of phonons and optical and structural characteristics is crucial for correlating them to their different microscopic traits. To the best of our knowledge, no experimental and/or theoretical studies of phonons exist for $X_{1-x}Y_xC$ alloys, especially ones examining the contributions of lattice dynamical properties to phonon-free energy, entropy, specific heat, and Debye temperatures, etc. Accurate RIM simulations of $\omega_j(\vec{q})$, and $g(\omega)$ for the binary XC materials are employed (cf. Section 2.1) to comprehend the lattice dynamics and thermodynamic properties of $X_{1-x}Y_xC$ alloys using a GF methodology (cf. Section 2.2).

2.1. Phonons in XC Materials

Earlier, the IXS [129] and RSS results [105] of phonon dispersions $\omega_j(\vec{q})$, for the zb 3C-SiC were analyzed by exploiting the ab initio methods [120–126]. Recent calculations of lattice dynamics for GeC and SnC using plane–wave pseudopotential DFT approaches in the local density approximation [124,125] have provided different results. By using a realistic RIM [127], we have probed the experimental and theoretical data [105,120–126,129] of phonon dispersions for binary XC (3C-SiC, GeC and SnC) materials. Our choice of RIM, briefly described in Section 2.1.1, is based on the following facts: (*a*) a truncated RIM [139–144], where the number of IFCs is limited, a priori, failed to give accurate $\omega_j(\vec{q})$, and it is not even less complicated than the more realistic scheme [127] adopted here; and (*b*) with regard to $\omega_j(\vec{q})$, our simulated results in the high-symmetry direction are comparable (cf. Section 3) with experimental (3C-SiC) [105,129] and ab initio methods (GeC, SnC) [120–126].

2.1.1. Rigid-Ion Model

Here, we have briefly outlined the RIM [127] to comprehend the vibrational and thermodynamic properties of perfect binary XC materials. In this approach, the atomic displacements $u_\alpha(l\kappa|\vec{q},j)$ of the jth vibrational modes $\omega_j(\vec{q})$, are expressed as plane waves with the wave vector $\vec{q}$ as [127]:

$$u_\alpha(l\kappa|\vec{q},j) = \frac{1}{\sqrt{M_\kappa}} e_\alpha(\kappa|\vec{q},j) e^{i[\vec{q}\,\vec{x}(l\kappa) - \omega_j(\vec{q})t]} \quad (1)$$

where t is the time; the $\vec{x}(l\kappa)$ and M_κ are, respectively, the position and mass of the $(l\kappa)$ atom. The term l represents the position and κ denotes the types of atoms with κ = 1, 2 to specify the C and X, respectively, in the XC materials (see: the crystal structure, in Figure 1a, and its BZ in Figure 1b).

In the harmonic approximation, the equations of motion can be expressed either in terms of the force constant $\Phi^{sC}_{\alpha\beta}(\kappa\kappa'|\vec{q})$ matrix elements [127]:

$$M_\kappa \omega_j^2(\vec{q}) e_\alpha(\kappa|\vec{q}j) = \sum_{\kappa'\beta} \Phi^{sC}_{\alpha\beta}(\kappa\kappa'|\vec{q}) e_\beta(\kappa\prime|\vec{q}j) \quad (2a)$$

or in terms of the dynamical $D^{sC}_{\alpha\beta}(\kappa\kappa'|\vec{q})$ matrix elements:

$$\omega_j^2(\vec{q}) e_\alpha(\kappa|\vec{q},j) = \sum_{\kappa'\beta} D^{sC}_{\alpha\beta}\left(\kappa\kappa'\middle|\vec{q}\right) e_\beta\left(\kappa'\middle|\vec{q},j\right); \text{ with } \kappa,\kappa' = 1,2 \quad (2b)$$

where,

$$D^{sC}_{\alpha\beta}(\kappa\kappa'|\vec{q}) = D^{s}_{\alpha\beta}(\kappa\kappa'|\vec{q}) - \frac{Z_\kappa Z_{\kappa'} e^2}{(M_\kappa M_{\kappa'})^{1/2}} D^{C}_{\alpha\beta}(\kappa\kappa'|\vec{q}) \quad (2c)$$

The term $\Phi^{sC}_{\alpha\beta}(\kappa\kappa'|\vec{q})$ $[D^{sC}_{\alpha\beta}(\kappa\kappa'|\vec{q})]$ in Equations (2a)–(2c) represents the force constant [dynamical] matrix elements involving components of both the short-range $\Phi^{s}_{\alpha\beta}(\kappa\kappa'|\vec{q})[D^{s}_{\alpha\beta}(\kappa\kappa'|\vec{q})]$ and long-range $\Phi^{C}_{\alpha\beta}(\kappa\kappa'|\vec{q})$ $[D^{C}_{\alpha\beta}(\kappa\kappa'|\vec{q})]$ Coulomb interactions [127, 139–142,145–147]. For simulating the $\omega_j(\vec{q})$ of binary materials, the short- (A, B, C_κ, D_κ, E_κ and F_κ) and long-range (Z_{eff}) interactions involved in $D^{s}_{\alpha\beta}(\kappa\kappa'|\vec{q})$ and $D^{C}_{\alpha\beta}(\kappa\kappa'|\vec{q})$ matrices are optimized (Table 2) following successive non-linear least square fitting procedures.

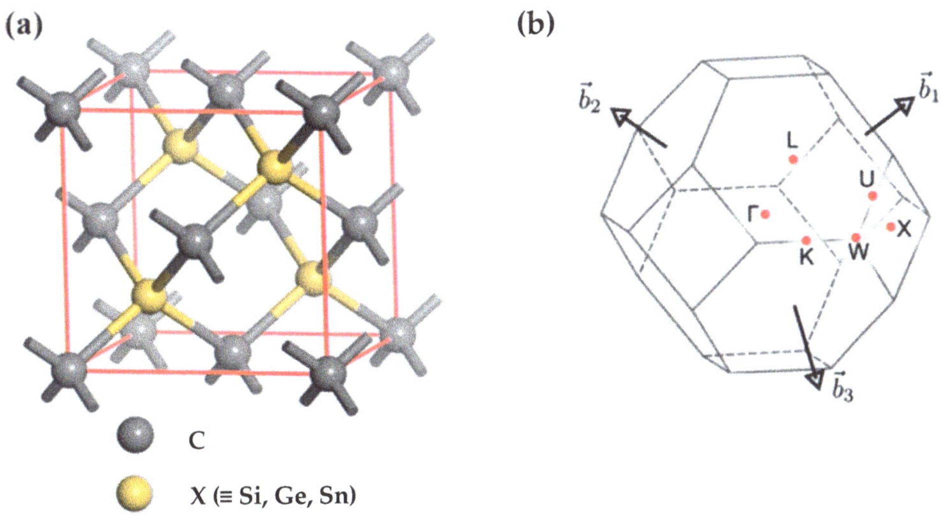

Figure 1. (**a**) The lattice structure of novel zinc-blende (zb) XC binary materials. The yellow color circles are represented for X (≡Si, Ge and Sn) atoms, while the grey color circles symbolize C atoms arranged in the tetrahedral (T_d^2 : $F\bar{4}3m$) point group symmetry. (**b**) The Brillouin zone of face-centered cubic material is labeled with high symmetry points (see: text).

Table 2. In the notations of Ref. [127], we have reported the optimized sets of rigid-ion model (RIM) parameters [$A, B, C_1, C_2, D_1, D_2, E_1, E_2, F_1, F_2$ (10^5 dyn/cm)] at ambient pressure for novel zb XC (X = Si, Ge and Sn) materials. The term Z_{eff} is the effective charge (see text).

Parameters [(a)]	3C-SiC	GeC	SnC
A	−0.91723	−0.68066	−0.39000
B	−0.44500	−0.66000	−0.37000
C_1	−0.04050	−0.02300	−0.00400
C_2	−0.15900	−0.13200	−0.09300
D_1	0.06440	−0.02097	−0.00135
D_2	−0.33088	−0.38000	−0.07200
E_1	0.06200	−0.01000	−0.01000
E_2	0.10850	0.02000	0.02000
F_1	−0.04100	0.03550	0.01050
F_2	0.28800	0.02780	0.01480
Z_{eff}	1.05300	0.92200	0.88030

[(a)] Ref. [127].

In the primitive unit cell of the zb lattice having two atoms per unit cell, the solutions of Equations (2a) and (2b) lead to an eigenvalue problem of size (6 × 6) with the values of wave vectors $\vec{q}$ restricted to the fcc lattice:

$$|\vec{q}| = \frac{\pi}{a_o}(q_1, q_2, q_3); -1 \leq q_1, q_2, q_3 \leq 1; \ (q_1 + q_2 + q_3) \leq \frac{3}{2} \tag{3}$$

with the triplets (q_1, q_2, q_3) distributed uniformly throughout the volume of the BZ. In our calculations of phonon dispersions $\omega_j(\vec{q})$, for the XC materials, we have used a mesh of 64,000 $|\vec{q}|$ points in the reduced BZ. At each $\vec{q}$, there are six [127] vibrational modes—three of them are acoustic (i.e., a longitudinal ω_{LA} and a doubly degenerate transverse ω_{TA}) and the remaining modes are optical, comprising a doubly degenerate transverse—ω_{TO} and a longitudinal ω_{LO} optical phonons (cf. Section 3.1).

2.2. Lattice Dynamics of Ternary $X_xY_{1-x}C$ Alloys

Composition-dependent phonon dispersions $\omega_j(\vec{q})$ for mixed zb $X_xY_{1-x}C$ crystals are simulated as a function of x ($0 \leq x \leq 1$) (cf. Section 3.2) by adopting a generalized GF theory in the VCA. For one dimensional mixed $X_xY_{1-x}C$ alloy with two interpenetrating sublattices, Kutty [131] has developed a GF approach and derived the dynamical matrix equation:

$$|(\omega^2 M_\alpha^\mu - \Phi_{\alpha\alpha}^{\mu\mu})\delta_{\alpha\beta}\delta_{\mu\nu} - C_\alpha^\mu \Phi_{\alpha\beta}^{\mu\nu}(\vec{q})| = 0 \tag{4}$$

by postulating that the sublattice 1 occupies C atoms while the sublattice 2 is randomly acquired by atoms of type X and Y having concentrations (x) and (1 − x), respectively. In Equation (4), $C_1^C = 1$, $C_2^Y = (1-x)$, $C_2^X = x$, $M_1^C = m_C$, $M_2^Y = m_Y$, $M_2^X = m_X$ and the term $\Phi_{\alpha\beta}^{\mu\nu}$ represents the IFCs. By substituting these parameters in Equation (4), we obtain the following [131]:

$$\begin{vmatrix} \omega^2 m_C - \Phi_{11}^{11} & -\Phi_{12}^{12}(\vec{q}) & -\Phi_{12}^{12}(\vec{q}) \\ -(1-x)\Phi_{21}^{21}(\vec{q}) & \omega^2 m_Y - \Phi_{22}^{22} & -(1-x)\Phi_{22}^{23}(\vec{q}) \\ -x\Phi_{21}^{31}(\vec{q}) & -x\Phi_{22}^{32}(\vec{q}) & \omega^2 m_X - \Phi_{22}^{33} \end{vmatrix} = 0 \tag{5}$$

with the solution of Equation (5) leading to the vibrational mode frequencies.

Extension of the above one-dimensional approach to simulate the lattice dynamics of three-dimensional ideal zb random $X_xY_{1-x}C$ ternary alloys using GF theory [131] in the framework of a RIM [127] is trivial. Here, we have assumed that the mixed $X_xY_{1-x}C$ crystal lattice achieves the following: (*a*) forms the ideal pseudo-binary alloys in the entire composition range x, (probably in contrast to reality), (*b*) the cation sublattice 2 is structurally close to the virtual crystal lattice where the X and Y atoms are randomly distributed having the concentration of Y $(1-x)$ and of X as x, (*c*) the anion sublattice 1 with C atoms remains undistorted, and (*d*) the characteristic NN atomic distances follow Vegard's law. It is to be noted that in the RIM GF methodology, no additional IFCs are required for the descriptions of wave-vector dependent phonon dispersions $\omega_j(\vec{q})$ of the $X_xY_{1-x}C$ alloys in arbitrary crystallographic directions. To simulate the dynamical characteristics of the pseudo-binary alloys using RIM, we have deliberated the IFCs of binary XC materials (cf. Table 2) in a three-body framework. While this approach has allowed for the calculations of phonon spectra for the optical, acoustical as well as disorder-induced modes, we have neglected here the disorder-related broadening of the spectral lines.

2.3. *Thermodynamic Properties*

To comprehend the effects of temperature on the structural stability of XC materials, we have calculated the thermodynamic properties (up to 1500 K) in the QHA by adopting a RIM and including appropriate values of their $g(\omega)$ and ω_j $(\vec{q})$ [132–138]. In the numerical computation of T-dependent lattice heat capacity ($C_V(T)$), we have used the following:

$$E(T) = E_{tot} + E_{zp} + \int \frac{\hbar\omega}{\exp\left(\frac{\hbar\omega}{k_B T}\right) - 1} g(\omega) d\omega \quad (6)$$

where E_{tot} is the total static energy at 0 K which can be calculated by first-principles methods, E_{zp} is the zero-point vibrational energy, k_B is Boltzmann's constant, ħ is Planck's constant. The term E_{zp} can be expressed as follows:

$$E_{zp} = \frac{1}{2} \int g(\omega) \hbar\omega d\omega \quad (7)$$

The lattice contribution to $C_V(T)$ is obtained by using [138] the following:

$$C_V(T) - k_B \int \frac{\left(\frac{\hbar\omega}{k_B T}\right)^2 \exp\left(\frac{\hbar\omega}{k_B T}\right)}{\left[\exp\left(\frac{\hbar\omega}{k_B T}\right) - 1\right]^2} g(\omega) d\omega \quad (8)$$

From Debye's equation,

$$C_V(T) = 9 r k_B \left(\frac{T}{\Theta_D(T)}\right)^3 \int_0^{\Theta_D(T)} \frac{\left(\frac{\hbar\omega}{k_B T}\right)^4 e^{\left(\frac{\hbar\omega}{k_B T}\right)}}{\left[\exp\left(\frac{\hbar\omega}{k_B T}\right) - 1\right]^2} d\omega \quad (9)$$

it is possible to calculate $\Theta_D(T)$. In Equation (9), the term r signifies the number of atoms per unit cell.

2.4. *Interaction of Photons with Solids*

In polar materials, the electrical polarization induced by the displacement of atoms offers the key quantities to understand how the deformations and electric fields are coupled, insulating them. The interaction of photons with solids comprises both ionic and electronic oscillations where the dielectric polarization is linked to their atomic polarizability [130]. The dynamic response of dielectric function on electromagnetic radiation is expressed in terms of elementary oscillators, where the strong interaction of photons with ω_{TO} phonons cause a large Reststrahlen absorption in the IR region. In polar crystals, the ω_{LO} modes produce a macroscopic electric field that interacts with electrons, resulting in a quasi-particle known as 'polaron' [130]. This long-range coupling instigates the Fröhlich interaction α_F (cf. Section 2.4.2). Born's transverse effective charge e_T^* (cf. Section 2.4.1) is an equally important quantity for studying the lattice dynamics of polar XC materials.

2.4.1. Born's Effective Charge e_T^*

In binary crystals, Born's transverse effective charge e_T^* is linked to the splitting of their optical phonon frequencies $\Delta\omega$ $[\equiv \omega_{LO(\Gamma)} - \omega_{TO(\Gamma)}]$. The screening of the Coulomb interaction depends on the electronic part of the dielectric function and volume of the unit cell [130]. In many compound semiconductors, the phonon splitting $\Delta\omega$ is assessed by measuring the pressure (P)-dependent optical modes [148] by Raman scattering. This optical phonon splitting is then related to the bonding characteristics by being scaled to the lattice constants a_o. The P-dependent Raman studies are difficult, and, in some cases, the results have been contentious [149]. Except for 3C-SiC [107], no P-dependent measurements are available for GeC and SnC. In Section 3.2, we have reported our RIM calculations of x-dependent phonon splitting for $X_xY_{1-x}C$ alloys and predicted the qualitative behavior of e_T^* by using [149] the following:

$$e_T^{*2} = \frac{\varepsilon_\infty \mu a_0^3}{16\pi}\left(\omega_{LO(\Gamma)}^2 - \omega_{TO(\Gamma)}^2\right) \tag{10}$$

where ε_∞ is the high-frequency dielectric constant and μ is the reduced mass of the anion–cation pair.

2.4.2. Fröhlich Coefficients α_F

The theory of the Fröhlich interaction in ternary alloys is very complex. In binary materials, the strength of electron–phonon interaction is expressed by a dimensionless Fröhlich coupling constant α_F [130]:

$$\alpha_F = \frac{1}{2}\frac{e^2/\sqrt{\hbar/2m_e^*\omega_{LO}}}{\hbar\omega_{LO}}\left(\frac{1}{\varepsilon_\infty} - \frac{1}{\varepsilon_o}\right) \tag{11}$$

where e is the electron charge, m_e^* the effective electron mass and ε_o is the static dielectric constant. Except for 3C-SiC, no systematic calculations of band structures or the effective electron masses m_e^* are known for GeC and SnC.

3. Numerical Computations and Results

Following standard practices, we have computed the lattice dynamical, as well as the thermodynamic properties of both zb binary XC and ideal ternary $X_xY_{1-x}C$ alloys. For the alloys, we have adopted (cf. Section 2.2) a generalized GF method in the framework of a RIM by incorporating the necessary IFCs (see Table 2) of the binary materials. The RIM results are analyzed by comparing/contrasting them with the existing experimental [105,129] and theoretical data [120–126].

3.1. Lattice Dynamics of Binary XC Materials

3.1.1. Phonon Characteristics

For binary XC materials, the simulated RIM phonon dispersions $\omega_j(\vec{q})$, and one phonon DOS $g(\omega)$, are displayed in Figures 2a and 2b, respectively. At ambient pressure, the acoustic phonon branches of the three (3C-SiC, GeC and SnC) compounds exhibited positive values—demonstrating their stability in the zb phases. From the DOS $g(\omega)$, one can clearly see two features, each with noticeable intensities in the low- (acoustic) as well as the high- (optical) frequency regions. Obviously, these characteristics are linked to the average ω_{TA}, ω_{LA} and ω_{TO}, ω_{LO} modes caused by the vibrations of heavier X and lighter C atomic masses, respectively.

For 3C-SiC, GeC and SnC materials, the simulated results of phonon frequencies at high critical points Γ, X, L reported in Table 3A–C are compared against the existing experimental [105,129] and/or theoretical [120–126] data. Our lattice dynamical results for 3C SiC contrasted reasonably well with the available phonon data from both the experimental (RSS [105], IXS [129]) and first-principles calculations [120–126]. For GeC and SnC, the results are also seen in very good accord with the phonon simulations reported by Zhang et al. [125] who used a plane–wave pseudopotential method in the density functional perturbation theory within the local density approximation.

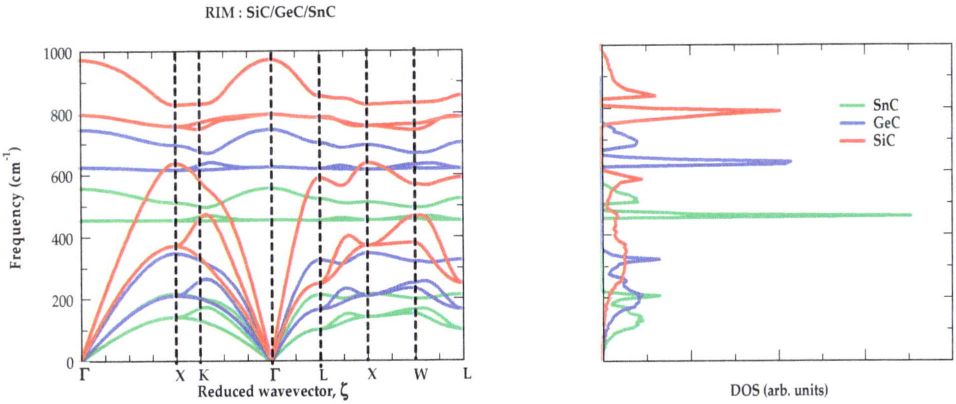

(**a**) (**b**)

Figure 2. (**a**) Simulated phonon dispersions of novel zinc-blende XC binary materials using a rigid-ion-model (RIM). The red color lines represent SiC, blue lines GeC and green lines SnC. Results are compared well with the experimental [105,129] and first-principles [120–126] data. (**b**) The RIM results of one phonon density of states for SiC (red color lines), GeC (blue lines) and SnC (green lines).

Table 3. Comparison of the RIM calculated phonon frequencies (cm^{-1}) at Γ, X, L critical points with the experimental and theoretical calculations (A) 3C-SiC, (B) GeC and (C) SnC, respectively (see text).

(A)

3C-SiC	$\omega_{LO(\Gamma)}$	$\omega_{TO(\Gamma)}$	$\omega_{LO(X)}$	$\omega_{TO(X)}$	$\omega_{LA(X)}$	$\omega_{TA(X)}$	$\omega_{LO(L)}$	$\omega_{TO(L)}$	$\omega_{LA(L)}$	$\omega_{TA(L)}$	$\Delta\omega$
Our [(a)]	974	797	828	760	639	373	857	787	591	250	177
Expt. [(b)]	974	793	830	759	644	373	850	770	605	260	181
Expt. [(c)]	972	796	829	761	640	373	838	766	610	266	176
Calc. [(d)]	953	783	811	749	623	364	832	755	608	260	170
Calc. [(e)]	956	783	829	755	629	366	838	766	610	261	173
Calc. [(f)]	945	774	807	741	622	361	817	747	601	257	171

[(a)] Our; [(b)] Ref. [105]; [(c)] Ref. [129]; [(d)] Ref. [125]; [(e)] Ref. [122]; [(f)] Ref. [123].

(B)

GeC	$\omega_{LO(\Gamma)}$	$\omega_{TO(\Gamma)}$	$\omega_{LO(X)}$	$\omega_{TO(X)}$	$\omega_{LA(X)}$	$\omega_{TA(X)}$	$\omega_{LO(L)}$	$\omega_{TO(L)}$	$\omega_{LA(L)}$	$\omega_{TA(L)}$	$\Delta\omega$
Our [(a)]	749	626	697	617	348	211	705	621	326	166	123
Calc. [(b)]	812	682	785	695	378	222	789	683	366	161	130
Calc. [(c)]	748	626	697	617	348	214	705	612	331	162	122

[(a)] Our; [(b)] Ref. [124], [(c)] Ref. [125].

(C)

SnC	$\omega_{LO(\Gamma)}$	$\omega_{TO(\Gamma)}$	$\omega_{LO(X)}$	$\omega_{TO(X)}$	$\omega_{LA(X)}$	$\omega_{TA(X)}$	$\omega_{LO(L)}$	$\omega_{TO(L)}$	$\omega_{LA(L)}$	$\omega_{TA(L)}$	$\Delta\omega$
Our [(a)]	558	456	512	454	216	141	524	454	214	102	102
Calc. [(b)]	711	590	689	621	268	150	694	600	262	106	121
Calc. [(c)]	558	456	503	450	216	134	516	440	199	109	102

[(a)] Our; [(b)] Ref. [124], [(c)] Ref. [125].

3.1.2. Thermodynamic Characteristics

To test the accuracy of our RIM $\omega_j(\vec{q})$, and $g(\omega)$, we have followed the methodology outlined in Section 2.3 and simulated the T-dependent Debye temperature and specific heat for the binary XC materials. The results of our calculations for $\Theta_D(T)$ and $C_V(T)$ displayed in Figure 3a,b are compared with the experimental (3C-SiC) [149–151] and/or theoretical data [152,153]. While the shapes of these thermodynamic quantities for the XC materials are typical to those of the group IV elemental and III–V, II–VI, I–VII compound semiconductors, the values of C-based materials, however, differ significantly [120,124,125,152].

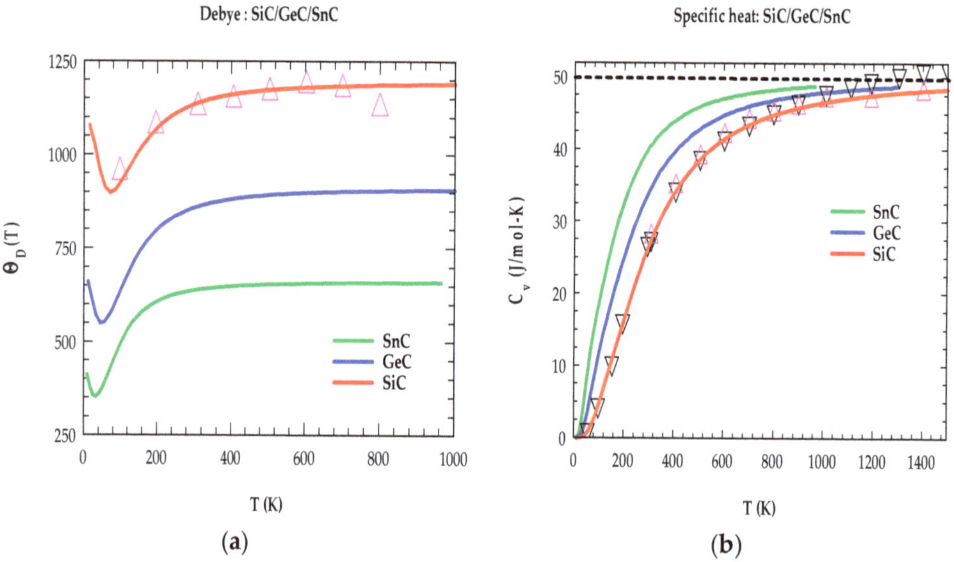

Figure 3. (**a**) Rigid-ion model simulations of temperature-dependent Debye temperature $\Theta_D(T)$ for the zinc-blende SiC (red color lines), GeC (blue lines) and SnC (green lines). For SiC, the results are compared with the experimental data (magenta color triangles). (**b**) Rigid-ion model calculations of $C_V(T)$ (in J/mol-K) as a function of T for SiC (red color lines), GeC (blue lines) and SnC (green lines). The calculations of 3C-SiC are compared with the experimental (magenta color triangles and black colored inverted triangles) (see text).

At a single temperature T, the outcomes of our results (cf. Figure 3a,b) for $\Theta_D(T)$ [$C_V(T)$] have shown considerable decrease [increase] in the simulated values with the increase of cation X atomic masses (i.e., from Si $\rightarrow$ Ge $\rightarrow$ Sn). Again, at low T, our results of $C_v(T)$ for the binary materials have not only confirmed the correct trends with appropriate shapes replicating the T^3 law but also provided a reasonably good match with the existing experimental/theoretical data [149–153]. For 3C-SiC, the experimental values of specific heat at constant volume (C_V) [149] and constant pressure (C_P) [150] are included in Figure 3b using different colored symbols. As expected, the results of $C_v(T)$ for XC binary materials at highest temperatures approached ~50 (J/mol-K) [see (Figure 3b)], in excellent agreement with the Dulong–Petit rule. Clearly, these results are justified because at higher T, one would anticipate all the excited phonon modes contributing to the thermodynamic characteristics.

3.2. Phonon and Thermodynamic Properties of $X_xY_{1-x}C$ Ternary Alloys

Recent developments in the crystal growth of ultrathin $X_xY_{1-x}C$ films on different substrates have given excellent opportunities to many researchers to prepare novel C-based semiconducting alloys with the expected crystal structures and compositions [42–46]. For future device designs and their applications using zb $X_xY_{1-x}C$ alloys, we have reported our comprehensive calculations and predicted their x-dependent phonon $\omega_j(\vec{q}), g(\omega)$, and thermodynamic $\Theta_D(T)$, $C_V(T)$ traits by using a RIM GF methodology in the VCA (see Table 4 and employing the results of binary XC materials).

Table 4. (A) Comparison of the temperature-dependent RIM calculations for the Debye temperature $\Theta_D(T)$, and (B) the specific heat $C_v(T)$ in (J/mol-K) for the zb SiC, GeC and SnC materials with the existing experimental Refs. [149,150] and/or theoretical Refs. [118–120,123,125,152] data. Please note that the reported results in Ref. [150] are for $C_P(T)$.

(A)

			Debye Temperature $\Theta_D(T)$			
T	**3C-SiC RIM** [(a)]	**Others**	**GeC RIM** [(a)]	**Others**	**SnC RIM** [(a)]	**Others**
0	1090 [(a)]		673 [(a)]		419 [(a)]	
100	923 [(a)]		635 [(a)]		490 [(a)]	
300	1134 [(a)]	960.61 [(b)], 1130 [(c)], 611.6 [(d)], 1151 [(e)], 1080 [(f)]	857 [(a)]	759.6 [(g)], 616 [(e)], 831 [(f)]	619 [(a)]	506.7 [(g)], 472 [(e)]
600	1180 [(a)]	951.54 [(b)]	899 [(a)]		651 [(a)]	
900	1188 [(a)]	942.52 [(b)]	906 [(a)]		657 [(a)]	
1200	1192 [(a)]	934.94 [(b)]	908 [(a)]		659 [(a)]	
1500	1194 [(a)]	930.02 [(b)]	910 [(a)]		660 [(a)]	

[(a)] Our; [(b)] Refs. [149,150]; [(c)] Ref. [152]; [(d)] Ref. [120]; [(e)] Ref. [118]; [(f)] Ref. [119]; [(g)] Ref. [123].

(B)

			Specific heat $C_v(T)$			
T	**3C-SiC RIM** [(a)]	**Others**	**GeC RIM** [(a)]	**Others**	**SnC RIM** [(a)]	**Others**
300	26.24	27.09 [(b)], 31.39 [(c)], 26.8 [(d)]	34.31	32.35 [(e)]	40.01	35.70 [(e)]
600	41.49	42.09 [(b)], 44.14 [(c)], 41.3 [(d)]	44.58	44.6 [(e*)]	46.96	46.10 [(e*)]
900	45.77	47.60 [(b)], 47.26 [(c)], 46.7 [(d)]	47.41	47.1 [(e*)]	48.56	48.32 [(e*)]
1200	47.51	50.19 [(b)], 48.41 [(c)], 49.6 [(d)]	48.46	48.2 [(e*)]	-	-
1500	48.34	51,77 [(b)], 48.94 [(c)], 51.2 [(d)]	-	-	-	-

[(a)] Our; [(b)] Refs. [149,150]; [(c)] Ref. [152]; [(d)] Ref. [120]; [(e)] Ref. [125] [(e*)] estimated from the graph.

3.2.1. Phonon Characteristics of $X_{1-x}Y_xC$ Alloys

In the framework of a generalized RIM GF formalism (cf. Section 2.2), systematic simulations are performed for predicting the results of x-dependent phonon dispersions $\omega_j(\vec{q})$, and DOS, $g(\omega)$ for the ideal mixed $X_{1-x}Y_xC$ ternary alloys. For instance, the calculated results of $Si_{1-x}Ge_xC$ alloys displayed in Figure 4a,b have clearly validated that the values of $\omega_j(\vec{q})$, and $g(\omega)$ steadily shifting towards the lower frequency regions as x increases (i.e., from ≡0.0, 0.2, 0.4, 0.6, 0.8 and 1.0). Moreover, in the extreme situations $x \to 0.0$, and $x \to 1.0$, the results are seen transforming to those of the binary SiC and GeC materials.

From the phonon characteristics of $Si_{1-x}Ge_xC$ alloys (cf. Figure 4a,b), we have also noticed a few interesting features: (*a*) the *x*-dependent curves of $\omega_j(\vec{q})$, and $g(\omega)$, exhibit very similar traits, (*b*) the ω_{TO} modes show nearly flat dispersions in the $X \to \Gamma \to L$ directions which are responsible for triggering appropriate strong peaks in the $g(\omega)$, (*c*) the ω_{LO} modes have demonstrated almost flat dispersions in the $L \to X \to W$ directions which initiated the correct weak peaks in the $g(\omega)$, and (*d*) the splitting of optical phonon modes $\Delta\omega$ ($\equiv \omega_{LO(\Gamma)} - \omega_{TO(\Gamma)}$) decreased from 177 cm^{-1} → 161 cm^{-1} → 149 cm^{-1} → 139 cm^{-1} → 130 cm^{-1} → 123 cm^{-1} with the increase of composition, x (≡0.0 → 0.2 → 0.4 → 0.6 → 0.8 → 1.0). Very similar phonon characteristics are also noticed in the $Si_{1-x}Sn_xC$ (cf. Figure 5a,b) and $Ge_{1-x}Sn_xC$ (cf. Figure 6a,b) ternary alloys.

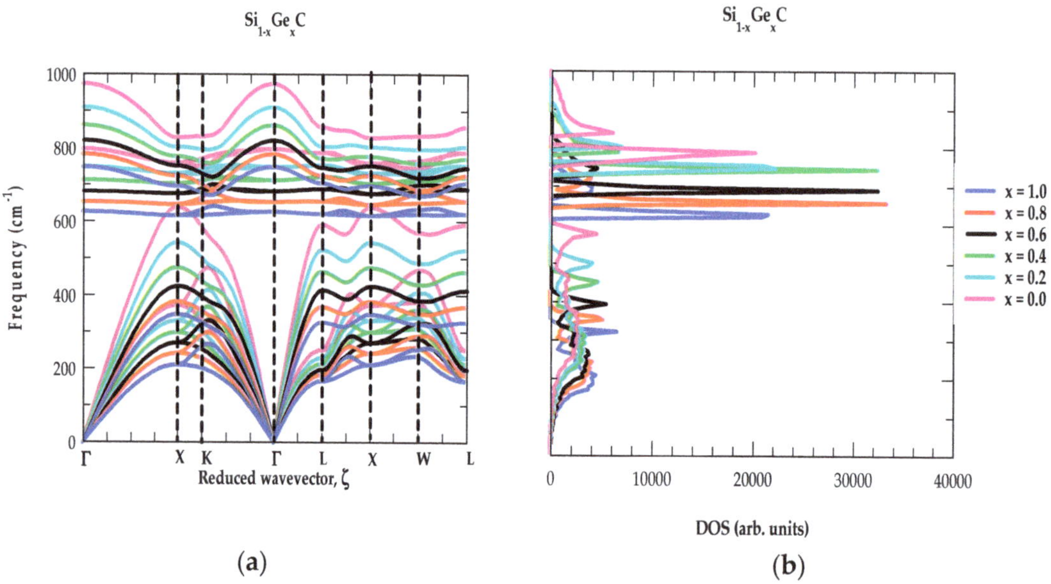

Figure 4. (**a**) Composition-dependent rigid-ion model (RIM) calculations of phonon dispersions $\omega_j(\vec{q})$ for $Si_{1-x}Ge_xC$ with x = 0.0, 0.2, 0.4, 0.6, 0.8, and 1.0. (**b**) Composition-dependent RIM calculations of one phonon density of states $g(\omega)$ for $Si_{1-x}Ge_xC$ with x = 0.0, 0.2, 0.4, 0.6, 0.8, and 1.0.

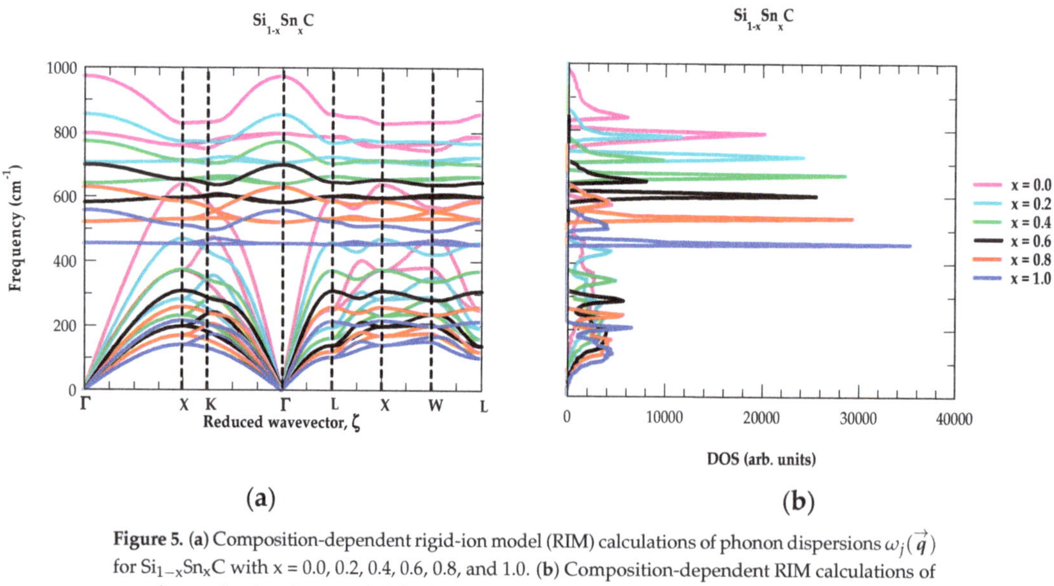

Figure 5. (**a**) Composition-dependent rigid-ion model (RIM) calculations of phonon dispersions $\omega_j(\vec{q})$ for $Si_{1-x}Sn_xC$ with x = 0.0, 0.2, 0.4, 0.6, 0.8, and 1.0. (**b**) Composition-dependent RIM calculations of one phonon density of states $g(\omega)$ for $Si_{1-x}Sn_xC$ with x = 0.0, 0.2, 0.4, 0.6, 0.8, and 1.0.

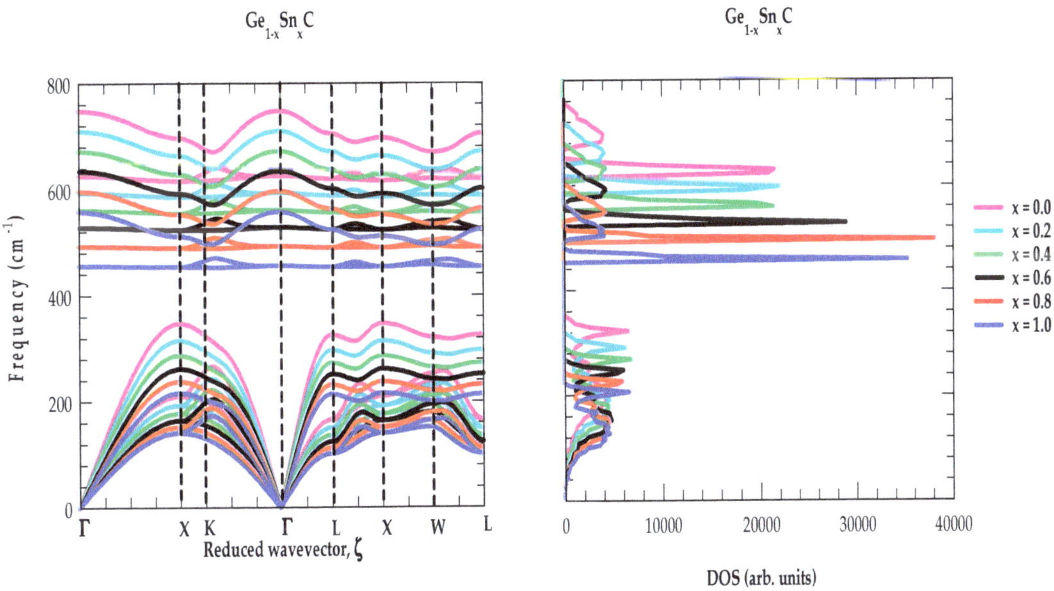

(**a**) (**b**)

Figure 6. (**a**) Composition-dependent rigid-ion model (RIM) calculations of phonon dispersions $\omega_j(\vec{q})$ for $Ge_{1-x}Sn_xC$ with x = 0.0, 0.2, 0.4, 0.6, 0.8, and 1.0. (**b**) Composition-dependent RIM calculations of one phonon density of states $g(\omega)$ for $Ge_{1-x}Sn_xC$ with x = 0.0, 0.2, 0.4, 0.6, 0.8, and 1.0.

3.2.2. Thermodynamic Characteristics of $X_{1-x}Y_xC$ Alloys

By using appropriate phonon dispersions and DOS, the x-dependent thermodynamic characteristics are simulated in the QHA for $X_{1-x}Y_xC$ ternary alloys between 0 < T < 1500 K. As an example, we have reported our results of $\Theta_D(T)$ and $C_v(T)$ for the $Si_{1-x}Ge_xC$ alloys in Figures 7a and 7b, respectively. Similar calculations have also been performed for other $Si_{1-x}Sn_xC$ (cf. Figure 8a,b) and $Ge_{1-x}Sn_xC$ (cf. Figure 9a,b) ternary alloys. From Figure 7a,b, some important noticeable facts can be justified. As x increases (from ≡0, 0.2, 0.4, 0.6, 0.8 and 1.0), the values of $\Theta_D(T)$ [$C_v(T)$] decrease [increase] and in the limiting situations $x \to 0.0$, and $x \to 1.0$, the results transform to those of SiC and GeC materials. Based on our simulations, the binary 3C-SiC has exhibited the highest $\Theta_D(T)$ and lowest $C_v(T)$, $\alpha(T)$ values. These results, related to its shorter bond-length and larger bond strength (see Table 2), can exhibit strong radiation tolerance with excellent resistance. Obviously, these characteristics have led to 3C-SiC being quite robust at higher T with less likelihood of breakdown in extreme conditions. We, therefore, feel that 3C-SiC is an ideal compound to be employed as a fuel-cladding material in nuclear reactors and high-temperature environments.

On the other hand, both GeC and SnC revealed significantly weaker bonding [149–153] (see Table 2) which instigated lower $\Theta_D(T)$ and higher $C_v(T)$, $\alpha(T)$ values. Thus, we anticipate that these materials may not be suitable for fuel-cladding layers in nuclear reactors and/or in higher temperature environments. However, with a very small composition (x), the ternary $Si_{1-x}Ge_xC$ alloys can still be deliberated. Both XC and $X_{1-x}Y_xC$ alloys have already been used to grow multilayer (viz., SiC/SiGe(Sn)C, GeC/Si(Ge)SnC) heterostructures. Thus, we feel that these structures may help engineers to design MQW/SL-based micro-/nanodevices for different strategic and civilian application needs.

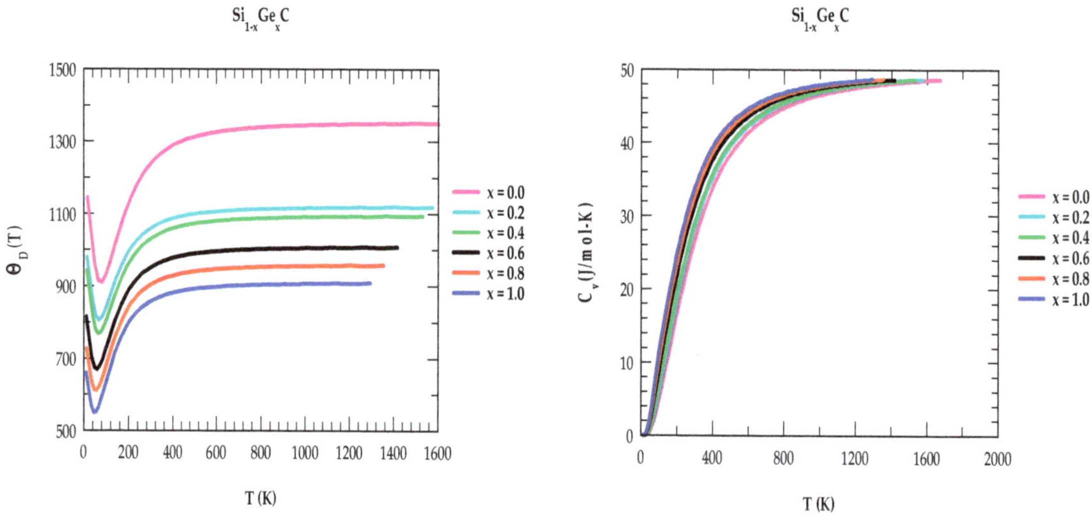

(a) (b)

Figure 7. (**a**) Composition-dependent rigid-ion model calculations of Debye temperatures $\Theta_D(T)$ for $Si_{1-x}Ge_xC$ with x = 0.0, 0.2, 0.4, 0.6, 0.8, and 1.0. (**b**) Composition-dependent RIM calculations of one specific heat Cv(T) for $Si_{1-x}Ge_xC$ with x = 0.0, 0.2, 0.4, 0.6, 0.8, and 1.0.

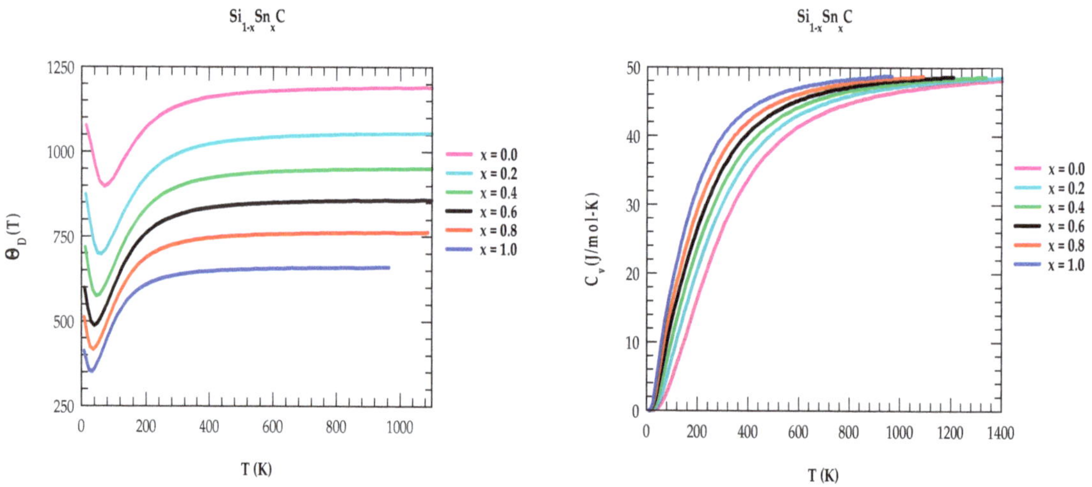

(a) (b)

Figure 8. (**a**) Composition-dependent rigid-ion model calculations of Debye temperatures $\Theta_D(T)$ for $Si_{1-x}Sn_xC$ with x = 0.0, 0.2, 0.4, 0.6, 0.8, and 1.0. (**b**) Composition-dependent RIM calculations of one specific heat Cv(T) for $Si_{1-x}Sn_xC$ with x = 0.0, 0.2, 0.4, 0.6, 0.8, and 1.0.

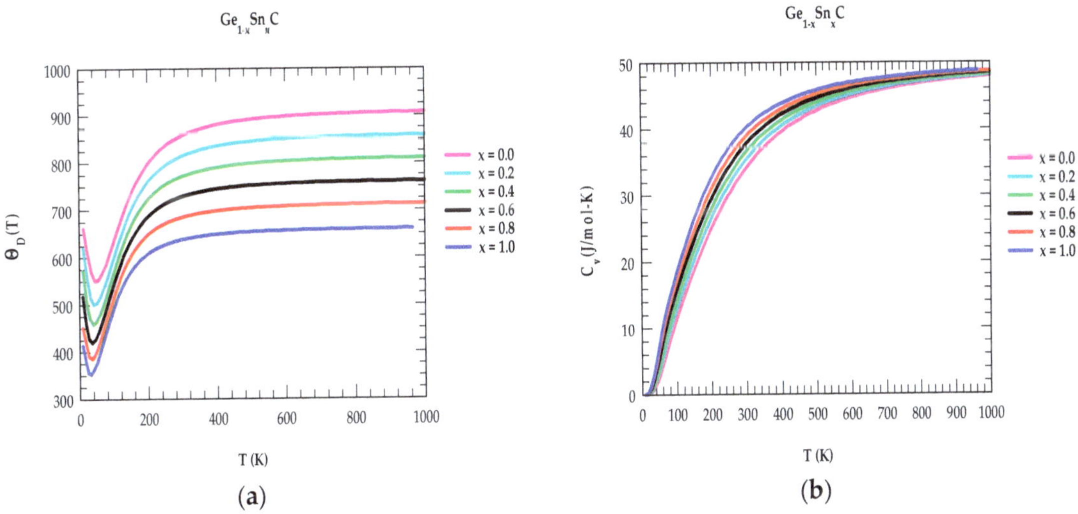

Figure 9. (**a**) Composition-dependent rigid-ion model calculations of Debye temperatures $\Theta_D(T)$ for $Ge_{1-x}Sn_xC$ with x = 0.0, 0.2, 0.4, 0.6, 0.8, and 1.0. (**b**) Composition-dependent RIM calculations of one specific heat Cv(T) for $Ge_{1-x}Sn_xC$ with x = 0.0, 0.2, 0.4, 0.6, 0.8, and 1.0.

3.2.3. Born Effective Charge for $X_{1-x}Y_xC$ Alloys

In polar materials, Born's effective charge e_T^* (also known as transverse or dynamic effective charge) manifests the coupling between lattice displacements and electrostatic fields. It is found that e_T^* remains insensitive to the isotropic volume change but strongly affected by changes in the atomic positions associated with phase transitions (P_t). From a theoretical standpoint, e_T^* in binary XC and ternary $X_{1-x}Y_xC$ alloys is important as P_t takes place due to the competition between long-range coulomb interactions and short-range forces. The long-range coulomb interactions are responsible for the observed splitting $\Delta\omega$ $[\equiv \omega_{LO(\Gamma)} - \omega_{TO(\Gamma)}]$ between $\omega_{LO(\Gamma)}$ and $\omega_{TO(\Gamma)}$ phonon frequencies. In Table 5, we have reported our simulated results of x-dependent $\Delta\omega$ and e_T^* for $X_{1-x}Y_xC$ alloys. In the absence of e_T^* data for ternary alloys, our theoretical results for the binary materials agree reasonably well with the experimental (for 3C-SiC) [107] and theoretical [123] results (for GeC and SnC) [122,124].

Table 5. Simulated composition-dependent Born's effective charges for zinc-blende ternary alloys: (A) $Si_{1-x}Ge_xC$, (B) $Si_{1-x}Sn_xC$, and (C) $Ge_{1-x}Sn_xC$.

(A)

			$Si_{1-x}Ge_xC$		
x	$\omega_{LO(\Gamma)}$	$\omega_{LO(\Gamma)}$	$\Delta\omega$	e_T^{*} [(a)]	e_T^* **others**
0.0	974	797	177	2.683	2.697 [(b)], 2.72 [(c)]
0.2	911	750	161	2.675	
0.4	862	713	149	2.659	
0.6	821	682	139	2.629	
0.8	783	653	130	2.611	
1.0	749	626	123	2.581	2.62 [(d)]

[(a)] Our; [(b)] Ref. [107]; [(c)] Refs. [122,123]; [(d)] Ref. [124].

Table 5. *Cont.*

(B)

x	$\omega_{LO(\Gamma)}$	$\omega_{LO(\Gamma)}$	$Si_{1-x}Sn_xC$ $\Delta\omega$	e_T^* (a)	e_T^* others
0.0	974	797	177	2.683	2.697 (b), 2.72 (c)
0.2	856	707	149	2.697	
0.4	773	642	131	2.695	
0.6	700	582	118	2.684	
0.8	630	522	108	2.658	
1.0	558	456	102	2.629	2.95 (d)

(a) Our; (b) Ref. [107]; (c) Refs. [122,123]; (d) Ref. [124].

(C)

x	$\omega_{LO(\Gamma)}$	$\omega_{LO(\Gamma)}$	$Ge_{1-x}Sn_xC$ $\Delta\omega$	e_T^* (a)	e_T^* others
0.0	749	626	177	2.581	2.62 (d)
0.2	710	594	116	2.595	
0.4	673	562	111	2.600	
0.6	635	528	107	2.615	
0.8	597	493	104	2.623	
1.0	558	456	102	2.629	2.95 (d)

(a) Our; (d) Ref. [124].

3.2.4. Fröhlich Coefficients

In semiconductors, the strength of the Fröhlich interaction is directly linked to the polar nature of its crystal lattice. In a highly polar material, the Coulomb field of a carrier (or exciton) couples more easily to the polar vibrations (i.e., ω_{LO} phonons) of the crystal lattice, resulting in a strong Fröhlich coupling coefficient, α_F . By using α_F (cf. Equation (11)), the polaron effective mass m_p^* can be calculated [154] in terms of the bare electron mass m_e using the following expression:

$$m_p^* = m_e \cdot \frac{1 - 0.0008\, \alpha_F{}^2}{1 - \left(\frac{\alpha_F}{6}\right) + 0.003\, \alpha_F{}^2} \quad (12)$$

where, the static ε_o and high-frequency ε_∞ values of the dielectric functions provide the means of quantifying the polar nature of materials, so that further insight into the Fröhlich interaction (see Equation (11)) can be gained via the dielectric characterizations.

It is to be noted that, except for 3C-SiC, no systematic calculations of the band structures and effective electron masses m_e^* are known for the GeC and SnC materials. With Equation (11), and using the existing parameters (ε_o, ε_∞, ω_{LO}, m_e^*) for 3C-SiC from the literature [105–107], our calculation of α_F (=0.576) has provided a value much higher than that reported by Adachi [130]. Interestingly, however, the calculation of polaron mass ($m_p^*/m_e \equiv 0.243$) using Equation (12) agrees very well with the theoretical result reported by Persson and Lindefelt [154]. Obviously, more experimental and theoretical efforts are needed for assessing the accurate values of the Fröhlich interaction coefficients for both the binary XC and ternary $X_{1-x}Y_xC$ alloys.

4. Discussions and Conclusions

In summary, we have used a realistic RIM and reported the results of our methodical simulations for comprehending the lattice dynamical and thermodynamic characteristics of zb (SiC, GeC, SnC) binary and ($S_{1-x}Ge_xC$, $S_{1-x}Sn_xC$, $Ge_{-x}Sn_xC$) ternary alloys. From a basic chemistry standpoint, one expects a gradual increase in the bond lengths (i.e., from Si-C → Ge-C → Sn-C) due to the differences in the sizes of the cations (Si, Ge, Sn) and (C) anion atoms. Accordingly, the increase in bond lengths will cause a decrease in their bond strengths. These facts are clearly revealed in our calculated IFCs of the RIM for XC materials (see Table 2), where the nearest-neighbor force constants [127] have shown steady decrease in their strength as one proceeds from SiC $\rightarrow$ GeC $\rightarrow$ SnC. Obviously, the atomic- and composition (x)-dependent variations in the bond strengths of mixed $X_{1-x}Y_xC$ alloys have instigated dramatic variations in the simulated RIM phonon dispersions $\omega_j(\vec{q})$, DOS$g(\omega)$, and thermodynamic [e.g., Θ_D(T), C_v(T), $\alpha(T)$] traits.

In recent years, consistent efforts have been made using group-IV carbides and III-nitrides to develop devices for achieving efficient operations in challenging environments (viz., radiation, high-power, extreme temperature) where the electronic systems based on Si material have indicated weaknesses of survival. In this quest, it is necessary to assess the electronic and thermodynamic characteristics of XC materials to see if they satisfy the necessary requirements for their use in the high temperature/high power settings. One must note that the devices based on wide bandgap GaN and SiC have recently emerged in the commercial market for slowly replacing the traditional Si-built electronic parts. Both GaN and SiC materials with wide bandgaps, high critical electric fields, and low dielectric constants have reflected on the lower on-state resistance for a given blocking voltage. In addition, these materials have exhibited high Debye temperatures Θ_D(T), low specific heats C_v(T) and low thermal expansion $\alpha(T)$ coefficients. As compared to SiC, there are a few disadvantages for the selection of GaN material. The main problems have been identified as follows: (*a*) the manufacturing complexity, cost, intrinsic defects, and reliability concerns about the integration of GaN into the existing processes with limited availability of substrates, and (*b*) the relatively lower [higher] values of Θ_D(T) [C_v(T), $\alpha(T)$] [108,136]. Again, with respect to Si ($E_g \equiv 1.12$ eV), the reported theoretical bandgap energies of GeC and SnC materials are 1.51 eV [118], and 0.75 eV [117], respectively. Obviously, compared to SiC, the binary GeC, SnC and/or ternary $Ge_{1-x}Sn_xC$ alloys with lower E_g and weaker bonding have exhibited different lattice dynamical and thermodynamic properties. Therefore, we strongly feel that devices based on binary GeC, SnC and/or $Ge_{1-x}Sn_xC$ materials may not be suitable for radiation detection in nuclear reactors or high-temperature, high-power settings. However, from recent successful efforts in the growth of ultrathin zb XC binary and $X_{1-x}Y_xC$ ternary alloys, along with their predicted results of phonon, structural, and thermodynamic traits, the materials can still be credible for the preparations of heterostructures in designing MQW and SL-based micro-/nanodevices for different strategic and civilian application needs.

Funding: This research has received no external funding.

Data Availability Statement: The data that support the findings of this study are available from the author upon reasonable request.

Acknowledgments: The author wishes to thank Deanne Snavely, Dean College of Natural Science and Mathematics at Indiana University of Pennsylvania (IUP), for the travel support and the Innovation Grant that he received from the School of Graduate Studies making this research possible.

Conflicts of Interest: The author declares having no financial and/or competing interests.

References

1. Bell Telephone Laboratories, Inc. *The Transistor: Selected Reference Material on Characteristics and Applications*; Western Electric Co., Inc.: New York, NY, USA, 1951.
2. Maram, R.; Kaushal, S.; Azaña, J.; Chen, L.R. Recent Trends and Advances of Silicon-Based Integrated Microwave Photonics. *Photonics* **2019**, *6*, 13. [CrossRef]
3. Bennett, H.S.; Brederlow, R.; Costa, J.C.; Cottrell, P.E.; Huang, W.M.; Immorlica, A.A., Jr.; Mueller, J.-E.; Racanelli, M.; Shichijo, H.; Weitzel, C.E.; et al. Device and Technology Evolution for Si-Based RF Integrated Circuits. *IEEE Trans. Electron Dev.* **2005**, *52*, 1235. [CrossRef]
4. Gutmann, R.J. Advanced Silicon IC Interconnect Technology and Design: Present Trends and RF Wireless Implications. *IEEE Trans. Microwave Theory Tech.* **1999**, *47*, 667–674. [CrossRef]
5. Burghartz, J.N. Silicon RF technology—The two generic approaches. In Proceedings of the 27th European Solid-State Device Research Conference, Stuttgart, Germany, 22–24 September 1997; pp. 143–153.

6. Razavi, B. CMOS technology characterization for analog and RF design. *IEEE J. Solid-State Circuits* **1999**, *34*, 268–276.
7. Harame, D.; Comfort, J.; Cressler, J.; Crabbé, E.; Sun, J.; Meyerson, B.; Tice, T. Si/SiGe epitaxial base transistors, Parts I and II. *IEEE Trans. Electron Dev.* **1995**, *42*, 455–482. [CrossRef]
8. Chen, D.; Li, R.; Xu, J.; Li, D.; Fei, C.; Yang, Y. Recent progress and development of radio frequency energy harvesting devices and circuits. *Nano Energy* **2023**, *117*, 108845. [CrossRef]
9. Raut, P.; Nanda, U.; Panda, D.K. Review—Recent Trends on Junction-Less Field Effect Transistors in Terms of Device Topology, Modeling, and Application. *ECS J. Solid Stat. Sci. Technol.* **2023**, *12*, 031010. [CrossRef]
10. Fu, Q.; Zhang, W.R.; Jin, D.Y.; Ding, C.B.; Zhao, Y.X.; Lu, D. Collector optimization for tradeoff between breakdown voltage and cut-off frequency in SiGe HBT. *Chin. Phys. B* **2014**, *23*, 114402. [CrossRef]
11. Shahbaz, M.; Butt, M.A.; Piramidowicz, R. Breakthrough in Silicon Photonics Technology in Telecommunications, Biosensing, and Gas Sensing. *Micromachines* **2023**, *14*, 1637. [CrossRef]
12. Kadri, E.; Messaoudi, O.; Krichen, M.; Dhahri, K.; Rasheed, M.; Dhahri, E.; Zouari, A.; Khirouni, K.; Barille, R. Optical and electrical properties of SiGe/Si solar cell heterostructures: Ellipsometric study. *J. Alloys Compd.* **2017**, *721*, 779–783. [CrossRef]
13. Sun, Z.; Wang, L.; Luo, H.; Hamer, P.; Ye, H.; Hallam, B. Study of the Hydrogen Passivation Effect of Low-Temperature Deposited Amorphous Silicon Layers on SiGe Solar Cells Grown on a Silicon Substrate. *ACS Appl. Energy Mater.* **2023**, *6*, 12064–12071. [CrossRef]
14. Ringel, S.A.; Carlin, J.; Andre, C.; Hudait, M.; Gonzalez, M.; Wilt, D.; Clark, E.; Jenkins, P.; Scheiman, D.; Allerman, A.; et al. Single-junction InGaP/GaAs solar cells grown on Si substrates with SiGe buffer layers. *Prog. Photovolt. Res. Appl.* **2002**, *10*, 417–426. [CrossRef]
15. Erdtmann, M.; Carroll, M.; Vineis, C.; Badawi, H.; Bulsara, M.; Ringel, S. Growth and characterization of high-Ge content SiGe virtual substrates. *Proc. Electrochem. Soc.* **2003**, *11*, 106–117.
16. Wang, L.; Conrad, B.; Soeriyadi, A.; Zhao, X.; Li, D.; Diaz, M.; Lochtefeld, A.; Gerger, A.; Perez-Wurfl, I.; Barnett, A. Current matched three-terminal dual junction GaAsP/SiGe tandem solar cell on Si. *Sol. Energy Mater. Sol. Cells* **2016**, *146*, 80–86. [CrossRef]
17. Sankarasubramaniam, Y.; Akyildiz, I.E.; Mchughlin, S.W. Energy Efficiency based Packet Size Optimization in Wireless Sensor Networks. In Proceedings of the First IEEE International Workshop on Sensor Network Protocols and Applications, Anchorage, AK, USA, 11 May 2003. [CrossRef]
18. Ojeda, F.; Mendez, D.; Fajardo, A.; Ellinger, F. On Wireless Sensor Network Models: A Cross-Layer Systematic Review. *J. Sens. Actuator Netw.* **2023**, *12*, 50. [CrossRef]
19. Xing, J.; Sun, C.; Xiong, B.; Wang, J.; Hao, Z.; Wang, L.; Han, Y.; Li, H.; Luo, Y. Membrane multiple quantum-well electro-optical modulator employing low loss high-k radio-frequency slot waveguides. *Opt. Express* **2022**, *30*, 23260. [CrossRef] [PubMed]
20. Zhao, X.; Xiong, B.; Sun, C.; Luo, Y. Low drive voltage optical phase modulator with novel InGaAlAs/InAlAs multiple-quantum-barrier based n-i-n heterostructure. *Opt. Express* **2013**, *21*, 24894–24903. [CrossRef] [PubMed]
21. Liu, X.; Xiong, B.; Sun, C.; Wang, J.; Hao, Z.; Wang, L.; Han, Y.; Li, H.; Yu, J.; Luo, Y. Wideband thin-film lithium niobate modulator with low half-wave-voltage length product. *Chin. Opt. Lett.* **2021**, *19*, 060016. [CrossRef]
22. Xing, J.; Sun, C.; Xiong, B.; Wang, J.; Hao, Z.; Wang, L.; Han, Y.; Li, H.; Yu, J.; Luo, Y. Low loss hybrid plasmon polariton Mach-Zehnder modulators. *OSA Contin.* **2021**, *4*, 2721–2733. [CrossRef]
23. Fujii, T.; Hiraki, T.; Aihara, T.; Nishi, H.; Takeda, K.; Sato, T.; Kakitsuka, T.; Tsuchizawa, T.; Matsuo, S. Development of an Epitaxial Growth Technique Using III-V on a Si Platform for Heterogeneous Integration of Membrane Photonic Devices on Si. *Appl. Sci.* **2021**, *11*, 1801. [CrossRef]
24. Rickman, A. The commercialization of silicon photonics. *Nat. Photonics* **2014**, *8*, 579–582. [CrossRef]
25. Nagarajan, R.; Joyner, C.H.; Schneider, R.P.; Bostak, J.S.; Butrie, T.; Dentai, A.G.; Dominic, V.G.; Evans, P.W.; Kato, M.; Kauffman, M.; et al. Large-scale photonic integrated circuits. *IEEE J. Sel. Top. Quantum Electron.* **2005**, *11*, 50–65. [CrossRef]
26. Komljenovic, T.; Davenport, M.; Hulme, J.; Liu, A.Y.; Santis, C.T.; Spott, A.; Srinivasan, S.; Stanton, E.J.; Zhang, C.; Bowers, J.E. Heterogeneous Silicon Photonic Integrated Circuits. *J. Light. Technol.* **2016**, *34*, 20–35. [CrossRef]
27. Dong, P.; Liu, X.; Chandrasekhar, S.; Buhl, L.L.; Aroca, R.; Chen, Y.-K. Monolithic Silicon Photonic Integrated Circuits for Compact 100+Gb/s Coherent Optical Receivers and Transmitters. *IEEE J. Sel. Top. Quantum Electron.* **2014**, *20*, 150–157. [CrossRef]
28. Horikawa, T.; Shimura, D.; Okayama, H.; Jeong, S.-H.; Takahashi, H.; Ushida, J.; Sobu, Y.; Shiina, A.; Tokushima, M.; Kinoshita, K.; et al. A 300-mm Silicon Photonics Platform for Large-Scale Device Integration. *IEEE J. Sel. Top. Quantum Electron.* **2018**, *24*, 1–15. [CrossRef]
29. Kukushkin, S.; Osipov, A.; Redkov, A. SiC/Si as a New Platform for Growth of Wide-Bandgap Semiconductors. In *Mechanics and Control of Solids and Structures*; Polyanskiy, V.A., Belyaev, A.K., Eds.; Advanced Structured Materials; Springer Nature: Cham, Switzerland, 2022; Chapter 18; p. 164.
30. Nguyen, T.-K.; Yadav, S.; Truong, T.-A.; Han, M.; Barton, M.; Leitch, M.; Guzman, P.; Dinh, T.; Ashok, A.; Vu, H.; et al. Integrated, Transparent Silicon Carbide Electronics and Sensors for Radio Frequency Biomedical Therapy. *ACS Nano* **2022**, *16*, 10890–10903. [CrossRef] [PubMed]
31. Pham, T.A.; Nguyen, T.K.; Vadivelu, R.K.; Dinh, T.; Qamar, A.; Yadav, S.; Yamauchi, Y.; Rogers, J.A.; Nguyen, N.T.; Phan, H.P. Stretchable Bioelectronics: A Versatile Sacrificial Layer for Transfer Printing of Wide Bandgap Materials for Implantable and Stretchable Bioelectronics. *Adv. Funct. Mater.* **2020**, *30*, 2070287. [CrossRef]

32. Nguyen, T.K.; Phan, H.P.; Kamble, H.; Vadivelu, R.; Dinh, T.; Iacopi, A.; Walker, G.; Hold, L.; Nguyen, N.T.; Dao, D.V. Superior Robust Ultrathin Single-Crystalline Silicon Carbide Membrane as a Versatile Platform for Biological Applications. *ACS Appl. Mater. Interfaces* **2017**, *9*, 41641–41647. [CrossRef] [PubMed]
33. Yang, H.; Ma, Y.; Da, Y. Progress of structural and electronic properties of diamond: A mini review. *Funct. Diam.* **2021**, *1*, 150–159. [CrossRef]
34. Yue, Y.; Gao, Y.; Hu, W.; Xu, B.; Wang, J.; Zhang, X.; Zhang, Q.; Wang, Y.; Ge, B.; Yang, Z.; et al. Hierarchically structured diamond composite with exceptional Toughness. *Nature* **2020**, *582*, 370–374. [CrossRef]
35. Huang, Q.; Yu, D.; Xu, B.; Hu, W.; Ma, Y.; Wang, Y.; Zhao, Z.; Wen, B.; He, J.; Liu, Z.; et al. Nanotwinned diamond with unprecedented hardness and stability. *Nature* **2014**, *510*, 250–253. [CrossRef]
36. Ekimov, E.A.; Sidorov, V.A.; Bauer, E.D.; Mel'Nik, N.N.; Curro, N.J.; Thompson, J.D.; Stishov, S.M. Superconductivity in diamond. *Nature* **2004**, *428*, 542–545. [CrossRef] [PubMed]
37. Bauer, M.R.; Tolle, J.; Bungay, C.; Chizmeshya, A.V.G.; Smith, D.J.; Menéndez, J.; Kouvetakis, J. Tunable band structure in diamond–cubic tin–germanium alloys grown on silicon substrates. *Solid State Commun.* **2003**, *127*, 355–359. [CrossRef]
38. Islam, M.S.; Mojumder, R.H.; Hassan, A.; Sohag, M.U.; Park, J. High-Efficiency Multi Quantum Well Blue LED Using 2D-SiC as an Active Material. In Proceedings of the 2021 5th International Conference on Electrical Engineering and Information & Communication Technology (ICEEICT), Dhaka, Bangladesh, 18–20 November 2021.
39. Zhang, L.; Cui, Z. Theoretical Study on Electronic, Magnetic and Optical Properties of Non-Metal Atoms Adsorbed onto Germanium Carbide. *Nanomaterials* **2022**, *12*, 1712. [CrossRef] [PubMed]
40. Mélinon, P.; Masenelli, B.; Tournus, F.; Perez, A. Playing with carbon and silicon at the nanoscale. *Nat. Mater.* **2007**, *6*, 479–490. [CrossRef] [PubMed]
41. Polyanskiy, V.A.; Belyaev, A.K. (Eds.) *Mechanics and Control of Solids and Structures, Advanced Structured Materials*; Springer Nature: Cham, Switzerland, 2022; p. 164. Available online: https://link.springer.com/bookseries/8611 (accessed on 27 February 2024).
42. Ikoma, Y.; Endo, T.; Watanabe, F.; Motooka, T. Growth of Ultrathin Epitaxial 3C-SiC Films on Si(100) by Pulsed Supersonic Free Jets of CH_3SiH_3. *Jpn. J. Appl. Phys.* **1999**, *38*, L301. [CrossRef]
43. Ikoma, Y.; Endo, T.; Watanabe, F.; Motooka, T. Growth of Si/3C–SiC/Si(100) heterostructures by pulsed supersonic free jets. *Appl. Phys. Lett.* **1999**, *75*, 3977–3979. [CrossRef]
44. Ohtani, R.; Ikoma, Y.; Motooka, T. Formation of Si/SiC heterostructures for silicon-based quantum devices using single CH_3SiH_3-gas source free jet. *Mater. Res. Soc. Symp. Proc.* **2004**, *815*, J5.11.1. [CrossRef]
45. Yoshimura, S.; Sugimoto, S.; Takeuchi, T.; Murai, K.; Kiuchi, M. Low energy Si+, SiCH5+, or C+ beam injections to silicon substrates during chemical vapor deposition with dimethyl silane. *Heliyon* **2023**, *9*, e19002. [CrossRef] [PubMed]
46. Gallagher, J.D.; Senaratne, C.L.; Kouvetakis, J.; Menéndez, J. Compositional dependence of the bowing parameter for the direct and indirect band gaps in Ge 1-y Sn y alloys. *Appl. Phys. Lett.* **2014**, *105*, 142102. [CrossRef]
47. Ghetmiri, S.A.; Du, W.; Margetis, J.; Mosleh, A.; Cousar, L.; Conley, B.R.; Domulevicz, L.; Nazzal, A.; Sun, G.; Soref, R.A.; et al. Direct-bandgap GeSn grown on silicon with 2230 nm photoluminescence. *Appl. Phys. Lett.* **2014**, *105*, 151109. [CrossRef]
48. Wirths, S.; Geiger, R.; von den Driesch, N.; Mussler, G.; Stoica, T.; Mantl, S.; Ikonic, Z.; Luysberg, M.; Chiussi, S.; Hartmann, J.M.; et al. Lasing in direct-bandgap GeSn alloy grown on Si. *Nat. Photonics* **2015**, *9*, 88–92. [CrossRef]
49. Xu, Z.; Li, Y.; Liu, Z. Controlling electronic and optical properties of layered SiC and GeC sheets by strain engineering. *Mater. Des.* **2016**, *108*, 333–342. [CrossRef]
50. Xua, Q.; Cai, W.; Li, W.; Sreeprasad, T.S.; He, Z.; Ong, W.-J.; Li, N. Two-dimensional quantum dots: Fundamentals, photoluminescence mechanism and their energy and environmental applications. *Mater. Today Energy* **2018**, *10*, 222–240. [CrossRef]
51. Schulte-Braucks, C.; Glass, S.; Hofmann, E.; Stange, D.; von den Driesch, N.; Hartmann, J.M.; Ikonic, Z.; Zhao, Q.T.; Buca, D.; Mantl, S. Process modules for GeSn nanoelectronics with high Sn-contents. *Solid-State Electron.* **2017**, *128*, 54–59. [CrossRef]
52. Shen, Z.; Chen, J.; Li, B.; Li, G.; Zheng, H.; Men, J.; Hou, X. Tunable fabrication and photoluminescence property of SiC nano wires with different microstructures. *Appl. Surf. Sci.* **2020**, *506*, 144979. [CrossRef]
53. Dey, T.; Reza, M.S.; Arbogast, A.W.; Holtz, M.; Droopad, R.; Bank, S.R.; Wistey, M.A. Molecular beam epitaxy of highly crystalline GeSnC using CBr4 at low temperatures. *Appl. Phys. Lett.* **2022**, *121*, 122104. [CrossRef]
54. Dey, T.; Arbogast, A.W.; Meng, Q.; Reza, M.S.; Muhowski, A.J.; Cooper, J.P.; Ozdemir, E.; Naab, F.U.; Borrely, T.; Anderson, J.; et al. Influence of H on Sn incorporation in GeSnC alloys grown using molecular beam epitaxy. *J. Appl. Phys.* **2023**, *134*, 193102. [CrossRef]
55. Giunto, A.; Morral, A.F.I. The GeSn Alloy and its Optoelectronic Properties: A Critical Review of the Current Understanding. *arXiv* **2023**, arXiv:2309.10584. [CrossRef]
56. Cardoux, C.; Casiez, L.; Pauc, N.; Calvo, V.; Coudurier, N.; Rodriguez, P.; Richy, J.; Barritault, P.; Lartigue, O.; Constancias, C.; et al. Room temperature spectral characterization of direct band gap $Ge_{0.85}Sn_{0.15}$ LEDs and photodiodes. In Proceedings of the 2022 Proceedings SPIE OPTO, Silicon Photonics XVII, San Francisco, CA, USA, 22–27 January 2022; p. 120060A. [CrossRef]
57. Moutanabbir, O.; Assali, S.; Gong, X.; O'Reilly, E.; Broderick, C.A.; Marzban, B.; Witzens, J.; Du, W.; Yu, S.-Q.; Chelnokov, A.; et al. Monolithic infrared silicon photonics: The rise of (Si)GeSn semiconductors. *Appl. Phys. Lett.* **2021**, *118*, 110502. [CrossRef]
58. Atalla, M.R.M.; Assali, S.; Koelling, S.; Attiaoui, A.; Moutanabbir, O. High-Bandwidth Extended-SWIR GeSn Photodetectors on Silicon Achieving Ultrafast Broadband Spectroscopic Response. *ACS Photonics* **2022**, *9*, 1425–1433. [CrossRef]

59. Simola, E.T.; Kiyek, V.; Ballabio, A.; Schlykow, V.; Frigerio, J.; Zucchetti, C.; De Iacovo, A.; Colace, L.; Yamamoto, Y.; Capellini, G.; et al. CMOS-Compatible Bias-Tunable Dual-Band Detector Based on GeSn/Ge/Si Coupled Photodiodes. *ACS Photonics* **2021**, *8*, 2166–2173. [CrossRef]
60. Tran, H.; Pham, T.; Margetis, J.; Zhou, Y.; Dou, W.; Grant, P.C.; Grant, J.M.; Al-Kabi, S.; Sun, G.; Soref, R.A.; et al. Si-Based GeSn Photodetectors toward MidInfrared Imaging Applications. *ACS Photonics* **2019**, *6*, 2807–2815. [CrossRef]
61. Zhang, D.; Hu, X.; Liu, D.; Lin, X.; Wang, W.; Ding, Z.; Wang, Z.; Cheng, B.; Xue, C. GeSn on Si avalanche photodiodes for short wave infrared detection. In *Optical Sensing and Imaging Technologies and Applications*; Liu, D., Gong, H., Guina, M., Lu, J., Eds.; SPIE: Bellingham, WA, USA, 2018; p. 54.
62. Kim, Y.; Assali, S.; Burt, D.; Jung, Y.; Joo, H.-J.; Chen, M.; Ikonic, Z.; Moutanabbir, O.; Nam, D. Improved GeSn microdisk lasers directly sitting on Si. In *Silicon Photonics XVII*; Reed, G.T., Knights, A.P., Eds.; SPIE: Bellingham, WA, USA, 2022; p. 21.
63. Ojo, S.; Zhou, Y.; Acharya, S.; Saunders, N.; Amoah, S.; Jheng, Y.-T.; Tran, H.; Du, W.; Chang, G.-E.; Li, B.; et al. Silicon-based electrically injected GeSn lasers. In *Physics and Simulation of Optoelectronic Devices XXX*; Osinski, M., Arakawa, Y., Witzigmann, B., Eds.; SPIE: Bellingham, WA, USA, 2022; p. 15.
64. Marzban, B.; Seidel, L.; Liu, T.; Wu, K.; Kiyek, V.; Zoellner, M.H.; Ikonic, Z.; Schulze, J.; Grutzmacher, D.; Capellini, G.; et al. Strain Engineered Electrically Pumped SiGeSn Microring Lasers on Si. *ACS Photonics* **2023**, *10*, 217–224. [CrossRef]
65. Kasper, E.; Herzog, H.J.; Kibbel, H. A one-dimensional SiGe superlattice grown by UHV epitaxy. *Appl. Phys.* **1975**, *8*, 199–205. [CrossRef]
66. Iyer, S.S.; Patton, G.L.; Delage, S.S.; Tiwari, S.; Stork, J.M.C. Silicon-germanium base heterojunction bipolar transistors by molecular beam epitaxy. In Proceedings of the International Electron Devices Meeting, IEDM Technical Digest, Washington, DC, USA, 6–9 December 1987; p. 874.
67. Okinaka, M.; Hamana, Y.; Tokuda, T.; Ohta, J.; Nunoshita, M. MBE growth mode and C incorporation of GeC epilayers on Si(0 0 1) substrates using an arc plasma gun as a novel C source. *J. Cryst. Growth* **2003**, *249*, 78–86. [CrossRef]
68. Lu, H.; Liu, W.; Wang, H.; Liu, X.; Zhang, Y.; Yang, D.; Pi, X. Molecular beam epitaxy growth and scanning tunneling microscopy study of 2D layered materials on epitaxial graphene/silicon carbide. *Nanotechnology* **2023**, *34*, 132001. [CrossRef] [PubMed]
69. Islam, M.R.; Islam, M.S.; Mitul, A.F.; Mojumder, M.R.H.; Jannatul, A.S.M.; Islam, C.S.; Park, J. Superior tunable photocatalytic properties for water splitting in two dimensional GeC/SiC van der Waals hetero bilayers. *Sci. Rep.* **2021**, *11*, 17739. [CrossRef] [PubMed]
70. Tayaba, S.; Sethi, H.; Shahid, H.; Malik, R.; Ikram, M.; Ali, S.; Khaliq, S.; Khan, Q.; Maqbool, M. Silicon-Germanium and carbon-based superconductors for electronic, industrial, and medical applications. *Mater. Sci. Eng. B* **2023**, *290*, 116332. [CrossRef]
71. Bean, J.C. Silicon based semiconductor heterostructures: Column IV bandgap engineering. *Proc. IEEE* **1992**, *80*, 571–587. [CrossRef]
72. Li, X.; Jacobson, H.; Boulle, A.; Chaussende, D.; Henry, A. Double-Position-Boundaries Free 3C-SiC Epitaxial Layers Grown on On-Axis 4H-SiC. *ECS J. Solid State Sci. Technol.* **2014**, *3*, P75. [CrossRef]
73. Xin, B.; Jia, R.X.; Hu, J.C.; Tsai, C.Y.; Lin, H.H.; Zhang, Y.M. A step-by-step experiment of 3C-SiC hetero-epitaxial growth on 4H-SiC by CVD. *Appl. Surf. Sci.* **2015**, *357*, 985–993. [CrossRef]
74. Wu, J.; Qian, S.T.; Huo, T.G.; Zheng, J.X.; Zhang, P.L.; Dai, Y.; Geng, D.S. Effect of PyC Inner Coating on Preparation of 3C-SiC Coating on Quartz Glass by Chemical Vapor Reaction. *Front. Mater.* **2022**, *9*, 897900. [CrossRef]
75. Kaloyeros, A.E.; Arkles, B. Silicon Carbide Thin Film Technologies: Recent Advances in Processing, Properties, and Applications—Part I Thermal and Plasma CVD. *ECS J. Solid State Sci. Technol.* **2023**, *12*, 103001. [CrossRef]
76. Hartmanna, J.M.; Abbadiea, A.; Vineta, M.; Claveliera, L.; Holligera, P.; Lafonda, D.; Semeria, M.N.; Gentile, P. Growth kinetics of Si on fullsheet, patterned and silicon-on-insulator substrates. *J. Cryst. Growth* **2003**, *257*, 19–30. [CrossRef]
77. Lukin, D.M.; Guidry, M.A.; Vučković, J. Silicon Carbide: From Abrasives to Quantum Photonics. *Opt. Photonics News* **2021**, *32*, 34–41. [CrossRef]
78. La Via, F.; Camarda, M.; La Magna, A. Mechanisms of growth and defect properties of epitaxial SiC. *Appl. Phys. Rev.* **2014**, *1*, 031301. [CrossRef]
79. Fraga, M.A.; Bosi, M.; Negri, N. Silicon carbide in microsystem technology—Thin film vs bulk material. In *Advanced Silicon Carbide Devices and Processing*; Saddow, S.E., La Via, F., Eds.; IntechOpen: London, UK, 2015; Chapter 1, p. 3.
80. Ou, H.; Shi, X.; Lu, Y.; Kollmuss, M.; Steiner, J.; Tabouret, V.; Syväjärvi, M.; Wellmann, P.; Chaussende, D. Novel photonic applications of silicon carbide. *Materials* **2023**, *16*, 1014. [CrossRef]
81. Kaloyeros, A.E.; Jové, F.A.; Goff, J.; Arkles, B. Silicon nitride and silicon nitride-rich thin film technologies: Trends in deposition techniques and related applications. *ECS J. Solid State Sci. Technol.* **2017**, *6*, P691. [CrossRef]
82. Kaloyeros, A.E.; Pan, Y.; Goff, J.; Arkles, B. Review—Silicon nitride and silicon nitride-rich thin film technologies: State-of-the-art processing technologies, properties, and applications. *ECS J. Solid State Sci. Technol.* **2020**, *9*, 062006. [CrossRef]
83. Marsi, N.; Majlis, B.Y.; Hamzah, A.; Mohd-Yasin, F. High reliability of MEMS packaged capacitive pressure sensor employing 3C-SiC for high temperature. *Energy Procedia* **2015**, *68*, 471–479. [CrossRef]
84. Feller, T.; Rosenfeldt, S.; Retsch, M. Carbothermal synthesis of micron-sized, uniform, spherical silicon carbide (SiC) particles. *Z. Anorg. Allg. Chem.* **2021**, *647*, 2172–2180. [CrossRef]
85. Kawanishi, S.; Daikoku, H.; Shibata, H.; Yoshikawa, T. Suppressing solvent compositional change during solution growth of SiC using SiC/C gradient crucible. *J. Cryst. Growth* **2021**, *576*, 126382. [CrossRef]

86. Sannodo, N.; Osumi, A.; Kaminaga, K.; Maruyama, S.; Matsumoto, Y. Vapor-liquid-solid-like growth of high-quality and uniform 3C-SiC heteroepitaxial films on alpha-Al_2O_3 (0001) substrates. *CrystEngComm* **2021**, *23*, 1709–1717. [CrossRef]
87. Kukushkin, S.A.; Osipov, A.V. Epitaxial silicon carbide on silicon. method of coordinated substitution of atoms (a review). *Russ. J. Gen. Chem.* **2022**, *92*, 584–610. [CrossRef]
88. Majid, A. A perspective on non-stoichiometry in silicon carbide (review article). *Ceram. Int.* **2018**, *44*, 1277–1283. [CrossRef]
89. Spera, M.; Greco, G.; Lo Nigro, R.; Bongiorno, C.; Giannazzo, F.; Zielinski, M.; La Via, F.; Roccaforte, F. Ohmic contacts on n-type and p-type cubic silicon carbide (3C-SiC) grown on silicon. *Mater. Sci. Semicond. Process.* **2019**, *93*, 295–298. [CrossRef]
90. Galashev, A.Y.; Abramova, K.A. Computer simulation of obtaining thin films of silicon carbide. *Phys. Chem. Chem. Phys.* **2023**, *25*, 3834–3847. [CrossRef] [PubMed]
91. Smith, D.J.; Todd, M.; McMurran, J.; Kouvetakis, J. Structural properties of heteroepitaxial germanium carbon alloys grown on (100) Si. *Philos. Mag. A* **2001**, *81*, 1613–1624. [CrossRef]
92. Schuh, P.; Scholer, M.; Wilhelm, M.; Syvajarvi, M.; Litrico, G.; La Via, F.; Mauceri, M.; Wellmann, P.J. Sublimation growth of bulk 3C-SiC using 3C-SiC-on-Si (100) seeding layers. *J. Cryst. Growth* **2017**, *478*, 159–162. [CrossRef]
93. Ha, C.V.; Ha, L.T.; Hue, D.T.; Nguyen, D.K.; Anh, D.T.; Sanchezd, J.G.; Hoat, D.M. First-principles study of SiC and GeC monolayers with adsorbed non-metal atoms. *RSC Adv.* **2023**, *13*, 14879–14886. [CrossRef]
94. Werninghaus, T.; Friedrich, M.; Cimalla, V.; Scheiner, J.; Goldhahn, R.; Zahn, D.R.T.; Pezold, J. Optical characterization of MBE grown cubic and hexagonal SiC films on Si(111). *Diam. Relat. Mater.* **1998**, *7*, 1385–1389.
95. Yasui, K.; Narita, Y.; Inubushi, T.; Akahane, T. In situ observation of reflection high-energy electron diffraction during the initial growth of SiC on Si using dimethylsilane. *J. Cryst. Growth* **2002**, *237–239*, 1254–1259. [CrossRef]
96. Kosiba, R.; Liday, J.; Ecke, G.; Ambacher, O.; Breza, J.; Vogrincic, P. Quantitative Auger electron spectroscopy of SiC. *Vacuum* **2006**, *80*, 990–995. [CrossRef]
97. Zhao, S.; Chen, J.; Yang, S.; Yan, G.; Shen, Z.; Zhao, W.; Wang, L.; Liu, X.; Sun, G.; Zeng, Y. Effect of temperature on growth of epitaxial layer on semi-insulating 4H-SiC substrate. *J. Cryst. Growth* **2023**, *603*, 127008. [CrossRef]
98. Tang, Z.; Gu, L.; Ma, H.; Dai, K.; Luo, Q.; Zhang, N.; Huang, J.; Fan, J. Study on the Surface Structure of N-Doped 4H-SiC Homoepitaxial Layer Dependence on the Growth Temperature and C/Si Ratio Deposited by CVD. *Crystals* **2023**, *13*, 193. [CrossRef]
99. Sreelakshmi, N.; Umapathy, G.R.; Abhaya, S.; David, C.; Ojha, S.; Amirthapandian, S. Ionization-induced annealing of defects in 3C–SiC: Ion channeling and positron annihilation spectroscopy investigations. *J. Mater. Res.* **2023**, *38*, 1349–1362. [CrossRef]
100. Bayu Aji, L.B.; Stavrou, E.; Wallace, J.B.; Boulle, A.; Debelle, A.; Kucheyev, S.O. Comparative study of radiation defect dynamics in 3C-SiC by X-ray diffraction, Raman scattering, and ion channeling. *Appl. Phys. A* **2019**, *125*, 28. [CrossRef]
101. Boulle, A.; Debelle, A.; Wallace, J.B.; Aji, L.B.; Kucheyev, S.O. The amorphization of 3C-SiC irradiated at moderately elevated temperatures as revealed by X-ray diffraction. *Acta Mater.* **2017**, *140*, 250–257. [CrossRef]
102. Zhang, L.; Jiang, W.; Pan, C.; Fadanelli, R.C.; Ai, W.; Chen, L.; Wang, T. Raman study of amorphization in nanocrystalline 3C–SiC irradiated with C+ and He+ ions. *J. Raman Spectrosc.* **2019**, *50*, 1197–1204. [CrossRef]
103. Lindner, J.K.N.; Tsang, W.M.; Wong, S.P.; Xu, J.B.; Wilson, I.H. XTEM characterization of tungsten implanted SiC thin films on silicon for field emission devices. *Thin Solid Films* **2003**, *427*, 417–421. [CrossRef]
104. Scuderi, V.; Calabretta, C.; Anzalone, R.; Mauceri, M.; La Via, F. Characterization of 4H- and 6H-Like Stacking Faults in Cross Section of 3C-SiC Epitaxial Layer by Room-Temperature µ-Photoluminescence and µ-Raman Analysis. *Materials* **2020**, *13*, 1837. [CrossRef]
105. Feldman, D.W.; Parker, J.H., Jr.; Choyke, W.J.; Patrick, L. Phonon Dispersion Curves by Raman Scattering in SiC, Polytypes 3C, 4H, 6H, 15R, and 21R. *Phys. Rev.* **1968**, *173*, 787. [CrossRef]
106. Yoshida, M.; Onodera, A.; Ueno, M.; Takemura, K.; Shimomura, O. Pressure-induced phase transition in SiC. *Phys. Rev. B* **1993**, *48*, 10587. [CrossRef] [PubMed]
107. Olego, D.; Cardona, M.; Vogl, P. Pressure dependence of the optical phonons and transverse effective charge in 3C-SiC. *Phys. Rev. B* **1982**, *25*, 3878. [CrossRef]
108. Talwar, D.N. Spectroscopic Investigations for the Dynamical Properties of Defects in Bulk and Epitaxially Grown 3C-SiC/Si (100). In *Handbook of Silicon Carbide Materials and Devices*; Feng, Z.C., Ed.; CRC Press: Boca Raton, FL, USA, 2023; Chapter 3.
109. Sahin, H.; Cahangirov, S.; Topsakal, M.; Bekaroglu, E.; Akturk, E.; Senge, R.T.; Ciraci, S. Monolayer honeycomb structures of group-IV elements and III-V binary compounds: First-principles calculations. *Phys. Rev. B* **2009**, *80*, 155453. [CrossRef]
110. Lin, S.S. Light-emitting two-dimensional ultrathin silicon carbide. *J. Phys. C* **2012**, *116*, 3951–3955. [CrossRef]
111. Jankousky, M.; Garrity, E.M.; Stevanovic, V. Polymorphism of group-IV carbides: Structures, (meta)stability, electronic, and transport properties. *Phys. Rev. Mater.* **2023**, *7*, 053606. [CrossRef]
112. Karch, K.; Bechstedt, F.; Pavone, P.; Strauch, D. Pressure-dependent properties of SiC polytypes. *Phys. Rev. B* **1996**, *53*, 13400. [CrossRef]
113. Miao, M.S.; Prikhodko, M.; Lambrecht, W.R.L. Changes of the geometry and band structure of SiC along the orthorhombic high-pressure transition path between the zinc-blende and rocksalt structures. *Phys. Rev. B* **2002**, *66*, 064107. [CrossRef]
114. Jiang, J.; Zheng, W.; Xiao, H.Y.; Liu, Z.J.; Zu, X.T. A comparative study of the mechanical and thermal properties of defective ZrC, TiC and SiC. *Sci. Rep.* **2017**, *7*, 9344. [CrossRef]

115. Lu, Y.P.; Wei, H.D.; Zhu, J.; Yang, X.D. First-principles study of pressure-induced phase transition in silicon carbide. *Physica B* **2008**, *403*, 3543–3546. [CrossRef]
116. Hao, A.; Yang, X.C.; Wang, X.M.; Zhu, Y.; Liu, X.; Liu, R.P. First-principles investigations on electronic, elastic, and optical properties of XC (X = Si, Ge, and Sn) under high pressure. *J. Appl. Phys.* **2010**, *108*, 063531. [CrossRef]
117. Pandey, R.; Rérat, M.; Darrigan, C.; Causà, M. A theoretical study of stability, electronic, and optical properties of GeC and SnC. *J. Appl. Phys.* **2000**, *88*, 6462–6466. [CrossRef]
118. Khenata, R.; Baltache, H.; Sahnoun, M.; Driz, M.; Rérat, M.; Abbar, B. Full potential linearized augmented plane wave calculations of structural and electronic properties of GeC, SnC and GeSn. *Physica B* **2003**, *336*, 321–328. [CrossRef]
119. Sekkal, W.; Zaoui, A. Predictive study of thermodynamic properties of GeC. *New J. Phys.* **2002**, *4*, 9. [CrossRef]
120. Varshney, D.; Shriya, S.; Varshney, M.; Singh, N.; Khenata, R. Elastic and thermo-dynamical properties of cubic (3C) silicon carbide under high pressure and high temperature. *J. Theor. Appl. Phys.* **2015**, *9*, 221–249. [CrossRef]
121. Karch, K.; Bechstedt, F.; Pavone, P.; Strauch, D. Pressure-dependent dynamical and dielectric properties of cubic SiC. *J. Phys. Condens. Matter* **1996**, *8*, 2945. [CrossRef]
122. Karch, K.; Pavone, P.; Windl, W.; Schütt, O.; Strauch, D. Ab initio calculation of structural and lattice-dynamical properties of silicon carbide. *Phys. Rev. B* **1994**, *50*, 17054. [CrossRef]
123. Wang, C.Z.; Yu, R.C.; Krakauer, H. Pressure dependence of Born effective charges, dielectric constant, and lattice dynamics in SiC. *Phys. Rev. B* **1996**, *53*, 5430. [CrossRef] [PubMed]
124. Souadkia, M.; Bennecer, B.; Kalarasse, F. Elastic, vibrational and thermodynamic properties of α-Sn based group IV semiconductors and GeC under pressure. *J. Phys. Chem. Solids* **2013**, *74*, 1615–1625. [CrossRef]
125. Zhang, X.; Quan, S.; Ying, C.; Li, Z. Theoretical investigations on the structural, lattice dynamical and thermodynamic properties of XC (X = Si, Ge, Sn). *Solid State Commun.* **2011**, *151*, 1545–1549. [CrossRef]
126. Zhang, P.; Crespi, V.H.; Chang, E.; Louie, S.G.; Cohen, M.L. Theory of metastable group-IV alloys formed from CVD precursors. *Phys. Rev. B* **2001**, *64*, 235201. [CrossRef]
127. Kunc, K. Dynamique de réseau de composés $A^{N}B^{8\text{-}N}$ présentant la structure de la blende. *Ann. Phys.* **1973**, *8*, 319–401.
128. Plumelle, P.; Vandevyver, M. Lattice dynamics of ZnTe and CdTe. *Phys. Stat. Sol.* **1976**, *73*, 271–281. [CrossRef]
129. Serrano, J.; Strempfer, J.; Cardona, M.; Schwoerer-Böhning, M.; Requardt, H.; Lorenzen, M.; Stojetz, B.; Pavone, P.; Choyke, W.J. Determination of the phonon dispersion of zinc blende (3C) silicon carbide by inelastic X-ray scattering. *Appl. Phys. Lett.* **2002**, *80*, 4360–4362. [CrossRef]
130. Adachi, S. *Properties of Semiconductor Alloys: Group-IV, III–V and II–VI, Semiconductors*; Wiley: Hoboken, NJ, USA, 2009.
131. Kutty, A.P.G. Phonons in mixed crystals. *Solid State Commun.* **1974**, *14*, 213–215. [CrossRef]
132. Talwar, D.N. On the pressure-dependent phonon characteristics and anomalous thermal expansion coefficient of 3C-SiC. *Mater. Sci. Eng. B* **2017**, *226*, 1–9. [CrossRef]
133. Talwar, D.N.; Vandevyver, M. Pressure-dependent phonon properties of III-V compound semiconductors. *Phys. Rev. B* **1990**, *41*, 12129–12139. [CrossRef]
134. Talwar, D.N.; Vandevyver, M.; Kunc, K.; Zigone, M. Lattice dynamics of zinc chalcogenides under compression: Phonon dispersion, mode Grüneisen, and thermal expansion. *Phys. Rev. B* **1981**, *24*, 741. [CrossRef]
135. Talwar, D.N.; Feng, Z.C.; Becla, P. Impurity-induced phonon disordering in $Cd_{1-x}Zn_xTe$ ternary alloys. *Phys. Rev. B* **1993**, *48*, 17064. [CrossRef]
136. Talwar, D.N.; Roughani, B.; Pellegrino, J.G.; Amirtharaj, P.; Qadri, S.B. Study of phonons in semiconductor superlattices by Raman scattering spectroscopy and microscopic model calculation. *Mater. Sci. Eng. B* **1997**, *44*, 143–146. [CrossRef]
137. Talwar, D.N. *Dilute III–V Nitride Semiconductors and Material Systems—Physics and Technology*; Erol, A., Ed.; Springer Series in Materials Science; Springer: Berlin/Heidelberg, Germany, 2008; Volume 105, Chapter 9; p. 222.
138. Boer, K.W.; Pohl, U.W. Phonon-induced thermal properties. In *Semiconductor Physics*; Springer International Publishing: Cham, Switzerland, 2014.
139. Grimm, A.; Maradudin, A.A.; Ipatova, I.P.; Subashiev, A.V. Impurity vibrations of copper defect complexes in gallium arsenide crystals. *J. Phys. Chem. Solids* **1972**, *33*, 775–796. [CrossRef]
140. Grimm, A. Infrared absorption by point defects in gallium arsenide. *J. Phys. C* **1972**, *5*, 1883. [CrossRef]
141. Vetelino, J.F.; Mitra, S.S. Lattice dynamics of cubic SiC. *Phys. Rev.* **1969**, *178*, 1349. [CrossRef]
142. Vetelino, J.F.; Mitra, S.S.; Namjoshi, K.V. Lattice Dynamics of ZnTe: Phonon Dispersion, Multi-phonon Infrared Spectrum, Mode Grüneisen Parameters, and Thermal Expansion. *Phys. Rev. B* **1970**, *2*, 967. [CrossRef]
143. Banerjee, R.; Varshni, Y.P. Lattice dynamics of III-V compounds. *Can. J. Phys.* **1969**, *47*, 451–462. [CrossRef]
144. Banerjee, R.; Varshni, Y.P. Lattice Dynamics and Thermodynamic Properties of β-ZnS, GaP and β-SiC. *J. Phys. Soc. Jpn.* **1971**, *30*, 1015–1021. [CrossRef]
145. Kellermann, E.W. Theory of the vibrations of the sodium chloride lattice. *Philos. Trans. R. Soc. Lond. A* **1944**, *238*, 513–548.
146. Cochran, W. Crystal stability and the theory of ferroelectricity. *Adv. Phys.* **1960**, *9*, 387–423. [CrossRef]
147. Merten, L. Zur Ultrarot-Dispersion zweiachsiger und einachsiger Kristalle II. Das außerordentliche Ultrarot-Spektrum von a-Quarz. *Z. Naturforsch A* **1958**, *13*, 662–679. [CrossRef]
148. Reparaz, J.S.; Muniz, L.R.; Wagner, M.R.; Goñi, A.R.; Alonso, M.I.; Hoffmann, A.; Meyer, B.K. Reduction of the transverse effective charge of optical phonons in ZnO under pressure. *Appl. Phys. Lett.* **2010**, *96*, 231906-03. [CrossRef]

149. Goldberg, Y.; Levinshtein, M.E.; Rumyantsev, S.L.; Levinshtein, M.E.; Rumyantsev, S.L.; Shur, M.S. (Eds.) *Properties of Advanced Semiconductor Materials GaN, AlN, SiC, BN, SiC, SiGe*; John Wiley & Sons, Inc.: New York, NY, USA, 2001; pp. 93–148.
150. Barin, I. *Thermochemical Data of Pure Substances Part II*; VCH: New York, NY, USA, 1989.
151. Hu, Y.-F.; Kong, F.-J.; Zhou, C. Structures and Thermodynamic Properties of 3C-SiC Compound. *Acta Phys. Chim. Sin.* **2008**, *24*, 1845–1849.
152. Pässler, R. Limiting Debye temperature behavior following from cryogenic heat capacity data for group-IV, III–V, and II–VI materials. *Phys. Status Solidi B* **2010**, *247*, 77–92. [CrossRef]
153. Pässler, R. Basic moments of phonon density of states spectra and characteristic phonon temperatures of group IV, III–V, and II–VI materials. *J. Appl. Phys.* **2007**, *101*, 093513-12. [CrossRef]
154. Persson, C.; Lindefelt, U. Relativistic band structure calculation of cubic and hexagonal SiC polytypes. *J. Appl. Phys.* **1997**, *82*, 5496–5508. [CrossRef]

Disclaimer/Publisher's Note: The statements, opinions and data contained in all publications are solely those of the individual author(s) and contributor(s) and not of MDPI and/or the editor(s). MDPI and/or the editor(s) disclaim responsibility for any injury to people or property resulting from any ideas, methods, instructions or products referred to in the content.

inorganics

MDPI

Review

Introductory Overview of Layer Formation Techniques of Ag Nanowires on Flexible Polymeric Substrates

Heebo Ha [1], Nadeem Qaiser [2] and Byungil Hwang [3,*]

[1] Department of Intelligent Semiconductor Engineering, Chung-Ang University, Seoul 06974, Republic of Korea
[2] Material Science and Engineering, Physical Science and Engineering Division, King Abdullah University of Science and Technology (KAUST), Thuwal 04700, Saudi Arabia
[3] School of Integrative Engineering, Chung-Ang University, Seoul 06974, Republic of Korea
* Correspondence: bihwang@cau.ac.kr

Abstract: Ag nanowire electrodes are promising substitutes for traditional indium tin oxide (ITO) electrodes in optoelectronic applications owing to their impressive conductivity, flexibility, and transparency. This review provides an overview of recent trends in Ag nanowire electrode layer formation, including key developments, challenges, and future prospects. It addresses several challenges in integrating Ag nanowires into practical applications, such as scalability, cost-effectiveness, substrate compatibility, and environmental considerations. Additionally, drawing from current trends and emerging technologies, this review explores potential avenues for improving Ag nanowire layer-forming technologies, such as material advancements, manufacturing scalability, and adaptability to evolving electronic device architectures. This review serves as a resource for researchers, engineers, and stakeholders in nanotechnology and optoelectronics, and underscores the relationship between advancements in patterning and the application of Ag nanowire electrodes. Through an examination of key developments, challenges, and future prospects, this review contributes to the collective knowledge base and encourages continued innovation in the ever-evolving realm of Ag nanowire-based optoelectronics.

Keywords: Ag nanowires; layer-forming technology; transparent conductive films; coating

1. Introduction

In the evolving fields of electronic engineering and optoelectronics, scientists and engineers are constantly searching for materials that can outperform indium tin oxide (ITO). The scarcity and cost fluctuations of indium, along with environmental concerns related to its extraction, challenge its long-term viability. Additionally, the adaptability of ITO to emerging technologies is limited by its brittleness, which reduces its durability, especially on flexible substrates, and its low conductivity at lower thicknesses [1–8]. As a result, there is increasing demand for electronic components that offer not only high performance but also environmental sustainability. In this context, Ag nanowires have emerged as a promising alternative owing to their excellent electrical conductivity, inherent flexibility, and outstanding transparency [9–17], which makes them a leading choice for next-generation electronics such as flexible displays and wearable devices [18–25]. Consequently, extensive research has been conducted on the synthesis, characterization, and applications of Ag nanowires (Figure 1) [26].

Layer formation technologies play a crucial role in unlocking the full potential of Ag nanowire electrodes, especially in the context of how form and function interact. By overcoming the technical challenges associated with Ag nanowire layer formation, the theoretical benefits of Ag nanowires can be translated into practical applications [5,12,27–32]. This review examines recent developments in the formation of Ag nanowire electrode layers [9,10,13,20,33–35] and highlights the associated challenges and opportunities, including scalability, cost-efficiency, material compatibility, and environmental considerations. The primary aim is to provide an overview of the multifaceted challenges faced by researchers and industry in utilizing Ag nanowires in actual devices. Furthermore, this review explores the future potential of Ag nanowire layers and their integration into evolving electronic designs and the opportunities they offer for improving the electrode potential of optoelectronic devices. It aims to provide accessible knowledge beneficial to researchers, engineers, industry professionals, and policymakers.

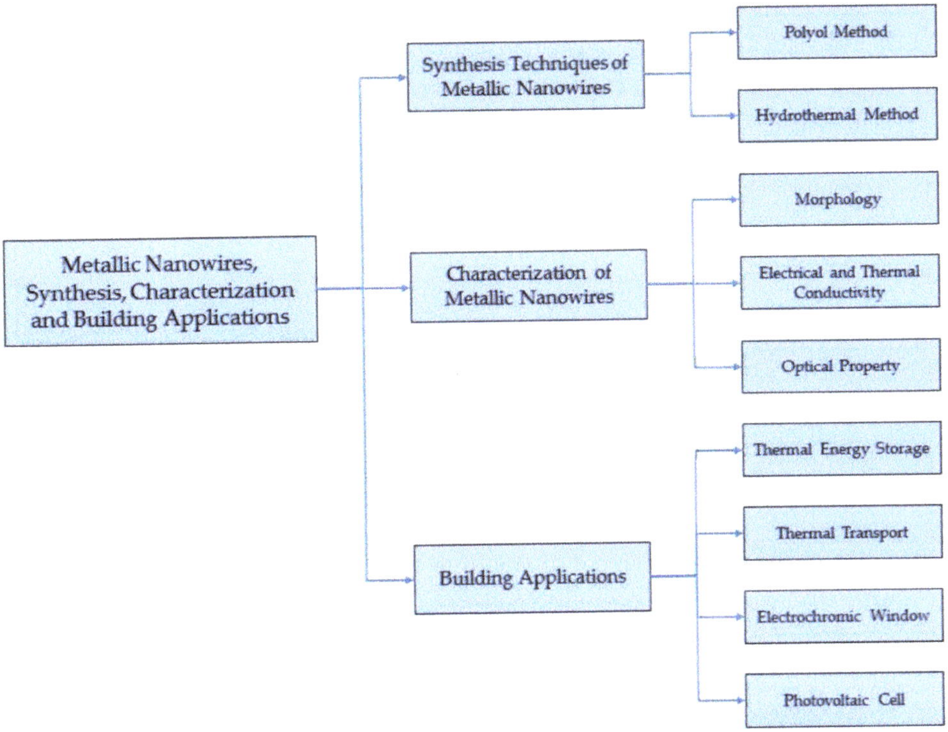

Figure 1. Synthesis, characterization, and applications of Ag nanowires [26]. Reproduced from Ref. [26] under the Creative Commons Attribution 4.0 International (CC BY 4.0) License.

2. Methods for Ag nanowire Layer Formation

The development of transparent conductive films on flexible polymeric substrates has led to significant advancements in coating techniques to improve the performance and applicability of the resulting films. Ag nanowires are increasingly recognized as candidates for transparent conductive layers in diverse applications, such as flexible electronics and solar cells, owing to their excellent electrical conductivity and optical transparency. Choosing an appropriate coating method is crucial, as it greatly affects the characteristics of the film and its suitability for a given application. The following sections explore different coating methods, including their operating principles, recent advancements and challenges, and the interaction between the coating method and the characteristics of the base material.

2.1. Spray Coating

Spray coating is a versatile and cost-effective method for depositing Ag nanowires on both rigid and flexible substrates [14,36–38]. This technique is simple and able to cover large areas uniformly. Recent advancements in spray coating have focused on refining precision, particularly with regard to nanowire density and orientation. Researchers are exploring novel approaches to optimize the spray parameters, such as the solvent composition and nozzle design, to enhance the homogeneity of the resultant films. In a notable example of these achievements, Zhu et al. developed a solvent-welding technique employing Ag nanowire spray coating (Figure 2) [36]. Ag nanowire networks combined with poly(vinylpyrrolidone) (PVP) and camphorquinone were spray coated onto a substrate containing ethylene glycol. By applying a current to the Ag nanowire network, the electrical resistance was reduced by a significant 96%. The versatility of spray-coating in terms of material selection and scalability makes it a valuable solution for wide-ranging applications, from small-scale laboratory research to large-scale industrial manufacturing. The ability to generate uniform and transparent conductive films through spray coating is particularly promising for the advancement of technologies such as flexible displays, touchscreens, and solar cells [14,36–38]. Furthermore, this method is cost-effective for the production of transparent conductive films. Advances in the spray-coating of Ag nanowire films on flexible polymeric substrates will facilitate the fabrication of next-generation transparent conductive films for diverse applications.

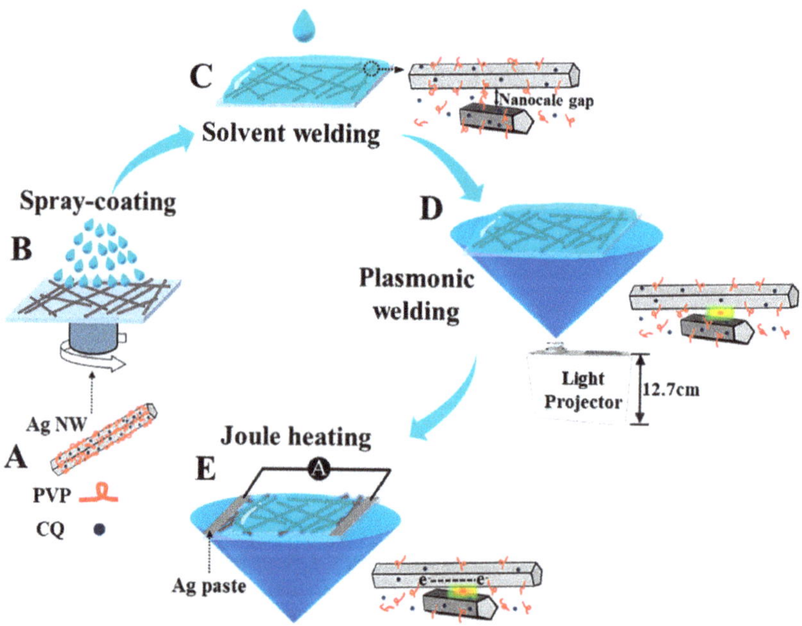

Figure 2. Ag nanowire layer formation using spray coating [36]. (**A**) Schematic of Ag nanowire synthesis with PVP (red lines) and camphorquinone (CQ; blue dots) capping; (**B**) fabrication of Ag nanowire network via spray coating; (**C**) solvent welding and corresponding state of Ag nanowire junctions; (**D**) solvent-based plasmonic welding; and (**E**) combined solvent-based plasmonic and Joule-heating welding. Reproduced from Ref. [36] under the Creative Commons Attribution 4.0 International (CC BY 4.0) License.

2.2. Dip Coating

Dip coating involves submerging a substrate in an Ag nanowire solution [37,39–42]. This method offers excellent control over the film thickness; however, achieving uniformity over large areas remains challenging. Therefore, recent research on dip-coating is focused on overcoming this limitation. In a recent study, Zeng et al. used dip coating to fabricate Ag nanowire electrodes on polyethylene terephthalate (PET) fabrics (Figure 3) [43], followed by hot pressing to improve the adhesion of the Ag nanowires to the fabric. This approach resulted in fabric with excellent electrical conductivity (of 464.2 S/m); strong electromagnetic shielding (17 dB); good strain sensing performance; minimal resistance changes (below −15%); and outstanding Joule-heating performance (110 °C at a current of 0.08 A). In the work by Choi et al., aligned Ag nanowire electrodes were fabricated with the dip-coating process [42]. They fabricated differently aligned Ag nanowire electrodes for flexible strain sensors, which were aligned longitudinally, parallel to the alignment direction, and the other aligned laterally, perpendicular to it (Figure 4). The sensor performance results indicated that the strain sensor with the longitudinally aligned Ag nanowire electrodes exhibited a gauge factor (GF) of 89.99 under 25% tensile strain, surpassing the GF of 12.71 of that with laterally aligned Ag nanowire electrodes. Other innovations such as advanced withdrawal techniques and modifications to the solution rheology have also been explored to enhance the uniformity of dip-coated films. Despite the challenges in achieving large-scale uniformity, dip coating remains a valuable tool, especially for applications and electronic components that require precise thickness control [37,39–42].

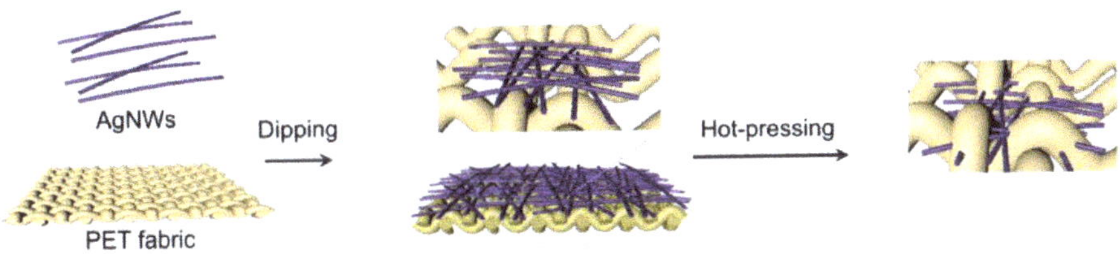

Figure 3. Formation of Ag nanowire (AgNW) layers using dip coating. Schematic diagram of dip-coating and hot-pressing processes for embedding Ag nanowires into PET fabric [43]. Reproduced from Ref. [43] under the Creative Commons Attribution 4.0 International (CC BY 4.0) License.

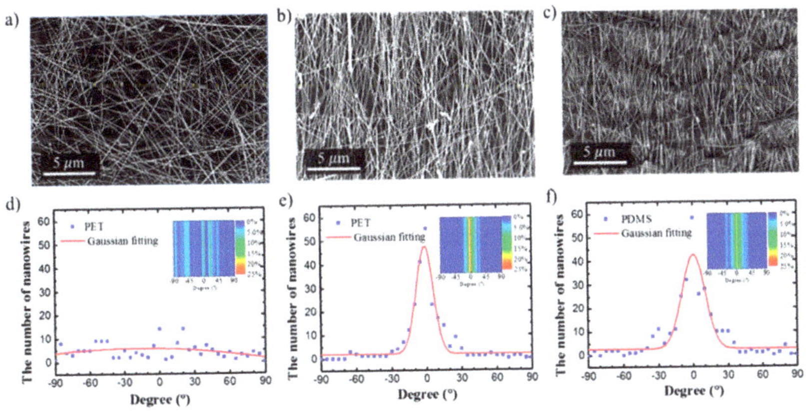

Figure 4. Comparison of the degree of alignment of silver nanowires [42]: (**a**) SEM image of unaligned silver nanowires; (**b**) SEM image of silver nanowires aligned using temperature-controlled dip coating process; (**c**) SEM image of silver nanowires transferred to PDMS; (**d**) analysis of degree of alignment of unaligned silver nanowires; (**e**) analysis of degree of alignment of silver nanowires aligned using temperature-controlled dip coating process; and (**f**) analysis of degree of alignment of transferred silver nanowires. Insets of (**d**–**f**) show the distribution of nanowires according to the angle, and the amount of silver nanowires is expressed in color. Reproduced from Ref. [42] under the Creative Commons Attribution 4.0 International (CC BY 4.0) License.

2.3. Roll-to-Roll Coating

Roll-to-roll (R2R) coating is a popular method of producing transparent conductive films, particularly in large-scale manufacturing [39,44–47]. This method involves continuously applying an Ag nanowire solution to a flexible substrate as it moves between rolls. The continuous nature of R2R coating makes it highly efficient for mass production. Consequently, it is particularly promising for products that require high-volume outputs, including flexible electronics and large-area photovoltaic cells. Furthermore, it is compatible with existing manufacturing processes, which offers significant advantages. Recently, Jeong et al. created a transparent conductive film made of an Ag nanowire-PVP composite through R2R coating (Figure 5) [48]. They achieved selective calendering by continuous R2R patterning using an embossed pattern roll, resulting in a pattern line width of 0.1 mm and spacing of 1 mm. R2R coating also has the ability to align the Ag nanowires by adjusting the coating conditions (Figure 6) [49]. Current research on R2R coating is focused on further innovations such as optimizing the drying process and improving the substrate-handling systems, which will enhance its efficiency and applicability to diverse industrial applications [39,44–47].

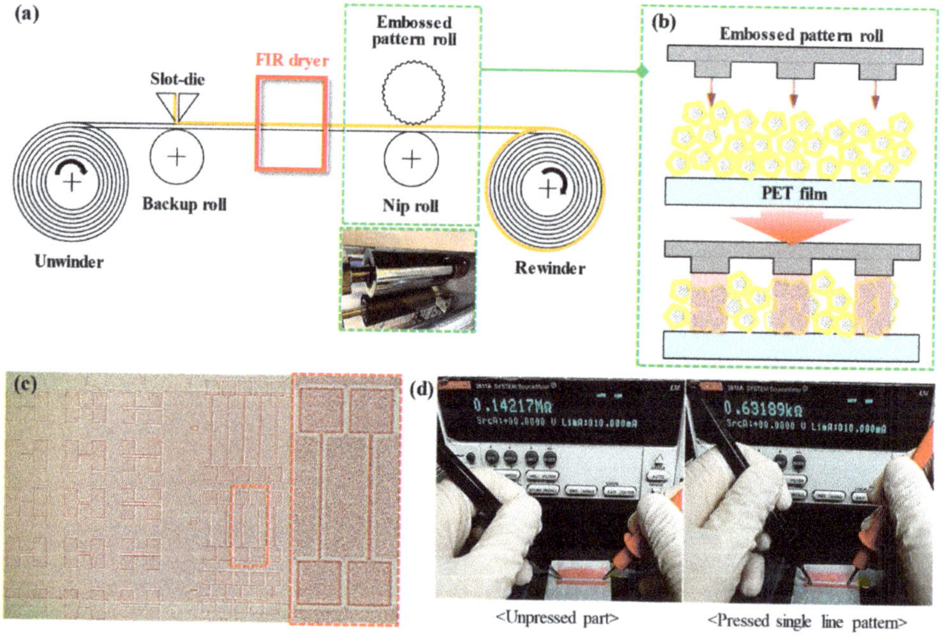

Figure 5. R2R-processed continuous patterning by selective calendering: (**a**) schematic illustration of patterning via R2R manufacturing; (**b**) selective calendering mechanism of Ag nanowire-PVP transparent conductive film using an embossed pattern roll; (**c**) pressure-sensitive paper pressed by an embossed pattern roll; and (**d**) comparison of the resistance between the unpressed and pressed part in a single-line pattern. Reproduced from Ref. [48] under the Creative Commons Attribution 4.0 International (CC BY 4.0) License.

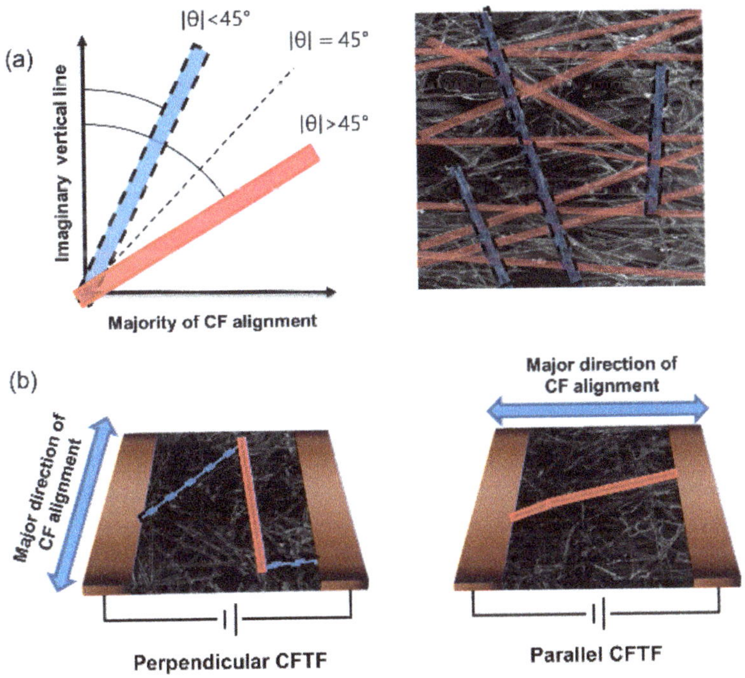

Figure 6. (**left**) (**a**) aligned angle of carbon fiber to an imaginary perpendicular line—carbon fibers with aligned angle less than 45° are colored red, and above 45° are colored blue; (**right**) (**a**) overlaid colored surface FESEM image of carbon fiber thin film [49]. (**left**) (**b**) schematic of electron pathway is overlaid on FESEM image of CF thin film, in which majority of CFs are aligned perpendicular to direction of applied voltage (perpendicular CFTF); (**right**) (**b**) majority of CFs aligned parallel to direction of applied voltage (parallel CFTF) [49]. Reproduced from Ref. [49] under the Creative Commons Attribution 4.0 International (CC BY 4.0) License.

2.4. Spin Coating

Spin coating is a prevalent method of applying thin films to substrates at laboratory scale [50–53]. This method involves spinning a substrate to evenly distribute an Ag nanowire solution over it, which offers precise control over the film thickness. However, its scalability is somewhat limited compared to other coating techniques. Current research on spin coating is focused on expanding its applicability by investigating new spinning techniques, exploring alternative solvents, and optimizing spin parameters to achieve larger-scale uniformity without compromising precision. For example, Zhang et al. prepared an Ag nanowire electrode by spin coating (Figure 7) [54]. They compared four common experimental methods, that is, Mayer rod coating, spin coating, spray coating, and vacuum filtration, to create transparent conductive films using a single type of Ag nanowire. Among the coating methods, spin coating was found to be well-suited for preparing small-sized Ag nanowire films that displayed excellent bending stability. By contrast, spray coating required precise control of the process parameters such as spray distance and traveling speed to obtain uniform Ag nanowire coatings; vacuum filtration required longer times to form Ag nanowire layers; and Mayer rod coating posed the risk of scratching the substrate owing to direct rod contact. Therefore, spin coating remains a valuable technique in research and development, especially for producing uniform films for in-depth electrical, optical, and mechanical analyses [50–53]. In addition, Lee et al. found that the spin coating can produce differently aligned Ag nanowire electrodes by adjusting the spin coating conditions (Figure 8) [55].

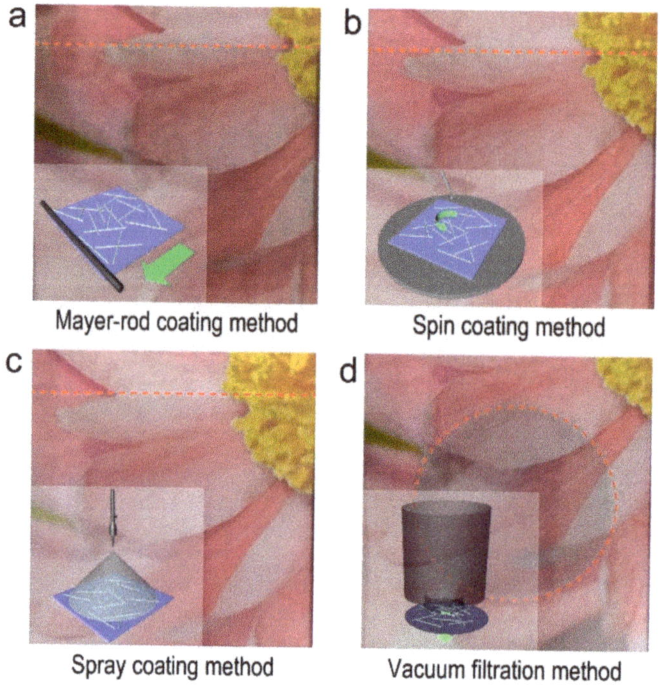

Figure 7. Photographs of Ag nanowire-based films prepared through (**a**) Mayer rod coating, (**b**) spin coating, (**c**) spray coating, and (**d**) vacuum-filtration methods. Insets are corresponding diagrams of these four methods. Reproduced from Ref. [54] under the Creative Commons Attribution 4.0 International (CC BY 4.0) License.

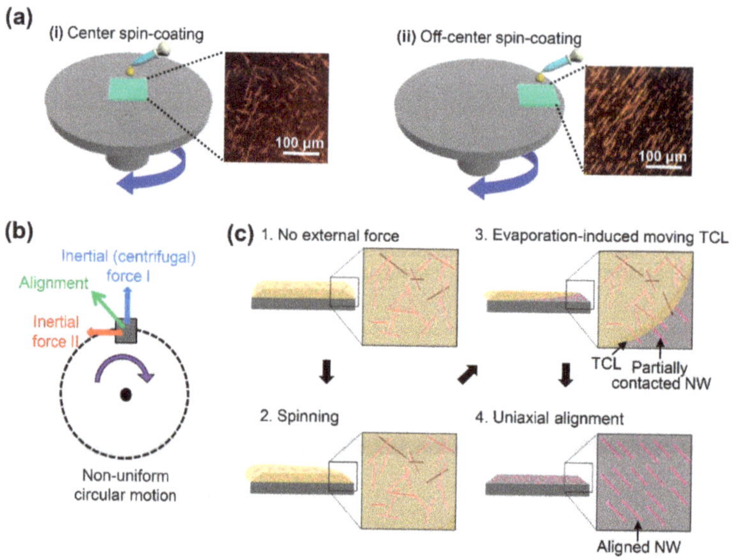

Figure 8. Schematic diagrams of the overall spin-coating processes and their effects upon NW alignment [55]: (**a**) the conventional spin-coating setup (i) and the proposed off-center spin-coating

setup (ii) with inset polarized optical microscope images of the as-deposited Si nanowires; (b) the forces involved in the off-center spin-coating mechanism, including Inertial (centrifugal) Force I, due to centripetal acceleration (blue), Inertial Force II (red), due to tangential acceleration, and the resultant force (green); (c) the sequential influence of the resultant force upon the uniaxial alignment of the NWs that are in partial contact with the substrate surface. Reproduced from Ref. [55] under the Creative Commons Attribution 4.0 International (CC BY 4.0) License.

2.5. Doctor-Blade Coating

Doctor-blade coating is a precise technique for spreading Ag nanowire solutions onto substrates using a specialized blade [56–58]. This method offers meticulous control over the film thickness and uniformity, making it particularly promising for applications where these characteristics are paramount. Recent developments in doctor-blade coating have focused on improving the blade design and exploring innovative materials for the blade itself to refine the coating process. Yoon et al. studied the effect of doctor-blading conditions on the electrical properties of Ag nanowire electrodes (Figure 9) [59]. The blade height and speed were the main parameters that determined the electrical properties of the Ag nanowire electrodes. They found that a lower blade height correlated with lower electrical resistance, and that the optimal blade speed was 20 mm/s. The ability to tailor the film properties using this technique makes it a valuable tool in both research and industrial settings. Consequently, doctor-blade coating is likely to see increasing use to meet the growing demand for tailored transparent conductive films [56–58].

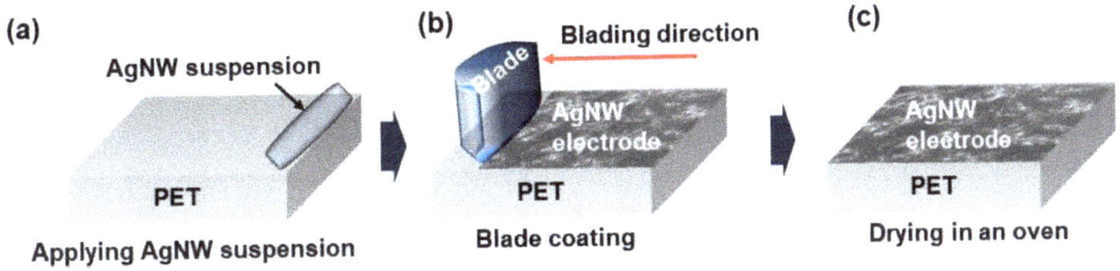

Figure 9. Schematic of the coating process for the production of transparent Ag nanowire (AgNW) electrodes using a doctor-blade system: (a) applying an AgNW suspension, (b) blade coating, and (c) drying in an oven. Reproduced from Ref. [59] under the Creative Commons Attribution 4.0 International (CC BY 4.0) License.

Doctor-blade coating shows excellent promise for the large-scale and cost-effective production of transparent conductive films on flexible polymeric substrates for applications including flexible electronics, touchscreens, and solar cells. This underscores the versatility and practicality of doctor-blade coating. Current research in this area is focused on exploring alternative materials, sustainable fabrication techniques, and integration with emerging technologies, highlighting the potential for continued innovation in doctor-blade-coated transparent conductive films.

2.6. Inkjet Printing

Inkjet printing is a digital deposition method that has gained popularity for its ability to precisely deposit Ag nanowire ink in controlled patterns on diverse substrates [60–63]. This technique enables intricate designs and patterns to be created, making it ideal for producing electronic circuits and displays. Recent innovations in inkjet printing technologies have focused on enhancing the printing resolution, exploring new ink formulations, and expanding substrate compatibility. For example, Wu et al. prepared Ag nanowire flexible transparent conductive films by inkjet printing (Figure 10) [64]. They tailored various factors, such as the surface tension and viscosity of the ink and the contact angle between the Ag nanowire ink droplet and PET substrate, and investigated the effect on the electrical properties of the prepared Ag nanowire layer. The best optical and electrical properties were achieved when using an Ag nanowire ink with a concentration of 0.38–0.57 mg/mL and a post-coating heat treatment at 60 °C for 10 min. In addition, Reenaers et al. showed that the final performance of electrodes fabricated by ink jet printing was mainly influenced by the post-sintering conditions (Figure 11) [65]. The adaptability of inkjet printing to different substrates, including flexible ones, makes it a promising method for applications requiring intricate designs and patterns, including advanced electronic devices and wearable technologies [60–63].

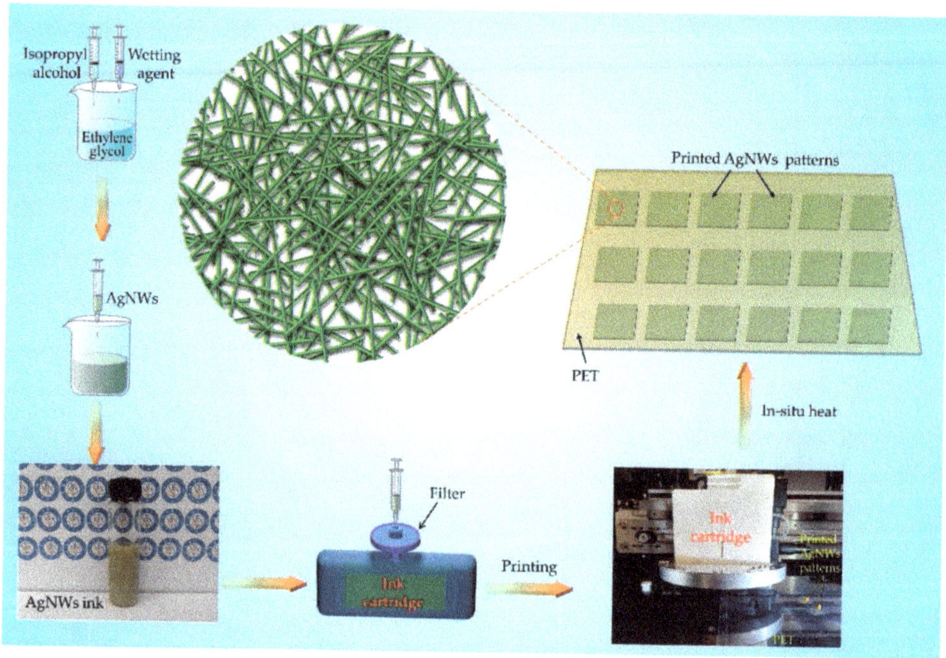

Figure 10. Schematic diagrams of Ag nanowire (AgNW)-based conductive ink fabrication and the inkjet printing process. A schematic of the deposited Ag nanowire flexible transparent conductive film is shown in the center. Reproduced from Ref. [64] under the Creative Commons Attribution 4.0 International (CC BY 4.0) License.

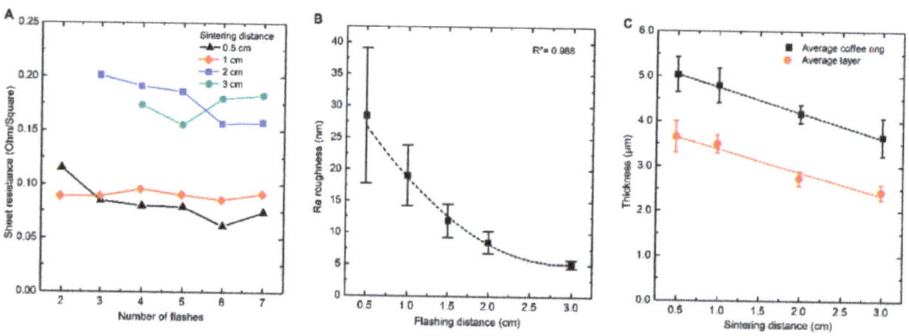

Figure 11. Sintering at shorter distances results in lower sheet resistance and requires a smaller number of flashes [65] (**A**). The higher light intensity at shorter distances causes a higher surface roughness; it exponentially decreases with increasing sintering distance according to the decrease of light intensity. (Every datapoint consists of three corresponding samples, with every sample measured on two locations with the profilometer (line measurement); the error bars represent the standard deviation

of each population). (**B**). The layer thickness decreases linearly when increasing the sintering distance for JS-A102A at a flashing intensity of 100% and a flash and cooling time of 2 s. (Each data point is representing the average of 8 to 12 measurements distributed over four to six different samples. The error bars represent the standard deviation of each population). (**C**). Reproduced from Ref. [65] under the Creative Commons Attribution 4.0 International (CC BY 4.0) License.

2.7. Electrospinning

Electrospinning is a less conventional yet innovative approach for preparing Ag nanowire films. This technique involves generating a nanofiber web on a substrate that captures Ag nanowires [66–70]. This method is particularly relevant for specialized applications that require the formation of nanofiber networks. Recent research in this area has focused on enhancing the nanofiber control, optimizing the solution parameters, and expanding the range of compatible substrates. Wang et al. deposited Ag nanowires on a stretchable thermoplastic polyurethane (TPU) substrate to fabricate a stretchable pressure sensor (Figure 12) [66]. The electrospun Ag nanowire layer exhibited high directionality, which was advantageous for directional sensing. The Ag nanowire/TPU electrode pressure sensor demonstrated excellent sensing performance, with a sensitivity of 7.24 kPa^{-1} within the range of 9.0×10^{-3} to 0.98 kPa. Notably, the versatility and potential of electrospinning for applications beyond traditional transparent conductive film applications, such as sensors and advanced filtration systems, demonstrates the impact of further research in this field [66–70].

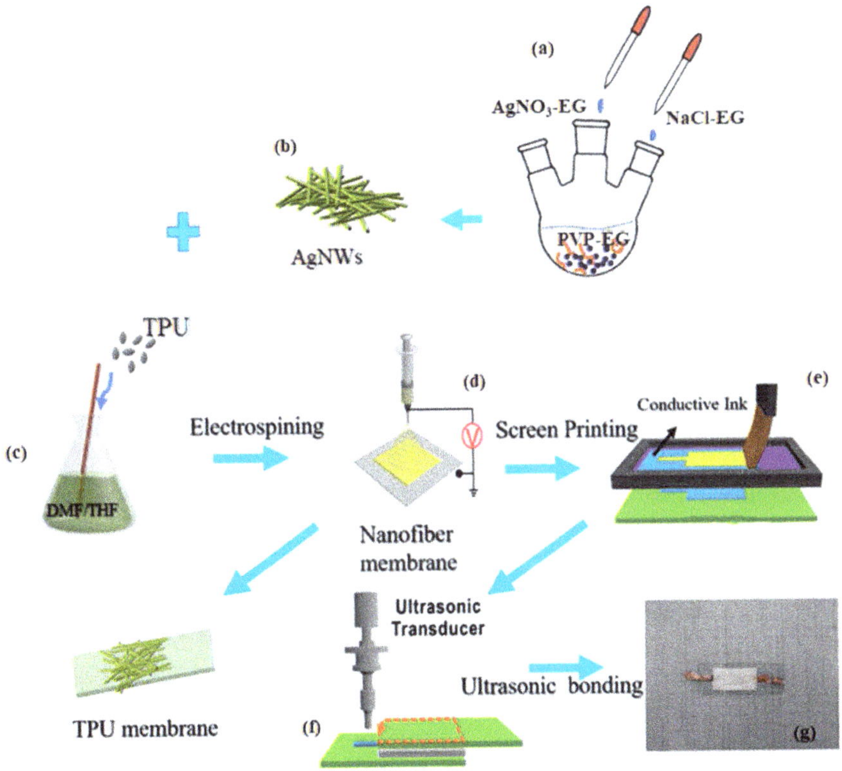

Figure 12. Fabrication of nanofiber membrane-based flexible capacitive pressure sensor: (**a**,**b**) preparation of Ag nanowires (AgNWs); (**c**) preparation of spinning solution; (**d**) electrospinning process; and (**e**–**g**) assembly of sensors—electrode printing and ultrasonic welding. Reproduced from Ref. [66] under the Creative Commons Attribution 4.0 International (CC BY 4.0) License.

2.8. Gravure Printing

Gravure printing, also known as rotogravure printing, is a high-throughput technique for transferring ink from engraved cells or depressions on a printing cylinder to a substrate [71–74]. The engraved cells are filled with ink, after which excess ink is removed from the surface, leaving ink only in the depressions. The substrate is then rolled over the cylinder to transfer the patterned ink. This technique is particularly beneficial in the field of printed electronics, as it offers an efficient and cost-effective method of producing conductive patterns, including electrodes, on various substrates. When applied to the deposition of Ag nanowires, gravure printing combines the advantages of traditional intaglio printing with the unique properties of Ag nanowires, resulting in the preparation of flexible and transparent conductive films suitable for diverse applications [75]. Notably, gravure printing allows precise control over the amount of ink transferred, ensuring uniformity of the electrode pattern. This is particularly important for applications such as electronic devices, because the performance is highly dependent on the consistency and reproducibility of the deposited film [71–74].

Ag nanowire ink for gravure printing is prepared by dispersing Ag nanowires in a solution along with suitable binders and solvents. The viscosity, surface tension, and other rheological properties of the ink must be optimized to achieve good transfer and pattern formation during printing. In addition, the use of stabilizing agents in the ink is essential to prevent nanowire agglomeration and maintain a stable dispersion for uniform printing. Li et al. fabricated Ag nanowire layers on glass substrates through gravure printing (Figure 13) [76]. They used an organic–inorganic nanohybrid ink prepared by incorporating an alkoxysilane-functionalized amphiphilic polymer precursor into a SiO_2–TiO_2 hybrid network (denoted as AGPTi) [76]. The gravure-printed Ag nanowire layers had uniform line widths of 490 ± 15 and 470 ± 12 μm, as well as excellent mechanical stability after 1000 bending cycles. Postprinting treatments such as thermal annealing or chemical treatments are often employed to enhance the conductivity and adhesion of Ag nanowire electrodes [71–74]. These treatments remove any residual binders and improve the interconnections between the nanowires, ultimately optimizing the electrical performance of the printed electrodes.

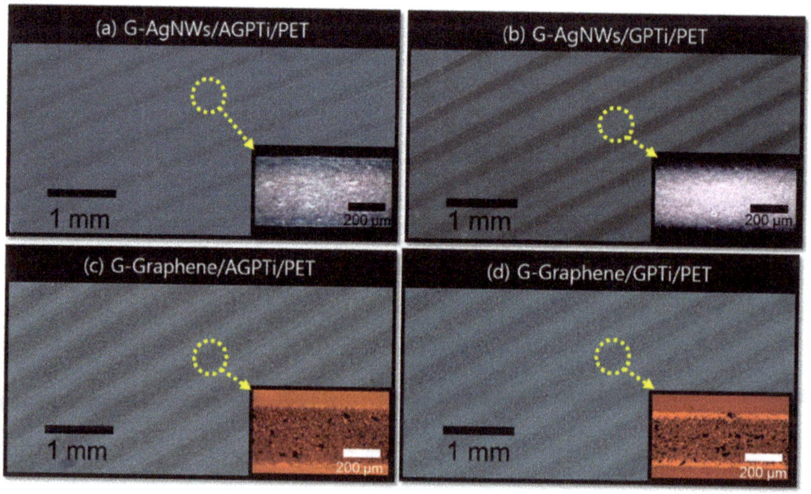

Figure 13. Gravure-printed lines of Ag nanowires and graphene on AGPTi: (**a**) G-Ag nanowires/AGPTi/PET; (**c**) G-graphene/AGPTi/PET) and GPTi; (**b**) G-Ag nanowires/GPTi/PET; and (**d**) G-graphene/GPTi/PET), previously deposited on PET. Reproduced from Ref. [76] under the Creative Commons Attribution 4.0 International (CC BY 4.0) License.

The transparency and conductivity of Ag nanowires make them well suited for use in optoelectronic devices, and the scalability of gravure printing is ideal for large-scale production [71–74]. Gravure printing is also versatile in terms of substrate compatibility, enabling the deposition of Ag nanowire electrodes on a wide range of materials, including flexible and transparent substrates. Consequently, the gravure printing of Ag nanowire electrodes is a promising technique for producing conductive patterns, including electrodes, on various substrates. Current research in this area is focused on maintaining ink stability, optimizing printing parameters, and ensuring the long-term stability of printed patterns. Future directions may involve the exploration of novel ink formulations, advances in printing cylinder technologies, and the development of in-line monitoring systems to enhance the precision and reproducibility of the printing process [71–74]. The ongoing exploration and refinement of Ag nanowire gravure printing in the field of printed electronics will contribute to the advancement of technologies that rely on high-performance, cost-effective, and scalable conductive patterns.

3. Additional Considerations for Ag Nanowire Layer Formation

There are several additional considerations for Ag nanowire layer formation. First, the choice of flexible polymeric substrate, like PET [77,78] or polyethylene naphthalate (PEN) [79,80], significantly impacts the mechanical, thermal, and adhesion properties of the prepared film. Recent advancements in coating techniques offer the possibility of tailoring these properties for specific applications, such as achieving mechanical flexibility for applications involving bending or stretching. Second, postcoating treatments such as drying and annealing are highly effective for enhancing the electrical conductivity and mechanical performance of a prepared film [14,15,20,30,31]. Researchers are continuously optimizing these treatments to improve film characteristics, including the morphology and Ag nanowire orientation. Advanced characterization techniques, including electrical conductivity measurements; optical transmittance analyses; and structural evaluations using scanning electron microscopy and atomic force microscopy, provide crucial insights for refining and optimizing post-fabrication processes [4,74,75,81]. Finally, the scalability of coating methods significantly influences their suitability for different applications [1,82–86]. R2R coating offers excellent scalability for mass production, whereas inkjet printing and spray coating offer advantages in terms of speed and efficiency.

Ag nanowire-coated films have diverse applications in flexible electronics, touchscreens, and solar cells. The exceptional properties of Ag nanowire-coated films, including transparency, flexibility, and conductivity, position them as essential components in the rapidly evolving field of flexible and transparent electronics.

Researchers are also actively exploring alternative materials and sustainable fabrication techniques, aiming to integrate these innovations with emerging technologies. This pursuit is driven by challenges including cost-effectiveness, scalability, and achieving uniformity over large areas. The complex interplay between coating methods, substrate choices, post-treatment processes, and characterization techniques should be more closely explored in future research to unlock new applications and breakthroughs in the field of transparent conductive films on flexible polymeric substrates.

4. Summary

Ag nanowire electrodes present a compelling alternative to traditional ITO electrodes for optoelectronic applications owing to their remarkable conductivity, flexibility, and transparency. This review explores recent trends in Ag nanowire electrode coating technologies. There are diverse coating methods for depositing Ag nanowires on flexible polymeric substrates, each with unique advantages for specific applications and production scales, as well as distinct challenges. Future research on these techniques will result in greater precision, uniformity, and substrate compatibility. The evolution of the field of Ag nanowire electrodes hinges on the development of these coating methods to meet the rapidly changing needs of diverse and advancing technologies.

Author Contributions: B.H. prepared the manuscript. H.H. and N.Q. revised the manuscript. All authors have read and agreed to the published version of the manuscript.

Funding: This research was supported by the GRDC (Global Research Development Center) Cooperative Hub Program through the National Research Foundation of Korea, funded by the Ministry of Science and ICT (MSIT) (RS-2023-00257595).

Data Availability Statement: Data sharing is not applicable.

Conflicts of Interest: The authors declare that they have no competing interests.

References

1. Ahn, Y.; Jeong, Y.; Lee, Y. Improved Thermal Oxidation Stability of Solution-Processable Silver Nanowire Transparent Electrode by Reduced Graphene Oxide. *ACS Appl. Mater. Interfaces* **2012**, *4*, 6410–6414. [CrossRef]
2. Bai, S.; Guo, X.; Chen, T.; Zhang, Y.; Zhang, X.; Yang, H.; Zhao, X. Solution processed fabrication of silver nanowire-MXene@PEDOT: PSS flexible transparent electrodes for flexible organic light-emitting diodes. *Compos. Part A Appl. Sci. Manuf.* **2020**, *139*, 106088. [CrossRef]
3. Bhadra, R.; Jana, T.; Mitra, A.; Sahoo, P. Effect of CNT radius on flattening contact behaviour of CNT-Al nanocomposite: A numerical approch. *Rep. Mech. Eng.* **2023**, *4*, 121–130. [CrossRef]
4. Basarir, F.; Madani, Z.; Vapaavuori, J. Recent Advances in Silver Nanowire Based Flexible Capacitive Pressure Sensors: From Structure, Fabrication to Emerging Applications. *Adv. Mater. Interfaces* **2022**, *9*, 2200866. [CrossRef]
5. Bobinger, M.; Hinterleuthner, S.; Becherer, M.; Keddis, S.; Schwesinger, N.; Lugli, P. Energy harvesting from ambient light using PVDF with highly conductive and transparent silver nanowire/PEDOT:PSS hybride electrodes. In Proceedings of the 2017 IEEE 17th International Conference on Nanotechnology (IEEE-NANO), Pittsburgh, PA, USA, 25–28 July 2017; pp. 426–429.
6. Hwang, B.; Han, Y.; Matteini, P. Bending Fatigue Behavior of Ag Nanowire/Cu Thin-Film Hybrid Interconnects for Wearable Electronics. *Facta Univ. Ser. Mech. Eng.* **2022**, *20*, 553–560. [CrossRef]
7. Milić, P.; Marinković, D.; Klinge, S.; Cojbašić, Ž. Reissner-Mindlin Based Isogeometric Finite Element Formulation for Piezoelectric Active Laminated Shells. *Teh. Vjesn.* **2023**, *30*, 416–425. [CrossRef]
8. Milić, P.; Marinković, D.; Ćojbašić, Ž. Geometrically Nonlinear Analysis of Piezoelectric Active Laminated Shells by Means of Isogeometric Fe Formulation. *Facta Univ. Ser. Mech. Eng.* 2023, *online first*. [CrossRef]
9. Bobinger, M.; Keddis, S.; Hinterleuthner, S.; Becherer, M.; Kluge, F.; Schwesinger, N.; Salmerón, J.F.; Lugli, P.; Rivadeneyra, A. Light and Pressure Sensors Based on PVDF with Sprayed and Transparent Electrodes for Self-Powered Wireless Sensor Nodes. *IEEE Sens. J.* **2019**, *19*, 1114–1126. [CrossRef]
10. Ayham, N.G.; Zuheir Fadhel, E.; Hashem Abbud, L. Investigation of the mechanical properties of nanocomposites with multi-wall carbon nanotube reinforcement and carbon fiber/epoxy. *Rep. Mech. Eng.* **2023**, *4*, 153–160. [CrossRef]
11. Chen, S.; Song, L.; Tao, Z.; Shao, X.; Huang, Y.; Cui, Q.; Guo, X. Neutral-pH PEDOT:PSS as over-coating layer for stable silver nanowire flexible transparent conductive films. *Org. Electron.* **2014**, *15*, 3654–3659. [CrossRef]
12. Cho, S.; Kang, S.; Pandya, A.; Shanker, R.; Khan, Z.; Lee, Y.; Park, J.; Craig, S.L.; Ko, H. Large-Area Cross-Aligned Silver Nanowire Electrodes for Flexible, Transparent, and Force-Sensitive Mechanochromic Touch Screens. *ACS Nano* **2017**, *11*, 4346–4357. [CrossRef] [PubMed]
13. Choi, J.H.; Lee, K.Y.; Kim, S.W. Ultra-bendable and durable Graphene–Urethane composite/silver nanowire film for flexible transparent electrodes and electromagnetic-interference shielding. *Compos. Part B Eng.* **2019**, *177*, 107406. [CrossRef]
14. Choi, Y.; Kim, C.S.; Jo, S. Spray Deposition of Ag Nanowire–Graphene Oxide Hybrid Electrodes for Flexible Polymer–Dispersed Liquid Crystal Displays. *Materials* **2018**, *11*, 2231. [CrossRef] [PubMed]
15. Fang, F.; Huang, G.-W.; Xiao, H.-M.; Li, Y.-Q.; Hu, N.; Fu, S.-Y. Largely enhanced electrical conductivity of layer-structured silver nanowire/polyimide composite films by polyaniline. *Compos. Sci. Technol.* **2018**, *156*, 144–150. [CrossRef]
16. Hwang, B.; Lund, A.; Tian, Y.; Darabi, S.; Müller, C. Machine-Washable Conductive Silk Yarns with a Composite Coating of Ag Nanowires and PEDOT:PSS. *ACS Appl. Mater. Interfaces* **2020**, *12*, 27537–27544. [CrossRef] [PubMed]
17. Kim, H.; Qaiser, N.; Hwang, B. Electro-Mechanical Response of Stretchable PDMS Composites with a Hybrid Filler System. *Facta Univ. Ser. Mech. Eng.* **2023**, *21*, 51–61. [CrossRef]
18. Ghosh, D.S.; Chen, T.L.; Mkhitaryan, V.; Pruneri, V. Ultrathin Transparent Conductive Polyimide Foil Embedding Silver Nanowires. *ACS Appl. Mater. Interfaces* **2014**, *6*, 20943–20948. [CrossRef]
19. Leroy, J.-E.; Popov, V.L. Stress tensor in the linear viscoelastic incompressible half-space beneath axisymmetric bodies in normal contact. *Rep. Mech. Eng.* **2023**, *4*, 310–316. [CrossRef]
20. Ha, H.; Müller, S.; Baumann, R.-P.; Hwang, B. PeakForce Quantitative Nanomechanical Mapping for Surface Energy Characterization on the Nanoscale: A Mini-Review. *Facta Univ. Ser. Mech. Eng.* 2023, *online first*. [CrossRef]
21. Gao, D.; Zhao, P.; Liu, J.; Zhou, Y.; Lyu, B.; Ma, J.; Shao, L. Polyaniline/silver nanowire cotton fiber: A flexible electrode material for supercapacitor. *Adv. Powder Technol.* **2021**, *32*, 3954–3963. [CrossRef]
22. Ha, H.; Amicucci, C.; Matteini, P.; Hwang, B. Mini review of synthesis strategies of silver nanowires and their applications. *Colloid Interface Sci. Commun.* **2022**, *50*, 100663. [CrossRef]

23. Bzinkowski, D.; Ryba, T.; Siemiatkowski, Z.; Rucki, M. Real-time monitoring of the rubber belt tension in an industrial conveyor. *Rep. Mech. Eng.* **2022**, *3*, 1–10. [CrossRef]
24. Ha, H.; Qaiser, N.; Yun, T.G.; Cheong, J.Y.; Lim, S.; Hwang, B. Sensing Mechanism and Application of Mechanical Strain Sensor: A Mini Review. *Facta Univ. Ser. Mech. Eng.* **2023**, *21*, 751–772. [CrossRef]
25. Seo, Y.; Hwang, B. Mulberry-paper-based composites for flexible electronics and energy storage devices. *Cellulose* **2019**, *26*, 8867–8875. [CrossRef]
26. Shah, K.W.; Xiong, T. Multifunctional Metallic Nanowires in Advanced Building Applications. *Materials* **2019**, *12*, 1731. [CrossRef]
27. Nam, V.B.; Lee, D. Copper Nanowires and Their Applications for Flexible, Transparent Conducting Films: A Review. *Nanomaterials* **2016**, *6*, 47. [CrossRef]
28. Xiong, W.; Liu, H.; Chen, Y.; Zheng, M.; Zhao, Y.; Kong, X.; Wang, Y.; Zhang, X.; Kong, X.; Wang, P.; et al. Highly Conductive, Air-Stable Silver Nanowire@Iongel Composite Films toward Flexible Transparent Electrodes. *Adv. Mater.* **2016**, *28*, 7167–7172. [CrossRef]
29. Lian, L.; Dong, D.; Feng, D.; He, G. Low roughness silver nanowire flexible transparent electrode by low temperature solution-processing for organic light emitting diodes. *Org. Electron.* **2017**, *49*, 9–18. [CrossRef]
30. Luo, M.; Liu, Y.; Huang, W.; Qiao, W.; Zhou, Y.; Ye, Y.; Chen, L.-S. Towards Flexible Transparent Electrodes Based on Carbon and Metallic Materials. *Micromachines* **2017**, *8*, 12. [CrossRef]
31. Morales-Masis, M.; De Wolf, S.; Woods-Robinson, R.; Ager, J.W.; Ballif, C. Transparent Electrodes for Efficient Optoelectronics. *Adv. Electron. Mater.* **2017**, *3*, 1600529. [CrossRef]
32. Hwang, B.; Yun, T.G. Stretchable and patchable composite electrode with trimethylolpropane formal acrylate-based polymer. *Compos. Part B Eng.* **2019**, *163*, 185–192. [CrossRef]
33. Xie, H.; Yang, X.; Du, D.; Zhao, Y.; Wang, Y. Flexible Transparent Conductive Film Based on Random Networks of Silver Nanowires. *Micromachines* **2018**, *9*, 295. [CrossRef]
34. He, X.; Shen, G.; Xu, R.; Yang, W.; Zhang, C.; Liu, Z.; Chen, B.; Liu, J.; Song, M. Hexagonal and Square Patterned Silver Nanowires/PEDOT:PSS Composite Grids by Screen Printing for Uniformly Transparent Heaters. *Polymers* **2019**, *11*, 468. [CrossRef]
35. Khadtare, S.; Ko, E.J.; Kim, Y.H.; Lee, H.S.; Moon, D.K. A flexible piezoelectric nanogenerator using conducting polymer and silver nanowire hybrid electrodes for its application in real-time muscular monitoring system. *Sens. Actuators A Phys.* **2019**, *299*, 111575. [CrossRef]
36. Zhu, Z.; Wang, X.; Li, D.; Yu, H.; Li, X.; Guo, F. Solvent Welding-Based Methods Gently and Effectively Enhance the Conductivity of a Silver Nanowire Network. *Nanomaterials* **2023**, *13*, 2865. [CrossRef]
37. Sohn, H.; Park, C.; Oh, J.-M.; Kang, S.W.; Kim, M.-J. Silver Nanowire Networks: Mechano-Electric Properties and Applications. *Materials* **2019**, *12*, 2526. [CrossRef]
38. Gorji, M.; Mazinani, S.; Faramarzi, A.-R.; Ghadimi, S.; Kalaee, M.; Sadeghianmaryan, A.; Wilson, L.D. Coating Cellulosic Material with Ag Nanowires to Fabricate Wearable IR-Reflective Device for Personal Thermal Management: The Role of Coating Method and Loading Level. *Molecules* **2021**, *26*, 3570. [CrossRef]
39. Kim, J.-H.; Ma, J.; Jo, S.; Lee, S.; Kim, C.S. Enhancement of Antibacterial Performance of Silver Nanowire Transparent Film by Post-Heat Treatment. *Nanomaterials* **2020**, *10*, 938. [CrossRef]
40. Jin, I.S.; Lee, H.D.; Hong, S.I.; Lee, W.; Jung, J.W. Facile Post Treatment of Ag Nanowire/Polymer Composites for Flexible Transparent Electrodes and Thin Film Heaters. *Polymers* **2021**, *13*, 586. [CrossRef]
41. Kumar, A.; Shaikh, M.O.; Chuang, C.-H. Silver Nanowire Synthesis and Strategies for Fabricating Transparent Conducting Electrodes. *Nanomaterials* **2021**, *11*, 693. [CrossRef]
42. Choi, J.H.; Shin, M.G.; Jung, Y.; Kim, D.H.; Ko, J.S. Fabrication and Performance Evaluation of Highly Sensitive Flexible Strain Sensors with Aligned Silver Nanowires. *Micromachines* **2020**, *11*, 156. [CrossRef]
43. Zeng, F.; Zheng, Y.; Wei, Y.; Li, H.; Wang, Q.; Shi, J.; Wang, Y.; Hong, X. Multifunctional Silver Nanowire Fabric Reinforced by Hot Pressing for Electromagnetic Interference Shielding, Electric Heating, and Sensing. *Polymers* **2023**, *15*, 4258. [CrossRef]
44. Lee, S.H.; Lee, S. Cantilever Type Acceleration Sensors Made by Roll-to-Roll Slot-Die Coating. *Sensors* **2020**, *20*, 3748. [CrossRef]
45. Liu, C.; Zhang, X.; Shan, J.; Li, Z.; Guo, X.; Zhao, X.; Yang, H. Large-Scale Preparation of Silver Nanowire-Based Flexible Transparent Film Heaters by Slot-Die Coating. *Materials* **2022**, *15*, 2634. [CrossRef]
46. Kim, Y.J.; Kim, G.; Kim, H.-K. Study of Brush-Painted Ag Nanowire Network on Flexible Invar Metal Substrate for Curved Thin Film Heater. *Metals* **2019**, *9*, 1073. [CrossRef]
47. Wu, X.; Zhou, Z.; Wang, Y.; Li, J. Syntheses of Silver Nanowires Ink and Printable Flexible Transparent Conductive Film: A Review. *Coatings* **2020**, *10*, 865. [CrossRef]
48. Jeong, H.; Lee, J.H.; Song, J.-Y.; Ghani, F.; Lee, D. Continuous Patterning of Silver Nanowire-Polyvinylpyrrolidone Composite Transparent Conductive Film by a Roll-to-Roll Selective Calendering Process. *Nanomaterials* **2023**, *13*, 32. [CrossRef]
49. Lim, S.-H.; Kim, H.-K. Thermal Profiles of Carbon Fiber Based Anisotropic Thin-Films: An Emerging Heat Management Solution for High-Current Flow Electrocatalysis and Electrochemical Applications. *Catalysts* **2020**, *10*, 1172. [CrossRef]
50. Xu, H.; Liu, P.; Huang, B.; Jiang, X.; Gao, Q.; Liu, L. Preparation of Double-Layer Crossed Silver Nanowire Film and Its Application to OLED. *Coatings* **2022**, *12*, 26. [CrossRef]

51. Camic, B.T.; Jeong, H.I.; Aslan, M.H.; Kosemen, A.; Kim, S.; Choi, H.; Basarir, F.; Lee, B.R. Preparation of Transparent Conductive Electrode via Layer-By-Layer Deposition of Silver Nanowires and Its Application in Organic Photovoltaic Device. *Nanomaterials* **2020**, *10*, 46. [CrossRef]
52. Heo, S.W. Ultra-Flexible Organic Photovoltaics with Nanograting Patterns Based on CYTOP/Ag Nanowires Substrate. *Nanomaterials* **2020**, *10*, 2185. [CrossRef]
53. Li, X.; Zhou, J.; Yan, D.; Peng, Y.; Wang, Y.; Zhou, Q.; Wang, K. Effects of Concentration and Spin Speed on the Optical and Electrical Properties of Silver Nanowire Transparent Electrodes. *Materials* **2021**, *14*, 2219. [CrossRef]
54. Zhang, J.; Zhu, X.; Xu, J.; Xu, R.; Yang, H.; Kan, C. Comparative Study on Preparation Methods for Transparent Conductive Films Based on Silver Nanowires. *Molecules* **2022**, *27*, 8907. [CrossRef]
55. Lee, G.; Kim, H.; Lee, S.B.; Kim, D.; Lee, E.; Lee, S.K.; Lee, S.G. Tailored Uniaxial Alignment of Nanowires Based on Off-Center Spin-Coating for Flexible and Transparent Field-Effect Transistors. *Nanomaterials* **2022**, *12*, 1116. [CrossRef]
56. Wang, J.; Yu, J.; Bai, D.; Li, Z.; Liu, H.; Li, Y.; Chen, S.; Cheng, J.; Li, L. Biodegradable, Flexible, and Transparent Conducting Silver Nanowires/Polylactide Film with High Performance for Optoelectronic Devices. *Polymers* **2020**, *12*, 604. [CrossRef]
57. Yang, X.; Du, D.; Wang, Y.; Zhao, Y. Silver Nanowires Inks for Flexible Circuit on Photographic Paper Substrate. *Micromachines* **2019**, *10*, 22. [CrossRef]
58. Kong, J.; Wang, Y.; Wu, Y.; Zhang, L.; Gong, M.; Lin, X.; Wang, D. Toward High-Energy-Density Aqueous Lithium-Ion Batteries Using Silver Nanowires as Current Collectors. *Molecules* **2022**, *27*, 8207. [CrossRef]
59. Yoon, H.; Matteini, P.; Hwang, B. Effect of the Blade-Coating Conditions on the Electrical and Optical Properties of Transparent Ag Nanowire Electrodes. *Micromachines* **2023**, *14*, 114. [CrossRef]
60. Du, D.; Yang, X.; Yang, Y.; Zhao, Y.; Wang, Y. Silver Nanowire Ink for Flexible Circuit on Textiles. *Micromachines* **2019**, *10*, 42. [CrossRef]
61. Wang, S.; Wu, X.; Lu, J.; Luo, Z.; Xie, H.; Zhang, X.; Lin, K.; Wang, Y. Inkjet-Printed Silver Nanowire Ink for Flexible Transparent Conductive Film Applications. *Nanomaterials* **2022**, *12*, 842. [CrossRef]
62. Wang, Y.; Wu, X.; Wang, K.; Lin, K.; Xie, H.; Zhang, X.; Li, J. Novel Insights into Inkjet Printed Silver Nanowires Flexible Transparent Conductive Films. *Int. J. Mol. Sci.* **2021**, *22*, 7719. [CrossRef]
63. Ke, S.-H.; Xue, Q.-W.; Pang, C.-Y.; Guo, P.-W.; Yao, W.-J.; Zhu, H.-P.; Wu, W. Printing the Ultra-Long Ag Nanowires Inks onto the Flexible Textile Substrate for Stretchable Electronics. *Nanomaterials* **2019**, *9*, 686. [CrossRef]
64. Wu, X.; Wang, S.; Luo, Z.; Lu, J.; Lin, K.; Xie, H.; Wang, Y.; Li, J.-Z. Inkjet Printing of Flexible Transparent Conductive Films with Silver Nanowires Ink. *Nanomaterials* **2021**, *11*, 1571. [CrossRef]
65. Reenaers, D.; Marchal, W.; Biesmans, I.; Nivelle, P.; D'Haen, J.; Deferme, W. Layer Morphology and Ink Compatibility of Silver Nanoparticle Inkjet Inks for Near-Infrared Sintering. *Nanomaterials* **2020**, *10*, 892. [CrossRef]
66. Wang, J.; Lou, Y.; Wang, B.; Sun, Q.; Zhou, M.; Li, X. Highly Sensitive, Breathable, and Flexible Pressure Sensor Based on Electrospun Membrane with Assistance of AgNW/TPU as Composite Dielectric Layer. *Sensors* **2020**, *20*, 2459. [CrossRef]
67. Li, B.; Xu, C.; Zheng, J.; Xu, C. Sensitivity of Pressure Sensors Enhanced by Doping Silver Nanowires. *Sensors* **2014**, *14*, 9889–9899. [CrossRef]
68. Chen, M.; Wang, Z.; Zheng, Y.; Zhang, Q.; He, B.; Yang, J.; Qi, M.; Wei, L. Flexible Tactile Sensor Based on Patterned Ag-Nanofiber Electrodes through Electrospinning. *Sensors* **2021**, *21*, 2413. [CrossRef]
69. Xiao, J.; Li, Y.; Wang, J.; Xu, Y.; Zhang, G.; Leng, C. Preparation and Antibiosis Investigation of Kaolinite Nanotubes and Silver Nanowires Co-Doped Electrospinning-Silk Fibroin/Gelatin Porous Fiber Films. *Metals* **2023**, *13*, 745. [CrossRef]
70. Wang, X.; Sun, F.; Yin, G.; Wang, Y.; Liu, B.; Dong, M. Tactile-Sensing Based on Flexible PVDF Nanofibers via Electrospinning: A Review. *Sensors* **2018**, *18*, 330. [CrossRef]
71. Park, S.; Kim, H.; Kim, J.-H.; Yeo, W.-H. Advanced Nanomaterials, Printing Processes, and Applications for Flexible Hybrid Electronics. *Materials* **2020**, *13*, 3587. [CrossRef]
72. Yang, J.; Zeng, W.; Li, Y.; Yi, Z.; Zhou, G. Fabrication of Screen Printing-Based AgNWs Flexible Transparent Conductive Film with High Stability. *Micromachines* **2020**, *11*, 1027. [CrossRef]
73. Garcia, A.J.L.; Sico, G.; Montanino, M.; Defoor, V.; Pusty, M.; Mescot, X.; Loffredo, F.; Villani, F.; Nenna, G.; Ardila, G. Low-Temperature Growth of ZnO Nanowires from Gravure-Printed ZnO Nanoparticle Seed Layers for Flexible Piezoelectric Devices. *Nanomaterials* **2021**, *11*, 1430. [CrossRef]
74. Giasafaki, D.; Mitzithra, C.; Belessi, V.; Filippakopoulou, T.; Koutsioukis, A.; Georgakilas, V.; Charalambopoulou, G.; Steriotis, T. Graphene-Based Composites with Silver Nanowires for Electronic Applications. *Nanomaterials* **2022**, *12*, 3443. [CrossRef]
75. Garcia, A.J.L.; Jalabert, T.; Pusty, M.; Defoor, V.; Mescot, X.; Montanino, M.; Sico, G.; Loffredo, F.; Villani, F.; Nenna, G.; et al. Size and Semiconducting Effects on the Piezoelectric Performances of ZnO Nanowires Grown onto Gravure-Printed Seed Layers on Flexible Substrates. *Nanoenergy Adv.* **2022**, *2*, 197–209. [CrossRef]
76. Li, X.; Kim, N.; Youn, S.; An, T.K.; Kim, J.; Lim, S.; Kim, S.H. Sol–Gel-Processed Organic–Inorganic Hybrid for Flexible Conductive Substrates Based on Gravure-Printed Silver Nanowires and Graphene. *Polymers* **2019**, *11*, 158. [CrossRef]
77. Huang, C.-H.; Wang, Y.-Y.; Lu, T.-H.; Li, Y.-C. Flexible Transparent Electrode of Hybrid Ag-Nanowire/Reduced-Graphene-Oxide Thin Film on PET Substrate Prepared Using H2/Ar Low-Damage Plasma. *Polymers* **2017**, *9*, 28. [CrossRef] [PubMed]
78. Xu, L.; Weng, W.-C.; Yeh, Y.-C. Continuous Wave Laser Nanowelding Process of Ag Nanowires on Flexible Polymer Substrates. *Nanomaterials* **2021**, *11*, 2511. [CrossRef] [PubMed]

79. Zacharatos, F.; Karvounis, P.; Theodorakos, I.; Hatziapostolou, A.; Zergioti, I. Single Step Laser Transfer and Laser Curing of Ag NanoWires: A Digital Process for the Fabrication of Flexible and Transparent Microelectrodes. *Materials* **2018**, *11*, 1036. [CrossRef] [PubMed]
80. Oh, J.; Wen, L.; Tak, H.; Kim, H.; Kim, G.; Hong, J.; Chang, W.; Kim, D.; Yeom, G. Radio Frequency Induction Welding of Silver Nanowire Networks for Transparent Heat Films. *Materials* **2021**, *14*, 4448. [CrossRef] [PubMed]
81. Zhu, Y.; Li, X.; Xu, Y.; Wu, L.; Yu, A.; Lai, G.; Wei, Q.; Chi, H.; Jiang, N.; Fu, L.; et al. Intertwined Carbon Nanotubes and Ag Nanowires Constructed by Simple Solution Blending as Sensitive and Stable Chloramphenicol Sensors. *Sensors* **2021**, *21*, 1220. [CrossRef] [PubMed]
82. Pang, S.; Hernandez, Y.; Feng, X.; Müllen, K. Graphene as Transparent Electrode Material for Organic Electronics. *Adv. Mater.* **2011**, *23*, 2779–2795. [CrossRef] [PubMed]
83. Tokuno, T.; Nogi, M.; Karakawa, M.; Jiu, J.; Nge, T.T.; Aso, Y.; Suganuma, K. Fabrication of silver nanowire transparent electrodes at room temperature. *Nano Res.* **2011**, *4*, 1215–1222. [CrossRef]
84. Hu, W.; Niu, X.; Zhao, R.; Pei, Q. Elastomeric transparent capacitive sensors based on an interpenetrating composite of silver nanowires and polyurethane. *Appl. Phys. Lett.* **2013**, *102*, 083303. [CrossRef]
85. Kim, A.; Won, Y.; Woo, K.; Kim, C.-H.; Moon, J. Highly Transparent Low Resistance ZnO/Ag Nanowire/ZnO Composite Electrode for Thin Film Solar Cells. *ACS Nano* **2013**, *7*, 1081–1091. [CrossRef] [PubMed]
86. Lee, D.; Lee, H.; Ahn, Y.; Jeong, Y.; Lee, D.-Y.; Lee, Y. Highly stable and flexible silver nanowire–graphene hybrid transparent conducting electrodes for emerging optoelectronic devices. *Nanoscale* **2013**, *5*, 7750–7755. [CrossRef] [PubMed]

Disclaimer/Publisher's Note: The statements, opinions and data contained in all publications are solely those of the individual author(s) and contributor(s) and not of MDPI and/or the editor(s). MDPI and/or the editor(s) disclaim responsibility for any injury to people or property resulting from any ideas, methods, instructions or products referred to in the content.

inorganics

MDPI

Review

Polymeric Protection for Silver Nanowire-Based Transparent Conductive Electrodes: Performance and Applications

Heebo Ha [1], Jun Young Cheong [2,*], Tae Gwang Yun [3,*] and Byungil Hwang [4,*]

[1] Department of Intelligent Semiconductor Engineering, Chung-Ang University, Seoul 06974, Republic of Korea
[2] Bavarian Center for Battery Technology (BayBatt) and Department of Chemistry, University of Bayreuth, Universitätsstraße 30, 95447 Bayreuth, Germany
[3] Department of Molecular Science and Technology, Ajou University, Suwon 16499, Republic of Korea
[4] School of Integrative Engineering, Chung-Ang University, Seoul 06974, Republic of Korea
* Correspondence: jun.cheong@uni-bayreuth.de (J.Y.C.); ytk0402@ajou.ac.kr (T.G.Y.); bihwang@cau.ac.kr (B.H.)

Abstract: Silver nanowires (AgNWs) are a potential alternative to conventional transparent conductive materials for various applications, such as flexible and transparent electrodes in optoelectronic devices, including touch screens, solar cells, and flexible displays. However, AgNW electrodes face degradation due to environmental factors, electrical instability, and mechanical stress. To overcome these challenges, strategies to protect AgNW-based electrodes via the incorporation of polymeric materials were widely investigated to improve the durability and stability of AgNW-based electrodes. This review paper gives a comprehensive overview of the incorporation of polymeric materials with AgNW electrodes, emphasizing their performance, and applications. We compare the different polymeric materials and their effect on the electrical, optical, and mechanical properties of AgNW electrodes. Furthermore, we evaluate the key factors affecting the choice of protective layers, such as their compatibility with AgNWs, and also we present current challenges and future opportunities for the development of polymeric materials for AgNW electrodes in emerging technologies.

Keywords: silver nanowire; protection layer; electrode; polymer; flexible

1. Introduction

The demand for advanced transparent conductive materials is increasing as flexible and high-performance electronic devices become more popular [1–8]. Silver nanowires (AgNWs) are one of the most promising candidates to replace the most widely used material, indium tin oxide (ITO) [5–7,9–14]. AgNWs possess a unique amalgamation of attributes, boasting exemplary electrical conductivity, remarkable optical transparency, and innate mechanical flexibility [1,8,15–21]. These distinctive properties render them exceptionally well-suited for a diverse array of applications, spanning from cutting-edge touchscreens to innovative photovoltaic systems, and beyond.

For a substantial duration, ITO has held sway as the predominant transparent conductor, lauded for its formidable combination of high conductivity and exceptional transparency [22]. However, ITO bears significant drawbacks, notably its brittleness, exorbitant production costs, and the scarcity of indium—a pivotal component within its composition. These inherent limitations have galvanized a fervent quest for alternative materials, thrusting AgNWs into the spotlight as a formidable contender.

AgNWs have undeniably had their strengths proven, excelling in the efficient conduction of electricity, often outperforming ITO in this crucial regard [23,24]. Moreover, their nanoscale dimensions, characterized by diameters typically ranging from tens to hundreds of nanometers and lengths extending across several micrometers, endow them with the remarkable ability to intricately weave complex networks [25,26]. These nanowire assemblies not only preserve an elevated level of electrical conductivity but also achieve superlative optical transparency—a prerequisite of paramount importance in applications where visual quality is of the essence [27,28]. In addition, the inherent flexibility of AgNWs has opened up new possibilities in electronics [29,30]. Devices that can bend, twist, and adapt to different shapes are now possible to be used, thanks to AgNW-based electrodes [27,28,30–33]. The potential of AgNWs as a key component in the field of flexible and stretchable electronics is an exciting prospect.

However, AgNWs have their own challenges that need to be overcome to fully exploit their potential, including their vulnerability to environmental factors such as oxidation and corrosion, which can degrade their performance over time [33,34]. Electrical instability is another issue, manifested in the disruption of electrical paths under mechanical stress or exposure to electric fields, leading to erratic device behavior. The limited mechanical stability of AgNW networks is another obstacle, requiring a delicate balance between flexibility and durability.

To address these formidable challenges, researchers have undertaken their own quest—to reinforce AgNW-based electrodes with protective materials, which can act as a shield against environmental threats, ensuring the long-term stability and reliability of AgNWs [35,36]. They also play an important role in improving the mechanical robustness of AgNW networks, enabling them to endure bending, stretching, and various forms of mechanical stress.

This review paper investigates the domain of polymeric protective materials for AgNW electrodes. We examine the various polymeric materials used to strengthen AgNWs against the challenges of the real world (Figure 1). Through a detailed analysis of protective layers, we aim to reveal their effect on the electrical, optical, and mechanical properties of AgNW-based electrodes. Moreover, we survey the diverse applications of these protected electrodes in different industries and the key factors influencing the choice of protective layers suited to specific use cases. Finally, we look ahead to the future, reflecting on the difficulties that remain and the infinite opportunities that await in the domain of AgNWs and their protection. Their incorporation with protective polymeric materials marks a crucial chapter in the story of AgNWs, one that has the potential to transform the field of transparent conductive materials and redefine the possibilities of modern technology

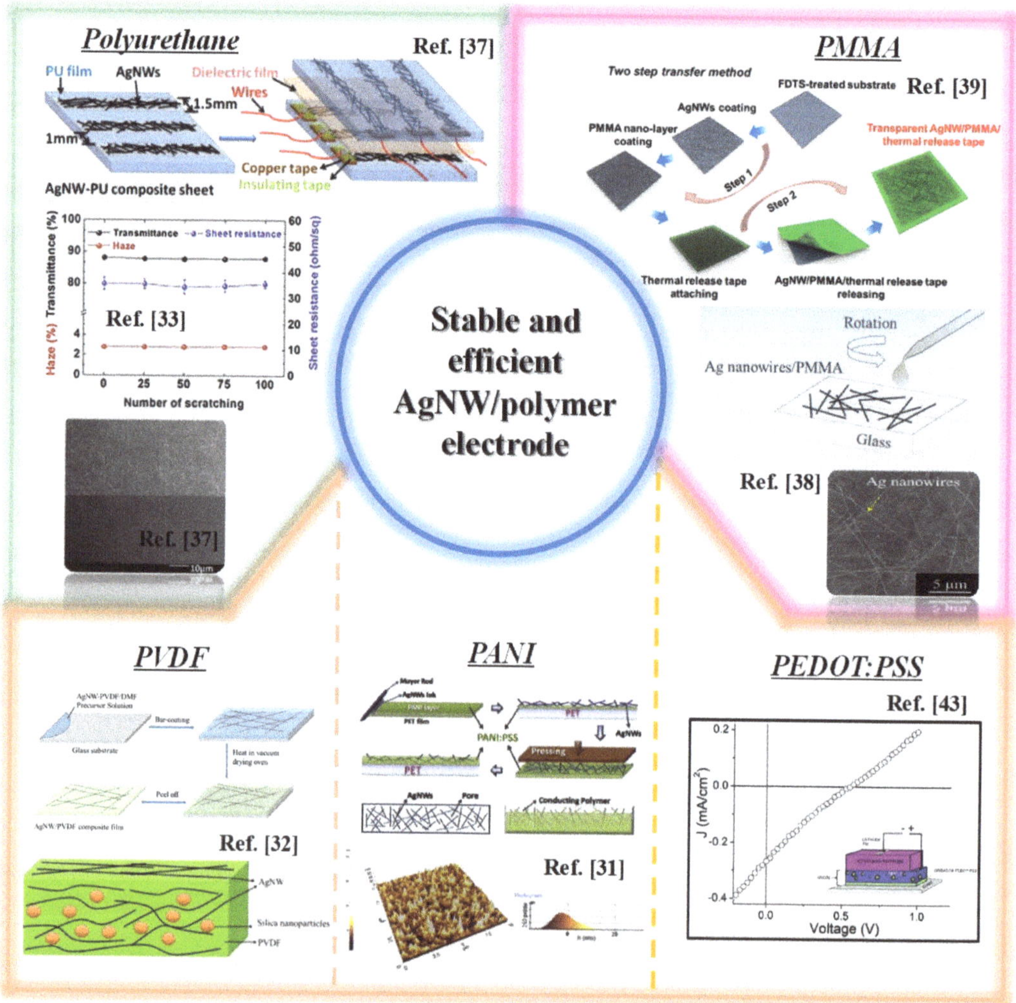

Figure 1. An overview of polymeric materials for AgNWs. Reproduced with permission from refs. [31–33,37–43].

2. Diverse Polymeric Overcoating

Polymeric materials are incorporated with AgNW electrodes using various methods, including spin coating, dip coating, spray coating, and inkjet printing [44–46]. The choice of method depends on the type of polymer, the desired thickness and uniformity of the layer, and compatibility with the substrate and the AgNW network. The details of each polymeric material for changing AgNW properties are discussed in following section. PDMS, as an elastomer, is excellent candidate for stretchable composites. There are numerous works on PDMS, and its applications are also diverse. Due to its exceptional mechanical properties, most of the works are focus on stretchable electrodes, and not on just flexible electrodes or on protection of AgNWs. Therefore, PDMS should be discussed and summarized as a separate review, which will be an interesting topic for future efforts in another review paper.

2.1. Polyurethane (PU)/AgNW

PU, a polymer comprising repeating urethane units characterized by the chemical formula NH_2-CO-O-CH_2, possesses an array of desirable attributes [33,47,48]. The strategic application of PU overcoating onto AgNW transparent electrodes yields multifaceted improvements in the performance and durability of AgNW-based devices, thereby addressing several pertinent challenges [33,47,48]. First, PU can reduce the surface roughness of AgNW electrodes by filling the gaps and smoothing the surface. This can enhance the optical transmittance and reduce the light scattering of the electrodes. Moreover, PU can improve the contact resistance between the AgNWs and the active materials of the devices by forming a uniform interface. This can increase the electrical conductivity and efficiency of the devices. Second, PU can protect AgNWs from external damages caused by mechanical deformation, chemical corrosion, or oxidation. This can increase the durability and reliability of the devices. PU can also prevent electrical shorting problems by isolating the AgNWs from each other and from the environment. This can reduce the power consumption and leakage current of the devices.

PU overcoating has shown to improve the performance and stability of AgNW-based devices without compromising their flexibility and transparency [33,47,48]. In the work by Hwang et al., a layer of PU was applied to AgNW electrodes, and various tests were conducted to assess their stability [33]. The incorporation of PU as an overcoating layer had a profound positive impact on the mechanical stability of AgNW electrodes. Even when subjected to 100 scratches (Figure 2a) and 60 wipes with isopropyl alcohol (IPA) (Figure 2b), the composite electrodes made of PU and AgNWs displayed no deterioration in their optical or electrical characteristics. Moreover, the AgNW electrodes coated with PU demonstrated remarkable reliability, maintaining their resistance even after enduring 300,000 bending cycles (Figure 2d). Additionally, PU served as a protective shield against oxidation for the Ag nanowires; even after exposure for 120 h at 85 °C, the stability of the AgNWs remained intact (Figure 2c). Notably, the application of a PU coating also enhanced the optical transmittance of the AgNW electrodes, as PU possesses a refractive index intermediate between that of air and the substrate. Furthermore, the use of UV-curable PU facilitated a straightforward and precise patterning process. This involved selectively curing the PU through a shadow mask and subsequently removing the AgNWs from the uncured regions by employing sonication in IPA.

Hu et al. presented PU/AgNW composites and used them for touch sensors (Figure 3) [37]. The AgNW electrodes coated with PU exhibited a low sheet resistance of 8 Ω/sq while maintaining a high optical transmittance of 74.6% at a wavelength of 550 nm. The integration of AgNWs into the PU layer resulted in exceptional mechanical stability, with the surface resistance remaining unchanged even after undergoing multiple scotch tape tests. Intriguingly, the PU/AgNW composite electrodes demonstrated a reduction in sheet resistance when subjected to a 60% strain. This phenomenon occurred because the external strain caused the AgNWs to establish tighter contact with each other, thereby reducing the contact resistance between individual AgNWs. Furthermore, the touch panel employing the PU/AgNW composite electrodes exhibited remarkable sensitivity to a pressing force of 30 kPa.

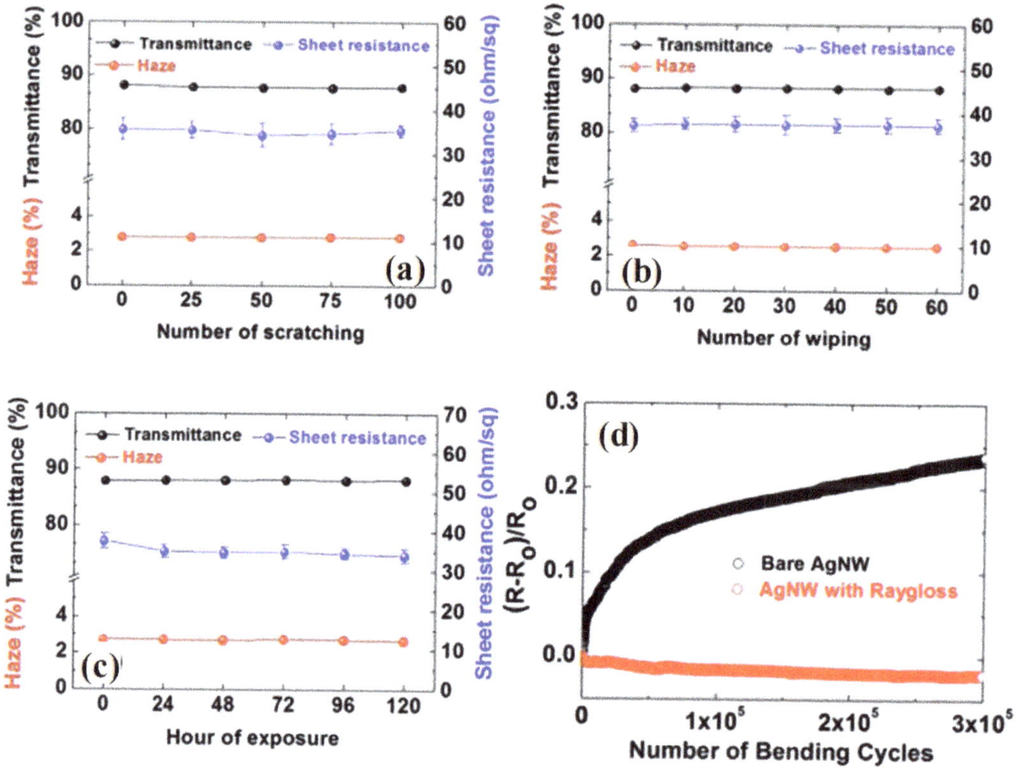

Figure 2. Change in optical transmittance, haze, and sheet resistance as a function of (**a**) the number of scratches made using a pen, (**b**) the number of wipes using IPA, (**c**) the hours of exposure to ambient air at 85 °C, and (**d**) the number of bending cycles of PU coating on the AgNW electrodes. Reproduced with permission from ref. [33].

2.2. Polymethyl Methacrylate (PMMA) on AgNW

PMMA is a polymer that consists of repeating units of methyl methacrylate, which is a chemical compound with the formula CH_2=C(CH_3)$_2$-C(O)-O-CH_2 [38,39,49,50]. PMMA has high transparency, good adhesion, low toxicity, and resistance to heat, moisture, and chemicals [38,39,49,50]. In the work by Sun et al., AgNWs with PMMA were utilized for two specific applications including a thin film heater and surface-enhanced Raman scattering (SERS) detector for biosensors [38]. The process involved dissolving PMMA in tetrahydrofuran (THF), which was then applied through spin-coating onto the AgNW electrodes (Figure 4a). The resulting PMMA/AgNW electrodes exhibited impressive characteristics, boasting a low sheet resistance of 8 Ω/sq alongside a high optical transmittance of approximately 85%. Importantly, these PMMA-coated AgNW electrodes demonstrated exceptional mechanical stability, even when subjected to bending deformations (Figure 4c,d). In the realm of thin film heating, devices employing PMMA/AgNW electrodes showcased robust joule heating performance, emphasizing their practical utility (Figure 4e). Furthermore, these PMMA/AgNW electrodes were effectively employed for the SERS detection of Rhodamine 6G (Rh6G) solutions, underscoring their versatility and efficacy in biosensing applications (Figure 4f).

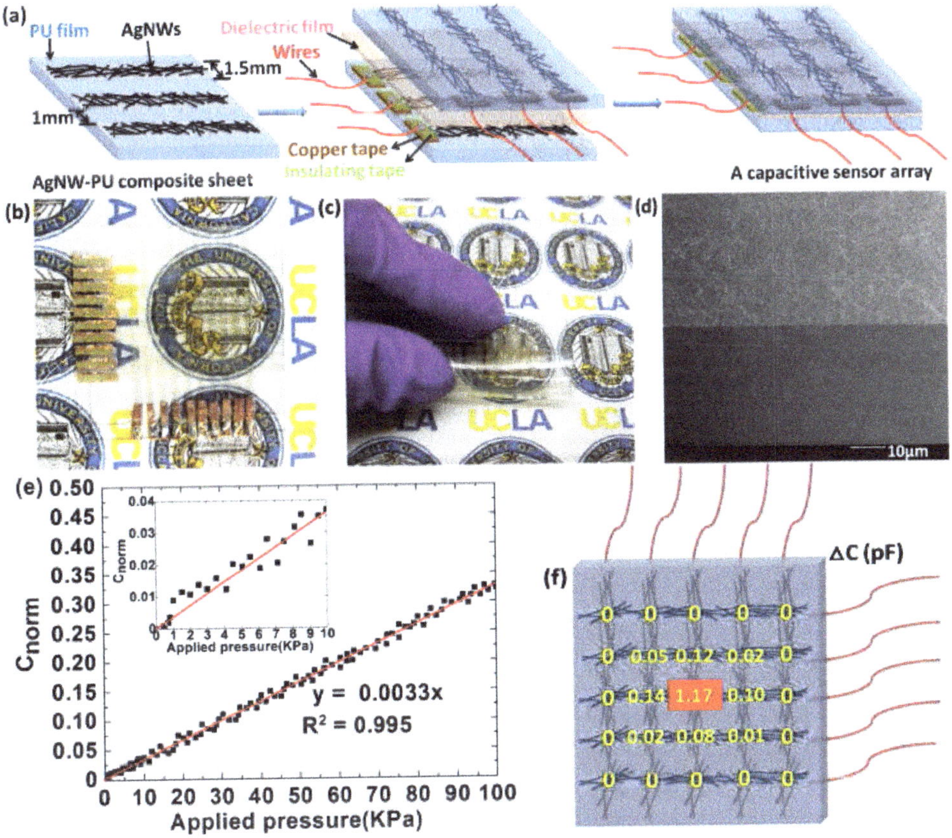

Figure 3. (**a**) A schematic illustration of the fabrication of a transparent capacitive array comprising an acrylic elastomer layer as the dielectric spacer between two transparent AgNW/PU composite electrodes. (**b**) Photograph of a pressure sensor array (10 × 10 pixels), each pixel representing a square area of 1.5 × 1.5 mm^2 separated by 1 mm from other areas. (**c**) Photograph of the sensor array bent at 180°. (**d**) SEM image of a surface area, half of which comprises patterned AgNW/PU electrodes. (**e**) Change in the capacitance, ΔC/C0, of one pixel with transversely applied pressure. (**f**) Mapping of the measured capacitance changes of pixels in the area where a pressure of 30 KPa was applied on the central pixel. Reproduced with permission from ref. [37].

In the work by Kim et al., a transparent quantum dot light-emitting diode (QLED) was demonstrated with PMMA/AgNW electrodes [39]. The PMMA/AgNW electrodes showed good chemical stability without degradation when exposed to ambient air. In addition, a low sheet resistance of 16.1 Ω/sq and a high optical transmittance of approximately ~87% were achieved (Figure 5a). The PMAM/AgNW electrodes, possessing optimized optoelectronic properties (with a figure of merit approximately equal to 3.3 × 10^{-2}), were coated with an ultrathin PMMA nanolayer (Figure 5b). These coated AgNWs were then transferred to QLEDs without causing significant damage to the adjacent active layer (Figure 5c,d). The resulting transparent QLEDs utilizing these transparent top electrodes demonstrated outstanding performance metrics (Figure 5e–g). They achieved a maximum total luminance of 27,310 cd·m^2 and a current efficiency of 45.99 cd·A^{-1}. These QLEDs, fabricated through entirely solution-based processes, create the potential for fabrication through the full roll-to-roll process.

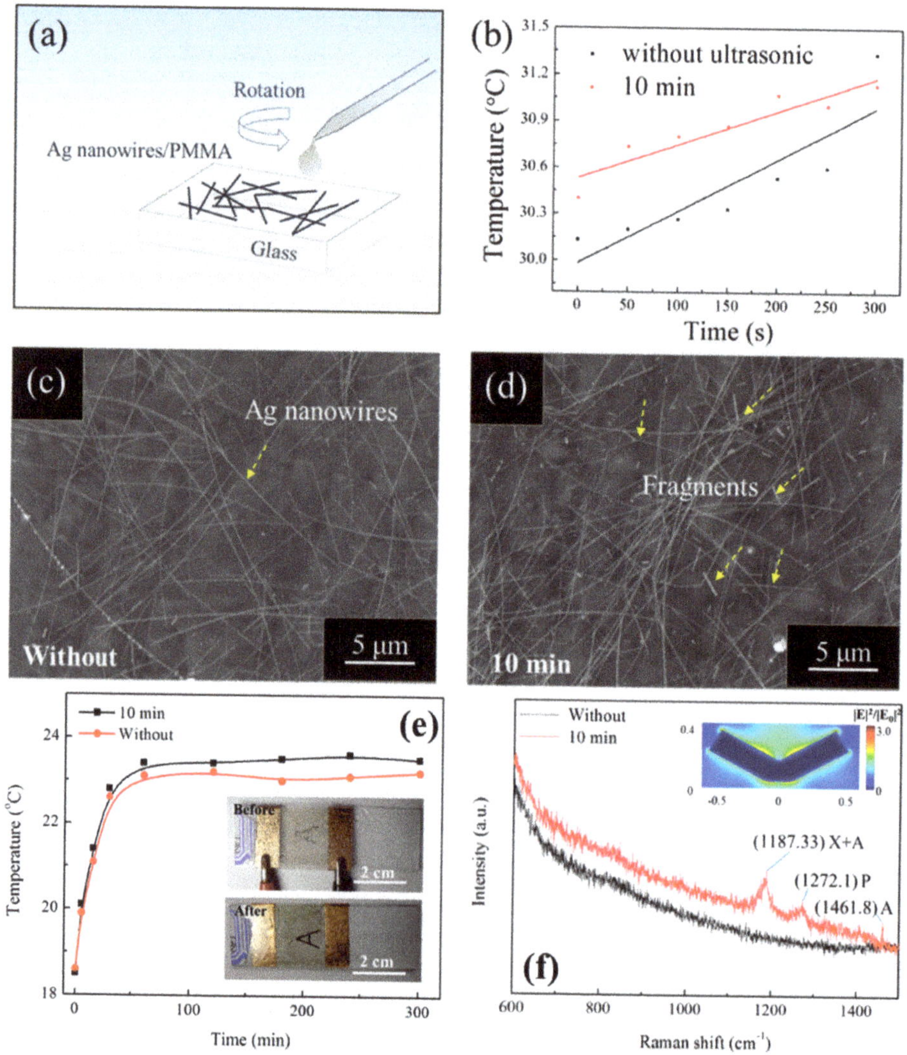

Figure 4. (**a**) A schematic of the structure of PMMA/AgNW electrodes; (**b**) temperature rise in the PMMA/AgNW electrode-based thin film heaters; (**c**,**d**) top views of AgNW electrodes after cyclic bendings without and with PMMA overcoating. Partial fragments of AgNWs are highlighted by yellow dot square frames. (**e**) Temperature increase curves of PMMA/AgNW electrode-based thin film heaters; the inset images are the optical photographs of the defogging process on the Ag-nanowire-based film heater. (**f**) SERS signals of PMMA/AgNW electrodes. Reproduced from ref. [38] under the Creative Commons Attribution 4.0 International (CC BY 4.0) License.

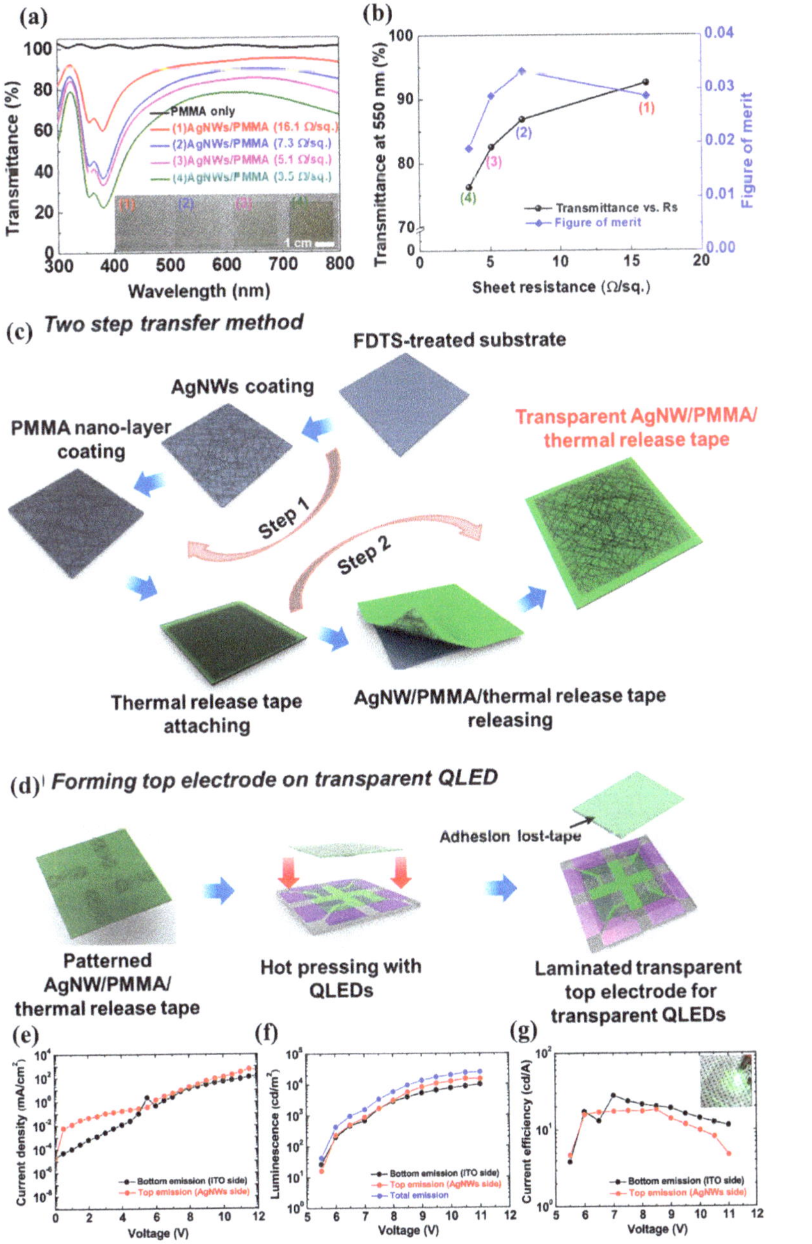

Figure 5. (**a**) UV–vis transmittance spectra of AgNW/PMMA as a function of the number of AgNW coatings (1–4 times). (**b**) Transmittance at 550 nm of (**a**) versus the sheet resistance measured for each sample with the different number of AgNW coatings. Figures of merit were calculated from the transmittance and sheet resistance values and are presented together in (**b**). The numbers in parentheses denote the number of AgNW coatings. Schematic illustrations of (**c**) the AgNW/PMMA TE fabrication process and (**d**) the applications in transparent QLEDs. Performance of the transparent QLEDs with the AgNW/PMMA top electrodes. (**e**) Current density–voltage, (**f**) luminance–voltage, and (**g**) current efficiency characteristics. Reproduced with permission from ref. [39].

2.3. *Polyvinylidene Difluoride (PVDF) on AgNW*

One of the methods is to use PVDF as an overcoating layer on Ag NWs [32,51–53]. PVDF is a piezoelectric polymer that has good thermal stability, chemical resistance, and mechanical properties [54,55]. Li et al. explored a highly thermally conductive film fabricated using a combination of AgNW and PVDF through a bar coating method (Figure 6a) [32]. The through-plane and in-plane thermal conductivity of the AgNW/PVDF composite film were measured at 0.31 and 1.61 W/mK, respectively, and significantly surpassed those of the pure PVDF film (Figure 6b,c). The experiment demonstrated the successful formation of thermally conductive pathways within the PVDF overcoating due to the incorporation of AgNW. Moreover, it was observed that heat preferentially transferred along these thermally conductive pathways rather than through the PVDF overcoating layer itself (Figure 6d,e).

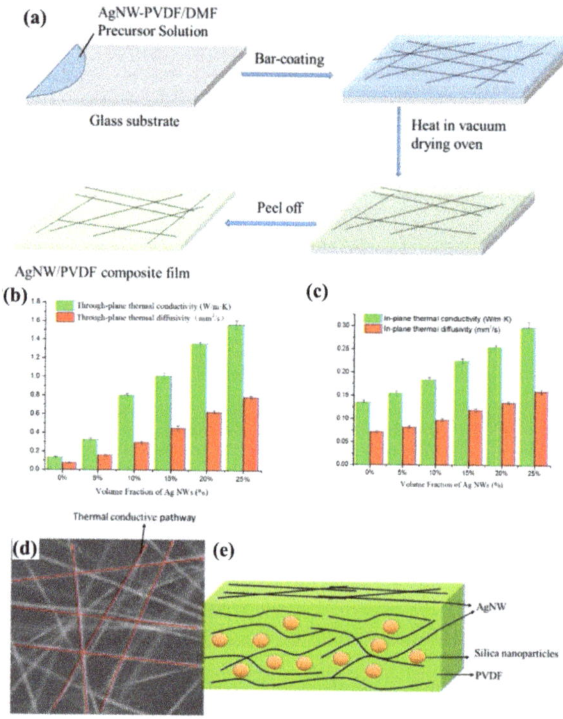

Figure 6. (**a**) Schematic illustration of the fabrication procedure of the AgNW/PVDF composite film. (**b**) Through-plane thermal conductivity and through-plane thermal diffusivity at different volume fractions. (**c**) In-plane thermal conductivity and in-plane thermal diffusivity at different volume fractions. (**d**) SEM image of the AgNW/PVDF composite film forming a thermally conductive pathway. (**e**) Dispersion principle diagram of AgNW in the AgNW/PVDF composite film. Reproduced with permission from ref. [32].

Zhang et al. developed transparent PVDF/AgNW electrodes, where AgNWs were partially embedded in PVDF layers that remarkably enhanced the electrical conductivity and adhesion of AgNWs (Figure 7a) [40]. The conductivity increased by more than two times during the formation of the PVDF/AgNW electrodes (Figure 7b,c). The adhesion of the AgNWs to the substrate was significantly enhanced, while the conductivity was persevered under multiple bending deformations (Figure 7d–f). Antioxidant ability and chemical resistance was also remarkably improved due to the excellent chemical stability of the PVDF substrate and the conductive structure of the PVDF/AgNW electrode. Therefore, the PVDF/AgNW electrode showed a much slower increase in sheet resistance than that of bare AgNWs during the 30 days of exposure to ambient air (Figure 7g).

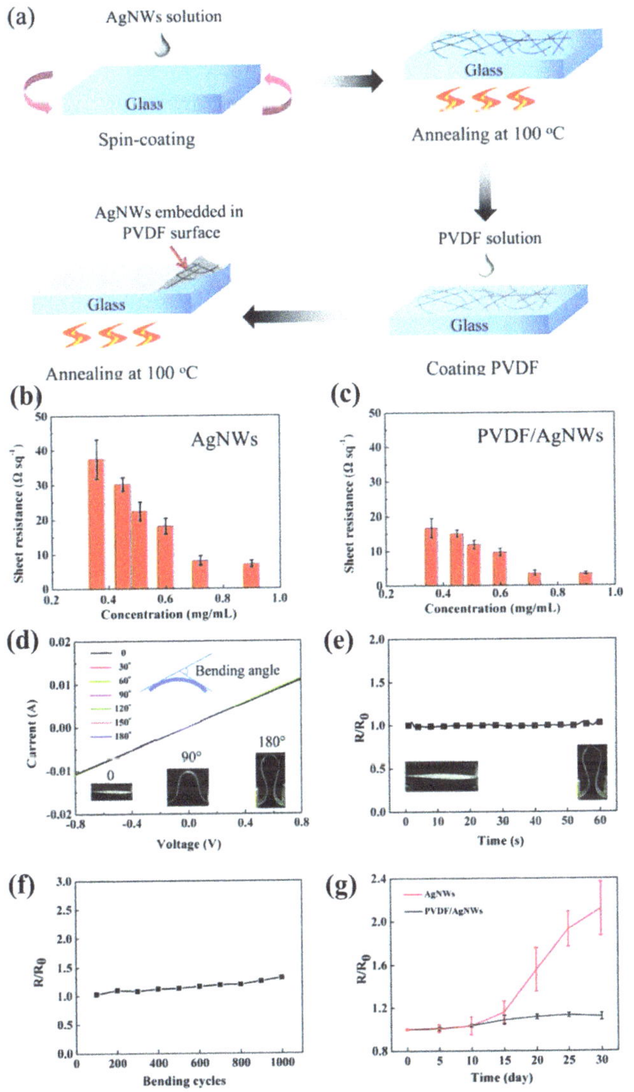

Figure 7. The fabrication and characteristics of PVDF/AgNW electrodes. (**a**) The fabrication of transparent PVDF/AgNW electrodes. (**b**) The sheet resistance of the AgNWs films decreasing with increasing concentrations. (**c**) The sheet resistance of PVDF/AgNW electrodes significantly decreasing by about half compared with that of AgNW films with the same concentrations. Flexibility tests and the long-term stability of the PVDF/AgNW electrodes are shown. The area of the electrode was 3 cm^2 (1 cm × 3 cm). (**d**) *I–V* curves of the PVDF/AgNW electrodes bent at different angles. The responses show that the bent electrodes did not change compared with those in the unbent state. (**e**) Dynamical resistance variation in the PVDF/AgNW electrodes with bending angles from 0° to 180° for 60 s. (**f**) The variation in the sheet resistance of the PVDF/AgNW electrodes after repeated bending from 0° to 180° for various cycles. (**g**) The long-term stability of the PVDF/AgNW electrodes after exposure to air for 30 days. Reproduced with permission from ref. [40].

2.4. Polyaniline (PANI)/AgNW

PANI is a conducting polymer renowned for its remarkable electrical and mechanical characteristics [41,56–58]. It can be produced through diverse synthesis methods, including chemical oxidation, electrochemical polymerization, and interfacial polymerization [41,56–58]. Polyaniline exhibits versatility through its existence in distinct oxidation states, namely leucoemeraldine, emeraldine, and pernigraniline, each characterized by unique colors and conductive properties. Moreover, the conductivity and stability of polyaniline can be significantly improved by doping it with acids or bases [41,56–58]. PANI can also improve the adhesion between AgNWs and substrates, and reduce the surface roughness of the electrodes.

Kumar et al. investigated AgNW/PANI transparent electrodes via layer-by-layer coating and mechanical pressing, resulting in excellent surface characteristics (Figure 8a,b) [41]. The empty spaces between the individual AgNWs were filled with a polyaniline:polystyrene-sulphonate (PANI:PSS) coating, which reduced the surface roughness of the AgNWs electrode to ~6.5 nm (Figure 8c,d). The transparent composite electrode achieved a reasonable sheet resistance of 25 Ω/sq and a high transmittance of 83.5%, highlighting the enhancement of the conducting properties of AgNWs when incorporated into a composite with PANI:PSS. In the work by Fang et al., AgNW/PANI composite electrodes were fabricated to have a free-standing layer structure [31]. The composite electrode showed a high electrical conductivity (EC) of 5300 S/cm, which is higher than that of bare AgNW electrodes (Figure 8e). The conductivity of the composite electrodes was retained even after cyclic bendings.

2.5. Poly(3,4-ethylenedioxythiophene) Polystyrene Sulfonate (PEDOT:PSS)/AgNW

PEDOT:PSS enhances the performance and stability of AgNW transparent electrodes [46,59–61]. PEDOT:PSS reduces the contact resistance between individual AgNWs, thereby increasing the overall conductivity of the network [44–46]. In addition, the optical performance of AgNW electrodes can be enhanced by filling the gaps between AgNWs and reducing the light reflection [44–46]. For example, PEDOT:PSS protects AgNWs from oxidation and corrosion by forming a passivation layer that prevents contact between oxygen and water molecules from the surface of the AgNWs [46,59–61]. In addition, the thermal degradation of AgNWs is suppressed by the overcoating acting as a thermal barrier that reduces the heat transfer between AgNWs and the substrate [46,59–61]. However, PEDOT:PSS is susceptible to high temperatures or humidity, resulting in the loss of conductivity or transparency. Therefore, the optimal thickness and composition of PEDOT:PSS overcoatings should be carefully controlled to balance the trade-off between protection and performance.

In the work by He et al., PEDOT:PSS-coated AgNW transparent electrodes were used for a flexible transparent heater [42]. PEDOT:PSS overcoating provides protection via enhancing the adhesion and flexibility of AgNW transparent electrodes, as well as preventing their oxidation and corrosion. Moreover, PEDOT:PSS overcoatings also improve heating efficiency by reducing the contact resistance and noise of AgNW transparent electrodes (Figure 9a,b).

Zappia et al. used PEDOT:PSS-coated AgNW transparent electrodes for perovskite solar cells (PSCs) [43]. PSCs require transparent electrodes that have high conductivity, transparency, work function, and stability. AgNW transparent electrodes meet these requirements, but they need to be compatible with the organic materials and the deposition techniques used in PSCs. PEDOT:PSS overcoating is compatible with the PSC system in that in increases the work function and surface energy of AgNW transparent electrodes, as well as preventing their aggregation or migration during the deposition process. The power conversion efficiency was 25% (Figure 9c). Furthermore, PEDOT:PSS overcoating improves the power conversion efficiency and lifetime of PSCs by enhancing the charge extraction and transport of AgNW transparent electrodes.

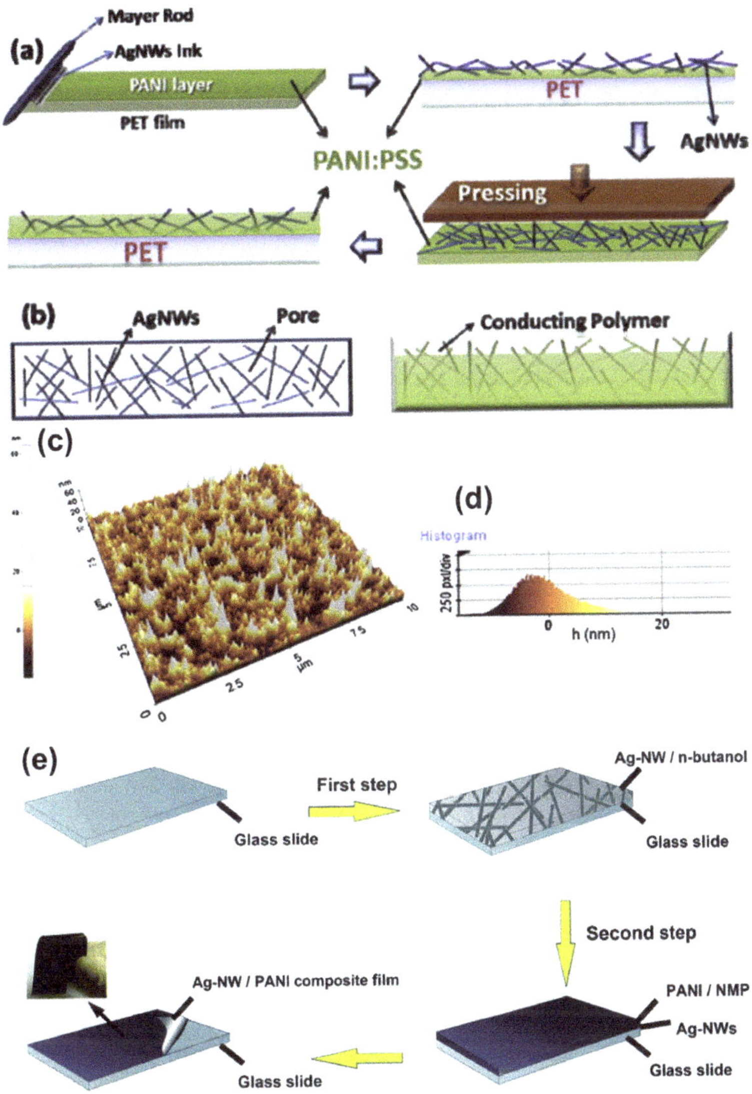

Figure 8. (**a**) Schematic representation of the preparation of the AgNW/PANI:PSS composite transparent conducting film via layer-by-layer coating. (**b**) Filling properties of the conducting film with and without PANI:PSS. (**c**) The AFM image and (**d**) height distribution of the AgNW/PANI transparent electrode after pressing. Reproduced with permission from ref. [41]. (**e**) Schematic showing the preparation of the layer-structured AgNW/PANI composite film. First step: an AgNW dispersion was cast on a glass slide to form a deposition layer; second step: a PANI solution was cast over the AgNW layer; third step: thermal evaporation was conducted and the AgNW/PANI composite film was peeled off from the glass substrate. Reproduced with permission from ref. [31].

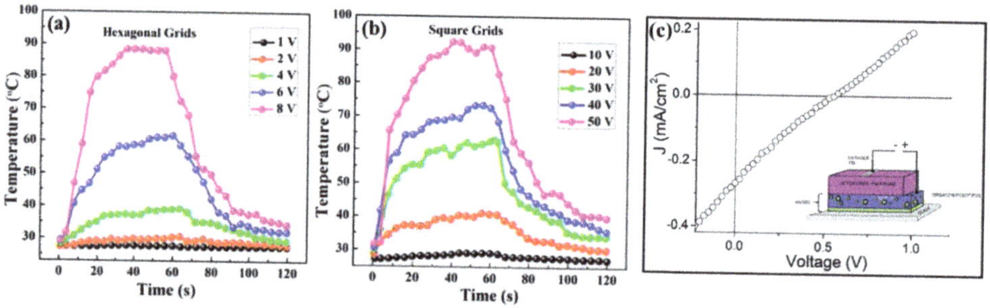

Figure 9. Evolution of generated temperature of the AgNW/PEDOT:PSS composite film with a (**a**) hexagonal pattern at varied voltages from 1 to 8 V, and (**b**) with a square pattern at varied voltages from 10 to 50 V [42]. Reproduced from ref. [42] under the Creative Commons Attribution 4.0 International (CC BY 4.0) License. (**c**) J–V curve for the solar cell using AgNW/PEDOT:PSS composites, the configuration of which is sketched in the inset [43]. Reproduced from ref. [43] under the Creative Commons Attribution 4.0 International (CC BY 4.0) License.

2.6. *Polyimide (PI)/AgNW*

Another important polymer material for the protection of AgNW is PI as it has high thermal and mechanical stabilities [62–64]. For example, Ghosh et al. presented a transparent conductor (TC) that consists of AgNWs embedded in a thin PI film (5 μm thick) [62]. The PI/AgNW TC showed outstanding optical transparency in the visible range, exceeding that of indium tin oxide (ITO) with more than 90% transmission. At the same time, it has a similar electrical sheet resistance of only 15 ohms per square. The PI film had two functions; it protected the AgNWs from environmental factors such as oxygen and moisture, and it supported the AgNW network mechanically. The TC showed excellent mechanical stability under bending with a radius of just 1 mm without any problems of adhesion. Moreover, this TC reduces the initial roughness of the AgNWs by about 15 times. In the work by Huang et al., PI/AgNW was also explored for a thin film heater [63]. The resistance of the PI/AgNW electrode under bending was excellent, showing no significant change even after 1000 outer bends. This amazing performance was due to the effective embedding of the AgNW network within the transparent PI film. The surface of PI/AgNW was very smooth (Rrms < 1 nm), and showed the electrode's resistance to oxidation and moisture, making it more durable. Wang et al. also presented PI/AgNW electrodes, wiere AgNWs were partially embedded in PI films [64]. The PI/AgNW electrodes exhibited exceptional performance across various parameters, including a low sheet resistance of ~12.7 Ω/sq, high optical transmittance of ~86.3%, and low RMS value of ~0.32 nm. Furthermore, they showed excellent stability when subjected to thermal, mechanical, and solution-based conditions.

3. Challenges and Future Perspectives

Polymeric materials applied to AgNW electrodes face several challenges and future prospects. One of the most significant challenges is ensuring scalability for mass production. To achieve this, researchers are optimizing deposition techniques such as roll-to-roll coating and inkjet printing to deliver uniform polymeric layers efficiently. The integration of these layers into emerging technologies such as foldable displays poses unique challenges, calling for materials that seamlessly combine robustness with flexibility. Ongoing research investigates advanced organic–inorganic composites and innovative manufacturing methods to address this challenge.

Long-term stability and reliability are paramount concerns, driving the establishment of standardized testing protocols, accelerated aging studies, and predictive models to ensure durability against environmental factors. Additionally, in an era prioritizing environmental sustainability, the assessment of the ecological footprint of polymeric materials is becoming indispensable. Researchers are spurring research into eco-friendly materials, sustainable fabrication techniques, and strategies for recycling and reusing AgNW electrodes with polymeric materials to minimize waste and environmental impacts.

4. Summary

AgNWs are susceptible to environmental factors such as oxidation and corrosion, which limit their long-term performance. Furthermore, electrical instability can manifest when they are subjected to mechanical stress and electric fields, resulting in unpredictable device behavior. The delicate balance between flexibility and durability also introduces a significant hurdle in the realization of robust AgNW networks.

To overcome these challenges, protective layers have emerged as indispensable preservers of AgNW-based electrodes. These layers serve as shields against environmental threats, ensuring the durability and stability of AgNWs even in the harshest conditions. Simultaneously, they fortify the mechanical integrity of AgNW networks, enabling them to withstand the rigors of flexibility and mechanical stress.

This review paper systematically examines various polymeric materials and their influence on the electrical, optical, and mechanical properties of AgNW-based electrodes. Table 1 provides the overview of various polymeric materials for AgNW electrodes. The paper also reviews the applications of these protected electrodes in different fields and demonstrates their adaptability and importance in various domains. Moreover, the paper clarifies the key factors that determine the choice of protective layers according to specific application needs.

Table 1. Summary of various polymeric materials for AgNW electrodes.

Polymeric Materials	Conductivity	Optical Transmittance	Mechanical Stability	Chemical Stability	Applications	Refs.
PU	~50 Ω/sq	~88%	Excellent	Excellent	Transparent electrode	[33]
	~8 Ω/sq	~74.6%	Excellent	N/A	Touch panel	[37]
PMMA	~8 Ω/sq	~85%	Excellent	N/A	Thin film heater/SERS detector	[38]
	~16.1 Ω/sq	~87%	Excellent	Excellent	QLED	[39]
PVDF	~8 Ω/sq	N/A	N/A	Excellent	Thermal conduction	[32]
	~5 Ω/sq	~73%	Excellent	Excellent	Touch sensor	[40]
PANI	~25 Ω/sq	~83.5%	N/A	Excellent	Solar cells	[41]
	~5300 S/cm	N/A	Excellent	N/A	EMI shielding	[31]
PEDOT:PSS	~2.3 Ω/sq	~70.5%	N/A	Excellent	Thin film heater	[42]
	~12 Ω/sq	~80%	N/A	N/A	Solar cells	[43]

As transparent conductive materials are advancing rapidly, the incorporation of protective materials is an essential step to enhance the performance and durability of AgNWs. These materials and their protection methods offer great potential and opportunities for various applications. However, there are still challenges and open questions that need to be addressed, and the research on AgNWs and their protection is ongoing, driven by the continuous pursuit of innovation and the exploration of the diverse possibilities in the dynamic field of AgNWs and their protection.

Author Contributions: H.H., J.Y.C., T.G.Y., and B.H. prepared the manuscript. All authors have read and agreed to the published version of the manuscript.

Funding: This research was supported by GRDC (Global Research Development Center) Cooperative Hub Program through the National Research Foundation of Korea(NRF) funded by the Ministry of Science and ICT(MSIT) (RS-2023-00257595).

Institutional Review Board Statement: Not applicable.

Informed Consent Statement: Not applicable.

Data Availability Statement: The datasets used and/or analyzed during the current study are available from the corresponding author upon reasonable request.

Acknowledgments: This research was supported by GRDC (Global Research Development Center) Cooperative Hub Program through the National Research Foundation of Korea(NRF) funded by the Ministry of Science and ICT(MSIT) (RS-2023-00257595).

Conflicts of Interest: The authors declare that they have no competing interest.

References

1. Wang, T.; Lu, K.; Xu, Z.; Lin, Z.; Ning, H.; Qiu, T.; Yang, Z.; Zheng, H.; Yao, R.; Peng, J. Recent Developments in Flexible Transparent Electrode. *Crystals* **2021**, *11*, 511. [CrossRef]
2. Pang, S.; Hernandez, Y.; Feng, X.; Müllen, K. Graphene as Transparent Electrode Material for Organic Electronics. *Adv. Mater.* **2011**, *23*, 2779–2795. [CrossRef]
3. Morales-Masis, M.; De Wolf, S.; Woods-Robinson, R.; Ager, J.W.; Ballif, C. Transparent Electrodes for Efficient Optoelectronics. *Adv. Electron. Mater.* **2017**, *3*, 1600529. [CrossRef]
4. Kim, H.; Qaiser, N.; Hwang, B. Electro-mechanical response of stretchable pdms composites with a hybrid filler system. *Facta Univ. Ser. Mech. Eng.* **2023**, *21*, 51–61. [CrossRef]
5. Jin, I.S.; Lee, H.D.; Hong, S.I.; Lee, W.; Jung, J.W. Facile Post Treatment of Ag Nanowire/Polymer Composites for Flexible Transparent Electrodes and Thin Film Heaters. *Polymers* **2021**, *13*, 586. [CrossRef] [PubMed]
6. Kim, W.-K.; Lee, S.; Hee Lee, D.; Hee Park, I.; Seong Bae, J.; Woo Lee, T.; Kim, J.-Y.; Hun Park, J.; Chan Cho, Y.; Ryong Cho, C.; et al. Cu Mesh for Flexible Transparent Conductive Electrodes. *Sci. Rep.* **2015**, *5*, 10715. [CrossRef] [PubMed]
7. Luo, M.; Liu, Y.; Huang, W.; Qiao, W.; Zhou, Y.; Ye, Y.; Chen, L.-S. Towards Flexible Transparent Electrodes Based on Carbon and Metallic Materials. *Micromachines* **2017**, *8*, 12. [CrossRef]
8. Xie, H.; Yang, X.; Du, D.; Zhao, Y.; Wang, Y. Flexible Transparent Conductive Film Based on Random Networks of Silver Nanowires. *Micromachines* **2018**, *9*, 295. [CrossRef]
9. Tokuno, T.; Nogi, M.; Karakawa, M.; Jiu, J.; Nge, T.T.; Aso, Y.; Suganuma, K. Fabrication of silver nanowire transparent electrodes at room temperature. *Nano Res.* **2011**, *4*, 1215–1222. [CrossRef]
10. Ahn, Y.; Jeong, Y.; Lee, Y. Improved Thermal Oxidation Stability of Solution-Processable Silver Nanowire Transparent Electrode by Reduced Graphene Oxide. *ACS Appl. Mater. Interfaces* **2012**, *4*, 6410–6414. [CrossRef]
11. Fang, Y.; Wu, Z.; Li, J.; Jiang, F.; Zhang, K.; Zhang, Y.; Zhou, Y.; Zhou, J.; Hu, B. High-Performance Hazy Silver Nanowire Transparent Electrodes through Diameter Tailoring for Semitransparent Photovoltaics. *Adv. Funct. Mater.* **2018**, *28*, 1705409. [CrossRef]
12. Choi, Y.; Kim, C.S.; Jo, S. Spray Deposition of Ag Nanowire–Graphene Oxide Hybrid Electrodes for Flexible Polymer–Dispersed Liquid Crystal Displays. *Materials* **2018**, *11*, 2231. [CrossRef] [PubMed]
13. Hajipour, P.; Bahrami, A.; Mehr, M.Y.; van Driel, W.D.; Zhang, K. Facile Synthesis of Ag Nanowire/TiO_2 and Ag Nanowire/TiO_2/GO Nanocomposites for Photocatalytic Degradation of Rhodamine B. *Materials* **2012**, *14*, 763. [CrossRef] [PubMed]
14. Nam, V.B.; Lee, D. Copper Nanowires and Their Applications for Flexible, Transparent Conducting Films: A Review. *Nanomaterials* **2016**, *6*, 47. [CrossRef] [PubMed]
15. Hwang, B.; Han, Y.; Matteini, P. Bending fatigue behavior of ag nanowire/cu thin-film hybrid interconnects for wearable electronics. *Facta Univ. Ser. Mech. Eng.* **2022**, *20*, 553–560. [CrossRef]
16. Hwang, B.; Matteini, P. Research Trends on Silk-Based Conductive Fibers with the Enhanced Machine Washa-bility by Adopting PEDOT:PSS. *J. Nat. Fibers* **2023**, *20*, 2148152. [CrossRef]
17. Limkatanyu, S.; Sae-Long, W.; Rungamornrat, J.; Buachart, C.; Sukontasukkul, P.; Keawsawasvong, S.; Chindaprasirt, P. Bending, buckling and free vibration analyses of nanobeam-substrate medium systems. *Facta Univ. Ser. Mech. Eng.* **2022**, *20*, 561–587. [CrossRef]
18. Ha, H.; Müller, S.; Baumann, R.-P.; Hwang, B. Peakforce quantitative nanomechanical mapping for surface energy characterization on the nanoscale: A mini-review. *Facta Univ. Ser. Mech. Eng.* **2023**. [CrossRef]
19. Seo, Y.; Ha, H.; Cheong, J.Y.; Leem, M.; Darabi, S.; Matteini, P.; Müller, C.; Yun, T.G.; Hwang, B. Highly Reliable Yarn-Type Supercapacitor Using Conductive Silk Yarns with Multilayered Active Materials. *J. Nat. Fibers* **2021**, *19*, 835–846. [CrossRef]
20. Shi, Y.; He, L.; Deng, Q.; Liu, Q.; Li, L.; Wang, W.; Xin, Z.; Liu, R. Synthesis and Applications of Silver Nanowires for Transparent Conductive Films. *Micromachines* **2019**, *10*, 330. [CrossRef]
21. Zhu, Y.; Li, X.; Xu, Y.; Wu, L.; Yu, A.; Lai, G.; Wei, Q.; Chi, H.; Jiang, N.; Fu, L.; et al. Intertwined Carbon Nanotubes and Ag Nanowires Constructed by Simple Solution Blending as Sensitive and Stable Chloramphenicol Sensors. *Sensors* **2021**, *21*, 1220. [CrossRef] [PubMed]

22. Im, H.-G.; Jin, J.; Ko, J.-H.; Lee, J.; Lee, J.-Y.; Bae, B.-S. Flexible transparent conducting composite films using a monolithically embedded AgNW electrode with robust performance stability. *Nanoscale* **2013**, *6*, 711–715. [CrossRef] [PubMed]
23. Kim, A.; Won, Y.; Woo, K.; Kim, C.-H.; Moon, J. Highly Transparent Low Resistance ZnO/Ag Nanowire/ZnO Composite Electrode for Thin Film Solar Cells. *ACS Nano* **2013**, *7*, 1081–1091. [CrossRef] [PubMed]
24. Kumar, A.; Shaikh, M.O.; Chuang, C.-H. Silver Nanowire Synthesis and Strategies for Fabricating Transparent Conducting Electrodes. *Nanomaterials* **2021**, *11*, 693. [CrossRef] [PubMed]
25. Wang, B.-Y.; Lee, E.-S.; Oh, Y.-J.; Kang, H.W. A silver nanowire mesh overcoated protection layer with graphene oxide as a transparent electrode for flexible organic solar cells. *RSC Adv.* **2017**, *7*, 52914–52922. [CrossRef]
26. Jang, J.; Choi, J.-Y.; Jeon, J.; Lee, J.; Im, J.; Lee, J.; Jin, S.-W.; Park, H.-J.; Lee, S.-H.; Kim, D.-B.; et al. Flexible Transparent Electrode Characteristics of Graphene Oxide/Cysteamine/AgNP/AgNW Structure. *Nanomaterials* **2020**, *10*, 2352. [CrossRef] [PubMed]
27. Cho, S.; Kang, S.; Pandya, A.; Shanker, R.; Khan, Z.; Lee, Y.; Park, J.; Craig, S.L.; Ko, H. Large-Area Cross-Aligned Silver Nanowire Electrodes for Flexible, Transparent, and Force-Sensitive Mechanochromic Touch Screens. *ACS Nano* **2017**, *11*, 4346–4357. [CrossRef] [PubMed]
28. Bari, B.; Lee, J.; Jang, T.; Won, P.; Ko, S.H.; Alamgir, K.; Arshad, M.; Guo, L.J. Simple hydrothermal synthesis of very-long and thin silver nanowires and their application in high quality transparent electrodes. *J. Mater. Chem. A* **2016**, *4*, 11365–11371. [CrossRef]
29. Lian, L.; Dong, D.; Feng, D.; He, G. Low roughness silver nanowire flexible transparent electrode by low temper-ature solution-processing for organic light emitting diodes. *Org. Electron.* **2017**, *49*, 9–18. [CrossRef]
30. Xiong, W.; Liu, H.; Chen, Y.; Zheng, M.; Zhao, Y.; Kong, X.; Wang, Y.; Zhang, X.; Kong, X.; Wang, P.; et al. Highly Conductive, Air-Stable Silver Nanowire@Iongel Composite Films toward Flexible Transparent Electrodes. *Adv. Mater.* **2016**, *28*, 7167–7172. [CrossRef]
31. Fang, F.; Li, Y.-Q.; Xiao, H.-M.; Hu, N.; Fu, S.-Y. Layer-structured silver nanowire/polyaniline composite film as a high performance X-band EMI shielding material. *J. Mater. Chem. C* **2016**, *4*, 4193–4203. [CrossRef]
32. Li, Z.; Zhang, L.; Qi, R.; Xie, F.; Qi, S. Improvement of the thermal transport performance of a poly(vinylidene fluoride) composite film including silver nanowire. *J. Appl. Polym. Sci.* **2016**, *133*. [CrossRef]
33. Hwang, B.; An, C.-H.; Becker, S. Highly robust Ag nanowire flexible transparent electrode with UV-curable polyurethane-based overcoating layer. *Mater. Des.* **2017**, *129*, 180–185. [CrossRef]
34. Hwang, B.; Qaiser, N.; Lee, C.; Matteini, P.; Yoo, S.J.; Kim, H. Effect of Al_2O_3/Alucone nanolayered composite overcoating on reliability of Ag nanowire electrodes under bending fatigue. *J. Alloys Compd.* **2020**, *846*, 156420. [CrossRef]
35. Lee, D.; Lee, H.; Ahn, Y.; Jeong, Y.; Lee, D.-Y.; Lee, Y. Highly stable and flexible silver nanowire–graphene hybrid transparent conducting electrodes for emerging optoelectronic devices. *Nanoscale* **2013**, *5*, 7750–7755. [CrossRef] [PubMed]
36. Wang, P.; Jian, M.; Zhang, C.; Wu, M.; Ling, X.; Zhang, J.; Wei, B.; Yang, L. Highly Stable Graphene-Based Flexible Hybrid Transparent Conductive Electrodes for Organic Solar Cells. *Adv. Mater. Interfaces* **2021**, *9*, 2101442. [CrossRef]
37. Hu, W.; Niu, X.; Zhao, R.; Pei, Q. Elastomeric transparent capacitive sensors based on an interpenetrating composite of silver nanowires and polyurethane. *Appl. Phys. Lett.* **2013**, *102*. [CrossRef]
38. Sun, J.; Yu, X.; Li, Z.; Zhao, J.; Zhu, P.; Dong, X.; Yu, Z.; Zhao, Z.; Shi, D.; Wang, J.; et al. Ultrasonic Modification of Ag Nanowires and Their Applications in Flexible Transparent Film Heaters and SERS Detectors. *Materials* **2019**, *12*, 893. [CrossRef] [PubMed]
39. Kim, S.; Kim, J.; Kim, D.; Kim, B.; Chae, H.; Yi, H.; Hwang, B. High-Performance Transparent Quantum Dot Light-Emitting Diode with Patchable Transparent Electrodes. *ACS Appl. Mater. Interfaces* **2019**, *11*, 26333–26338. [CrossRef]
40. Zhang, L.; Wang, Y.; Gui, J.; Wang, X.; Li, R.; Liu, W.; Sun, C.; Zhao, X.; Guo, S. Efficient Welding of Silver Nanowires embedded in a Poly(vinylidene fluoride) Film for Robust Wearable Electronics. *Adv. Mater. Technol.* **2018**, *4*, 1800438. [CrossRef]
41. Kumar, A.B.V.K.; Jiang, J.; Bae, C.W.; Seo, D.M.; Piao, L.; Kim, S.-H. Silver nanowire/polyaniline composite trans-parent electrode with improved surface properties. *Mater. Res. Bull.* **2014**, *57*, 52–57. [CrossRef]
42. He, X.; Shen, G.; Xu, R.; Yang, W.; Zhang, C.; Liu, Z.; Chen, B.; Liu, J.; Song, M. Hexagonal and Square Patterned Silver Nanowires/PEDOT:PSS Composite Grids by Screen Printing for Uniformly Transparent Heaters. *Polymers* **2019**, *11*, 468. [CrossRef] [PubMed]
43. Zappia, S.; Alloisio, M.; Valdivia, J.C.; Arias, E.; Moggio, I.; Scavia, G.; Destri, S. Silver Nanoparticle–PEDOT:PSS Composites as Water-Processable Anodes: Correlation between the Synthetic Parameters and the Optical/Morphological Properties. *Polymers* **2023**, *15*, 3675. [CrossRef]
44. Li, X.; Park, H.; Lee, M.H.; Hwang, B.; Kim, S.H.; Lim, S. High resolution patterning of Ag nanowire flexible trans-parent electrode via electrohydrodynamic jet printing of acrylic polymer-silicate nanoparticle composite overcoating layer. *Org. Electron.* **2018**, *62*, 400–406. [CrossRef]
45. Yang, H.; Bai, S.; Guo, X.; Wang, H. Robust and smooth UV-curable layer overcoated AgNW flexible transparent conductor for EMI shielding and film heater. *Appl. Surf. Sci.* **2019**, *483*, 888–894. [CrossRef]
46. Chen, S.; Song, L.; Tao, Z.; Shao, X.; Huang, Y.; Cui, Q.; Guo, X. Neutral-pH PEDOT:PSS as over-coating layer for stable silver nanowire flexible transparent conductive films. *Org. Electron.* **2014**, *15*, 3654–3659. [CrossRef]
47. Choi, J.H.; Lee, K.Y.; Kim, S.W. Ultra-bendable and durable Graphene–Urethane composite/silver nanowire film for flexible transparent electrodes and electromagnetic-interference shielding. *Compos. Part B Eng.* **2019**, *177*, 107406. [CrossRef]
48. Tiwari, N.; Ankit, A.; Rajput, M.; Kulkarni, M.R.; John, R.A.; Mathews, N. Healable and flexible transparent heaters. *Nanoscale* **2017**, *9*, 14990–14997. [CrossRef]

49. Yu, S.; Ma, X.; Li, X.; Li, J.; Gong, B.; Wang, X. Enhanced adhesion of Ag nanowire based transparent conducting electrodes for application in flexible electrochromic devices. *Opt. Mater.* **2021**, *120*, 111414. [CrossRef]
50. Zhu, Y.; Deng, Y.; Yi, P.; Peng, L.; Lai, X.; Lin, Z. Flexible Transparent Electrodes Based on Silver Nanowires: Mate-rial Synthesis, Fabrication, Performance, and Applications. *Adv. Mater. Technol.* **2019**, *4*, 1900413. [CrossRef]
51. Khadtare, S.; Ko, E.J.; Kim, Y.H.; Lee, H.S.; Moon, D.K. A flexible piezoelectric nanogenerator using conducting polymer and silver nanowire hybrid electrodes for its application in real-time muscular monitoring system. *Sens. Actuators A Phys.* **2019**, *299*, 111575. [CrossRef]
52. Bobinger, M.; Keddis, S.; Hinterleuthner, S.; Becherer, M.; Kluge, F.; Schwesinger, N.; Salmeron, J.F.; Lugli, P.; Rivadeneyra, A. Light and Pressure Sensors Based on PVDF With Sprayed and Transparent Electrodes for Self-Powered Wireless Sensor Nodes. *IEEE Sens. J.* **2018**, *19*, 1114–1126. [CrossRef]
53. Basarir, F.; Madani, Z.; Vapaavuori, J. Recent Advances in Silver Nanowire Based Flexible Capacitive Pressure Sensors: From Structure, Fabrication to Emerging Applications. *Adv. Mater. Interfaces* **2022**, *9*, 2200866. [CrossRef]
54. Tan, Q.; Yuan, L.; Liang, G.; Gu, A. Flexible, transparent, strong and high dielectric constant composite film based on polyionic liquid coated silver nanowire hybrid. *Appl. Surf. Sci.* **2021**, *576*, 151827. [CrossRef]
55. Bobinger, M.; Hinterleuthner, S.; Becherer, M.; Keddis, S.; Schwesinger, N.; Lugli, P. Energy harvesting from ambient light using PVDF with highly conductive and transparent silver nanowire/PEDOT:PSS hybride electrodes. In Proceedings of the 2017 IEEE 17th International Conference on Nanotechnology (IEEE-NANO), Pittsburgh, PA, USA, 25–28 July 2017; pp. 426–429.
56. Gao, D.; Zhao, P.; Liu, J.; Zhou, Y.; Lyu, B.; Ma, J.; Shao, L. Polyaniline/silver nanowire cotton fiber: A flexible electrode material for supercapacitor. *Adv. Powder Technol.* **2021**, *32*, 3954–3963. [CrossRef]
57. Fang, F.; Huang, G.-W.; Xiao, H.-M.; Li, Y.-Q.; Hu, N.; Fu, S.-Y. Largely enhanced electrical conductivity of layer-structured silver nanowire/polyimide composite films by polyaniline. *Compos. Sci. Technol.* **2018**, *156*, 144–150. [CrossRef]
58. Che, B.; Zhou, D.; Li, H.; He, C.; Liu, E.; Lu, X. A highly bendable transparent electrode for organic electrochromic devices. *Org. Electron.* **2018**, *66*, 86–93. [CrossRef]
59. Bai, S.; Guo, X.; Chen, T.; Zhang, Y.; Zhang, X.; Yang, H.; Zhao, X. Solution processed fabrication of silver nanowire-MXene@PEDOT: PSS flexible transparent electrodes for flexible organic light-emitting diodes. *Compos. Part A Appl. Sci. Manuf.* **2020**, *139*, 106088. [CrossRef]
60. Liu, Y.-S.; Feng, J.; Ou, X.-L.; Cui, H.-F.; Xu, M.; Sun, H.-B. Ultrasmooth, highly conductive and transparent PE-DOT:PSS/silver nanowire composite electrode for flexible organic light-emitting devices. *Org. Electron.* **2016**, *31*, 247–252. [CrossRef]
61. Nair, N.M.; Pakkathillam, J.K.; Kumar, K.; Arunachalam, K.; Ray, D.; Swaminathan, P. Printable Silver Nanowire and PEDOT:PSS Nanocomposite Ink for Flexible Transparent Conducting Applications. *ACS Appl. Electron. Mater.* **2020**, *2*, 1000–1010. [CrossRef]
62. Ghosh, D.S.; Chen, T.L.; Mkhitaryan, V.; Pruneri, V. Ultrathin Transparent Conductive Polyimide Foil Embedding Silver Nanowires. *ACS Appl. Mater. Interfaces* **2014**, *6*, 20943–20948. [CrossRef]
63. Huang, Q.; Shen, W.; Fang, X.; Chen, G.; Guo, J.; Xu, W.; Tan, R.; Song, W. Highly flexible and transparent film heaters based on polyimide films embedded with silver nanowires. *RSC Adv.* **2015**, *5*, 45836–45842. [CrossRef]
64. Wang, Y.; Chen, Q.; Zhang, G.; Xiao, C.; Wei, Y.; Li, W. Ultrathin Flexible Transparent Composite Electrode via Semi-embedding Silver Nanowires in a Colorless Polyimide for High-Performance Ultraflexible Organic Solar Cells. *ACS Appl. Mater. Interfaces* **2022**, *14*, 5699–5708. [CrossRef]

Disclaimer/Publisher's Note: The statements, opinions and data contained in all publications are solely those of the individual author(s) and contributor(s) and not of MDPI and/or the editor(s). MDPI and/or the editor(s) disclaim responsibility for any injury to people or property resulting from any ideas, methods, instructions or products referred to in the content.

MDPI AG
Grosspeteranlage 5
4052 Basel
Switzerland
Tel.: +41 61 683 77 34

Inorganics Editorial Office
E-mail: inorganics@mdpi.com
www.mdpi.com/journal/inorganics

Disclaimer/Publisher's Note: The title and front matter of this reprint are at the discretion of the Guest Editor. The publisher is not responsible for their content or any associated concerns. The statements, opinions and data contained in all individual articles are solely those of the individual Editor and contributors and not of MDPI. MDPI disclaims responsibility for any injury to people or property resulting from any ideas, methods, instructions or products referred to in the content.

www.ingramcontent.com/pod-product-compliance
Lightning Source LLC
Chambersburg PA
CBHW061616210926
51850CB00009B/2341
* 9 7 8 3 7 2 5 8 6 2 3 2 0 *